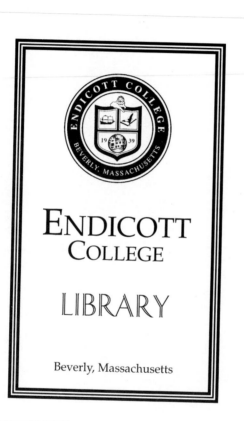

AIDS AND SEX

THE KINSEY INSTITUTE SERIES

June Machover Reinisch, *General Editor*

Volume I
MASCULINITY/FEMININITY: *Basic Perspectives*
Edited by
June Machover Reinisch, Leonard A. Rosenblum,
and Stephanie A. Sanders

Volume II
HOMOSEXUALITY/HETEROSEXUALITY:
Concepts of Sexual Orientation
Edited by
David P. McWhirter, Stephanie A. Sanders,
and June Machover Reinisch

Volume III
ADOLESCENCE AND PUBERTY
Edited by
John Bancroft and June Machover Reinisch

Volume IV
AIDS AND SEX: *An Integrated Biomedical and Biobehavioral Approach*
Edited by
Bruce Voeller, June Machover Reinisch,
and Michael Gottlieb

AIDS AND SEX

An Integrated Biomedical and Biobehavioral Approach

Edited by

Bruce Voeller, Ph.D.
June Machover Reinisch, Ph.D.
Michael Gottlieb, M.D.

New York   Oxford
OXFORD UNIVERSITY PRESS
1990

Oxford University Press

Oxford   New York   Toronto
Delhi   Bombay   Calcutta   Madras   Karachi
Petaling Jaya   Singapore   Hong Kong   Tokyo
Nairobi   Dar es Salaam   Cape Town
Melbourne   Auckland

and associated companies in
Berlin   Ibadan

Copyright © 1990 The Kinsey Institute for Research
in Sex, Gender, and Reproduction

Published by Oxford University Press, Inc.,
200 Madison Avenue, New York, New York 10016

Oxford is a registered trademark of Oxford University Press

Library of Congress Cataloging-in-Publication Data
AIDS and sex : an integrated biomedical and biobehavioral approach /
edited by Bruce Voeller, June Machover Reinisch, Michael Gottlieb.
p.   cm.—(Kinsey Institute series ; v. 4)
Based on a symposium held at the Kinsey Institute, Dec. 5–8, 1987,
cosponsored by the National Institute of Child Health and Human
Development, the National Institute of Allergy and Infectious Diseases,
and the Kinsey Institute for Research in Sex, Gender, and Reproduction.
Includes bibliographical references. Includes index.
ISBN 0-19-506909-9
1. AIDS (Disease)—Congresses 2. AIDS (Disease)—Social aspects—Congresses.
I. Voeller, Bruce R.   II. Reinisch, June Machover.   III. Gottlieb, Michael S.
IV. National Institute of Child Health and Human Development (U.S.)
V. National Institute of Allergy and Infectious Diseases (U.S.)
VI. Kinsey Institute for Research in Sex, Gender, and Reproduction.   VII. Series.
[DNLM: 1. Acquired Immunodeficiency Syndrome—congresses.
2. Sex Behavior—congresses. WD 308 A287785   1987]
RA644.A25A3527   1990   616.97'92—dc20   DNLM/DLC
for Library of Congress   90-7945

The Kinsey Institute "K" symbol stamped on the
front cover of this book was
designed by Enoch, Inc., New York, New York.

2 4 6 8 9 7 5 3 1

Printed in the United States of America
on acid-free paper

Preface

AIDS and Sex: An Integrated Biomedical and Biobehavioral Approach is the fourth volume in *The Kinsey Institute Series*. In each volume, researchers from a wide range of academic disciplines draw on their own data and on the viewpoints of their area of expertise to address the central issues in a specific arena of discourse. The chapters for each volume are written after the contributors participate in a *Kinsey Symposium*. As a result these contributions reflect the diverse perspectives that have emerged during sustained discussions among colleagues from many different fields. The editors of each volume provide an introduction based on the full range of discussions at the Symposium and the contents of the final contributions. This overview highlights the central themes and research findings of the volume as well as the major issues for future consideration.

The Kinsey Symposium on AIDS and Sex was co-sponsored by The Kinsey Institute for Research in Sex, Gender, and Reproduction at Indiana University, the National Institute of Allergy and Infectious Diseases and the National Institute of Child Health and Human Development. Additionally, we would like to acknowledge the support of Indiana University. Special thanks go to Dr. Stephanie Sanders of The Kinsey Institute for coordination of the meeting and compilation and preparation of this volume; Wendy Liffers of the National Institute of Allergy and Infectious Diseases for her assistance in the development and coordination of the Fourth Kinsey Symposium; and to Elizabeth Roberge, Janet Rowland, J. Susan Straub, Sandra Stewart Ham, Carolyn Kaufman, Terry Sare, Stacey Trainer, Mary Ziemba-Davis of the Kinsey

Institute staff for assistance from the planning of the conference to publication of this volume.

Bloomington, Indiana   J. M. R.
August 1990

Contents

PART I   AIDS Research and the Demand for Information on Human Sexuality

PART II   Contributions from Research on Human Sexuality

PART III   Historical Perspectives

PART IV   Cross-Cultural Perspectives

PART V   Perspectives from the Media

Contributors

DEBORAH J. ANDERSON, Ph.D.
Fearing Research Laboratory
Brigham and Women's Hospital
Harvard Medical School
Boston, Massachusetts

ROY M. ANDERSON, Ph.D., FRS
Head, Department of Biology
Imperial College
London, England

GERALD S. BERNSTEIN, Ph.D., M.D.
Professor, Department of Obstetrics
  and Gynecology
University of Southern California
School of Medicine
Los Angeles, California

JOHN E. BOSWELL, Ph.D.
Professor, Department of History
Yale University
New Haven, Connecticut

C. L. CASSITY
Hillcrest Biologicals, Inc.
Cypress, California

JOAN S. CHMIEL, Ph.D.
Principal Investigator—DCCWG
Northwestern University Medical
  School
Chicago, Illinois

MARLENE CIMONS
Correspondent, Washington Bureau,
  Los Angeles Times
Washington, D.C.

SUSAN D. COCHRAN, Ph.D.
Associate Professor, Department of
  Psychology
California State University at
  Northridge
Northridge, California
Associate Research Psychologist,
  Institute for Social Science
  Research
University of California at
  Los Angeles
Los Angeles, California

ANNE HERSEY COULSON
Senior Lecturer, Research
  Epidemiologist
Department of Epidemiology
School of Public Health
UCLA Center for the Health Sciences
Los Angeles, California

DON D. DES JARLAIS, Ph.D.
Formerly, New York State Division
  of Substance Abuse Services
Director of Research
Chemical Dependency Institute
Beth Israel Medical Center
New York, New York

ROGER DETELS, M.D., M.S.
Professor, Department of
  Epidemiology
UCLA School of Public Health
Los Angeles, California

LOIS J. ELDRED
Baltimore City Health Department
Baltimore, Maryland

PATRICIA ENGLISH
Multicenter AIDS Cohort Study
Department of Epidemiology
UCLA School of Public Health
Los Angeles, California

MARIA EUGENIA FERNANDES, M.D.
National Program for the Prevention
  of AIDS
São Paulo, Brazil

STEVEN FINDLAY
Medical Writer
Associate Editor, U.S. News &
  World Report
Washington, D.C.

SAMUEL R. FRIEDMAN, Ph.D.
Narcotic and Drug Research, Inc.
New York, New York

HAROLD GINZBURG, M.D., J.D., Ph.D.
Chief Medical Consultant
Bureau of Health Care Delivery and
  Assistance
Rockville, Maryland

DOUGLAS GOLDSMITH
Narcotic and Drug Research, Inc.
New York, New York

MARSHA F. GOLDSMITH
Associate Editor, Journal of the
  American Medical Association
JAMA Medical News
Chicago, Illinois

MICHAEL GOTTLIEB, M.D.
Assistant Clinical Professor of
  Medicine
UCLA School of Medicine
and Medical Director
Immune Suppressed Unit
Sherman Oaks Community Hospital
Los Angeles, California

DONALD R. HICKS, Ph.D.
Long Beach, California

WILLIAM HOPKINS
New York State Division of
  Substance Abuse Services
New York, New York

LISA P. JACOBSON, Sc.M.
Coordinator, CAMACS
School of Hygiene and Public Health
Johns Hopkins University
Baltimore, Maryland

ANNE M. JOHNSON, MB BS, MRCGP,
  MFCM
Senior Lecturer in Epidemiology
Academic Department of
  Genito-Urinary Medicine
University College and Middlesex
  School of Medicine
London, England

VIRGINIA JOHNSON, D.Sc. (Hon)
Director, Masters and Johnson
  Institute
St. Louis, Missouri

HOWARD B. KAPLAN, Ph.D.
Professor, Department of Sociology
Texas A & M University
College Station, Texas

LAWRENCE A. KINGSLEY, Dr.P.H.
Assistant Professor, Department of
  Infectious Diseases and
  Microbiology and Epidemiology
Pittsburgh Men's Study
Pittsburgh, Pennsylvania

DIANNE M. KRAFT, M.S.
Department of Psychology
University of Houston
Houston, Texas

JOSEPH F. LOVETT
President, Lovett Productions
Former Producer, ABC News 20/20
New York, New York

LEON McKUSICK, Ph.D.
Clinical Psychologist
Assistant Research Psychologist
Department of Medicine
University of California at
  San Francisco
San Francisco, California

DAVID P. McWHIRTER, M.D.
Associate Clinical Professor of
  Psychiatry
Medical Director, San Diego County
  Psychiatric Hospital
San Diego, California

WILLIAM H. MASTERS, M.D.
Consultant, Masters and Johnson
  Institute
St. Louis, Missouri

MARIA HELENA MATARAZZO, M.A.
President, Brazilian Association of
  Sex Education
UNESCO International Consultant
São Paulo, Brazil

ANDREW M. MATTISON, MSW, Ph.D.
Assistant Clinical Professor of
  Community Medicine (Family
  Medicine)
Director, Psychosocial Services
HIV Neurobehavioral Research
  Center
University of California, San Diego
  School of Medicine
San Diego, California

VICKIE M. MAYS, Ph.D.
Associate Professor, Department of
  Psychology
University of California at
  Los Angeles
Los Angeles, California

ROBERT M. NAKAMURA, Ph.D.
Associate Professor
Department of Obstetrics and
  Gynecology
The Women's Hospital
LAC/USC Medical Center
Los Angeles, California

J. O. NDINYA-ACHOLA, M.D.
Department of Microbiology
University of Nairobi
Nairobi, Kenya

MARGARET NICHOLS, Ph.D.
Executive Director, Institute for
  Personal Growth
New Brunswick, New Jersey
Project Director, Northern Brooklyn
  AIDS Partner Study
Brooklyn, New York

P. Piot, M.D.
Department of Microbiology
Institute of Tropical Medicine
Antwerp, Belgium

F. A. Plummer, M.D.
Department of Microbiology
University of Nairobi
Nairobi, Kenya

Malcolm Potts, M.B., B. Chir., Ph.D.
President, Family Health International
Research Triangle Park, North Carolina

Jeffrey Pudney, Ph.D.
Department of Biology
Boston University
Boston, Massachusetts

June Machover Reinisch, Ph.D.
Director, The Kinsey Institute for Research in Sex, Gender, and Reproduction
Professor, Departments of Psychology and Psychiatry
Indiana University
Bloomington, Indiana

Lionel Resnick, M.D.
Department of Dentistry
Mount Sinai Medical Center
Miami Beach, Florida

Santiago Rodriguez de Cordoba, Ph.D.
Centro de Investigaciones Biologicas
CSIC, Madrid, Spain

A. R. Ronald, M.D.
Institute of Tropical Medicine
Antwerp, Belgium

Pablo Rubinstein, M.D.
Director, The Fred Allen Laboratory of Immunogenetics of The New York Blood Center
New York, New York

Stephanie A. Sanders, Ph.D.
Assistant Director, The Kinsey Institute for Research in Sex, Gender and Reproduction
Indiana University
Bloomington, Indiana

Sara Silva
Hillcrest Biologicals, Inc.
Cypress, California

William Simon, Ph.D.
Department of Sociology
University of Houston
Houston, Texas

Cladd E. Stevens, M.D., M.P.H.
Senior Investigator
Head, Laboratory of Epidemiology
The Lindsley F. Kimball Research Institute of The New York Blood Center
New York, New York

The Rt. Rev. William E. Swing, D.D.
Diocese of California
San Francisco, California

Patricia E. Taylor, Ph.D.
Epidemiologist
The Lindsley F. Kimball Research Institute of The New York Blood Center
New York, New York

Contributors

BARBARA R. VISSCHER, M.D., Ph.D.
Professor, Department of
  Epidemiology
UCLA School of Public Health
Los Angeles, California

BRUCE VOELLER, Ph.D.
President, The Mariposa Education
  and Research Foundation
Topanga, California

CHRISTOPHER WEEKS
Hillcrest Biologicals, Inc.
Cypress, California

HANS WOLFF, M.D.
Fearing Research Laboratory
Brigham and Women's Hospital
Harvard Medical School
Boston, Massachusetts

GAIL ELIZABETH WYATT, Ph.D.
Associate Professor of Medical
  Psychology
Department of Psychiatry and
  Biobehavioral Sciences
University of California
Los Angeles, California

EDITH A. ZANG, Ph.D.
Consulting Biostatian
Laboratory of Epidemiology
The Lindsley F. Kimball
  Research Institute of
  The New York Blood Center
New York, New York

WENHAO ZHANG, M.D.
Fearing Research Laboratory
Brigham and Women's Hospital
Harvard Medical School
Boston, Massachusetts

MARY ZIEMBA-DAVIS
Graduate Assistant, The Kinsey
  Institute for Research in Sex,
  Gender, and Reproduction
Indiana University
Bloomington, Indiana

AIDS AND SEX

An Integrated Biomedical and Biobehavioral Approach to AIDS: An Introduction

Bruce Voeller, June Machover Reinisch, and Michael Gottlieb

As researchers active in the era of AIDS from the disease's first identification, we have had the opportunity to observe and participate in the
scientific community's growing awareness and active response to this
unprecedented epidemic. In the early days, it was difficult to generate
interest on the part of many legitimate scientists. Some understandably
resisted abandoning established lines of research; some hesitated because of the stigma associated with this specific disease, and others
because they thought that AIDS would be a brief episode—short lived
and quickly resolved—and would affect only a small number of people,
like the previous public health crises of legionnaires' disease and toxic
shock syndrome. The identification of the AIDS virus (human immunodeficiency virus type 1; HIV-1), an appreciation of its unique biological
behavior, its severe danger as a human pathogen, its worldwide impact,
and the numbers of people worldwide potentially and actually affected
by HIV-1 have changed these attitudes. Once some scientific successes
were achieved, the morale of the scientific community improved, and
this was enhanced considerably by the growing availability of research
funding.

While advances in immunology, epidemiology, and virology have
progressed quickly, there is a growing concern that bench science alone
cannot erase this epidemic. There is increasing appreciation that the
involvement of the behavioral sciences in collaboration with biomedical
research is essential to limit the disease's spread. It is only through the
integration of the "natural" and medical sciences with behavioral science
that we can arrive at new insights and strategies for controlling AIDS

3

and for conducting effective research that in turn will reduce suffering, loss of life, and the ruinous worldwide economic costs.

It is hoped, of course, that these strategies will permit continued expression of the most basic human feelings. There are moralistic forces in our society that would have us believe that human sexuality causes AIDS, and of course this is not true. It is important to distinguish between measures as strategies that encourage appropriate AIDS awareness and those that seek to limit sexual expression on religious or moral grounds.

Alfred C. Kinsey, the Institute's visionary founder, recognized that research on sexual phenomena demands multidisciplinary approaches, encompassing not only clinical medical science but also psychology, anthropology, and sociology, developmental physiology and genetics, the other behavioral sciences such as ethology, and related disciplines. Also, although he shaped new paradigms for research, Kinsey may be remembered even more for his leadership in uncovering and disseminating information he thought people must have to understand their society better—even when that information was controversial and, at times, unwelcome.

The editors of this volume believe that biomedical leaders at the cutting edge of the fight against AIDS and distinguished experts in human behavior and sexuality must collaborate if AIDS is to be controlled before the imminent end of the millennium. Too little reciprocal, interdisciplinary collaboration took place during the first half-decade of the AIDS crisis. Sex researchers had only a limited awareness of progress in AIDS research and planning. AIDS researchers were unfamiliar with the sexual practices and behaviors of the individuals and groups they studied and knew little of the research methodologies known to sex researchers.

To foster and encourage collaboration, The Kinsey Institute invited authorities from around the world and from a wide variety of disciplines to a 4-day conference. Designed to expand the participants' knowledge of research and scholarship in divergent fields relevant to the AIDS crisis, the conference's goals included a forum in which information would be shared, the nature and extent of the gaps in our knowledge would be revealed, and fuller awareness of the perspectives represented would be developed.

This book is the product of the interaction that occurred among the participants, and a notable consequence of this interaction has been the establishment of many lasting intellectual and collegial ties, which continue to evidence the "hybrid vigor" that derives from the interdisciplinary merging of open and creative minds sharing strong commitment to solving a common problem such as AIDS. Since the conference on AIDS and Sex, there has been continuing interaction among the participants, and, in turn, there has been an increase in interaction between behavioral and biomedical scientists in general. At recent international AIDS conferences, for example, there is an ever larger proportion of behavioral, cross-cultural, and subcultural data reported by participants than in earlier years.

The chapters in this volume, which resulted from the conference, contribute to this interdisciplinary effort. We believe they will be of value to all those working toward conquering AIDS and other sexually transmitted diseases (STDs). Because of the general limited awareness of either the breadth or of the limits of our scholarly knowledge of the biobehavioral aspects of human sexuality, this volume should serve a useful role. For this reason, we explore below in somewhat more detail how and why the need for such a conference and such a volume evolved.

Lack of Information on Human Sexuality

Researchers universally concur that AIDS is spread throughout the world chiefly as a result of sexual contact. Some sexual practices, such as anal intercourse and vaginal intercourse, carry markedly greater risk than others. In some regions or countries, contact with infected blood adds to the risk of infection (e.g., use of contaminated syringes, which includes intravenous drug users who share needles; use of transfusions or blood products that have not been properly screened out of distribution). Those who themselves have become infected through needle use or blood commonly then pass their infection on to others through sexual contact or *in utero*.

In the absence of a cure for AIDS or a vaccine to prevent it and, even when one or both are developed, in the absence of distribution routes and the resources to universally supply them, the central role of human sexuality in the dispersal of AIDS unavoidably forces us to focus on altering sexual behavior in order to control the epidemic. To accomplish this there is a need for solidly grounded knowledge of the diverse sexual practices found in the many cultures throughout the world, including our own. Detailed and comprehensive knowledge of sexual behavior is an essential prerequisite to predicting the temporal and geospatial patterns of the disease's migration and to controlling its spread effectively.

The ease of travel nowadays and the massive migration of innumerable, long-isolated populations from rural to dense urban environments threaten to transform previously localized and unknown, or little noted, endemic diseases into worldwide epidemics. Overpopulation, malnutrition, lack of medical measures, and disease, of course, add to the risk. The wildfire dispersal of AIDS that we have seen in the 1980s may well be a bellwether of future sexually transmitted scourges that are yet to surface.

For AIDS, and for subsequent such diseases, lack of basic, accurate, and reliable data describing human sexual behavior can only generate flawed research, detours into costly and unproductive cul-de-sacs, unreliable national and global planning, and tragic errors in prioritization and allocation of precious and limited resources.

In Western industrialized society, we believe ourselves to be the best informed and most enlightened peoples in human history. Yet of those things that influence, regulate, and dominate our lives in major ways, human sexuality is one of the most powerful, and the least illuminated. We are a nation beleaguered with AIDS, rampant STDs, and millions of unwanted teenage pregnancies on a planet suffering from worldwide starvation and possibly irreparable environmental damage, all ultimately linked to massive overpopulation—a problem intimately linked to sexual behavior. Historically and contemporaneously, we often censure those who undertake scholarly or scientific exploration of this basic aspect of human existence through denial of academic promotion, access to public or private research funds, and respect in the community of scholars.

To be sure, textbooks detail the anatomy and the embryonic organogenesis and pubertal transformation of the genitalia; the role of sex chromosomes in determining whether we are male or female; and the biochemistry, endocrinology, and physiology regulating gamete (sperm and egg) production and fertilization. We even know that the human female neonate already carries all the eggs she will ovulate during her entire adult life, whereas the male continues to generate his gametes throughout his reproductive life. Such laboratory findings are of great importance. Often, however, these reproductive data are all that is available, and a broader context that includes biobehavioral information about sexuality is lacking. We do not know accurately how many heterosexual women engage in anal intercourse, or what portion of them have bisexual male partners. We do not even know very precisely how many Americans engage only in homosexual behavior, and how many are behaviorally bisexual—essential statistics if models for predicting the future extent and directions of the spread of AIDS are to be useful. We do not know with any precision how many homosexual men engage in anal intercourse, or what predicts being the insertive or receptive partner. We are very ignorant about the sexual practices of our various subcultures, whether among diverse black, white, Asian, Native American, or Latino groups or among individuals of different age, educational, and socioeconomic groups.

We have few data regarding the sexual practices of any culture. This was recently illustrated when the National Institute of Medicine of the National Academy of Science relied on the original Kinsey data—data published as long ago as 1948—for their 1986 report on the AIDS crisis. Most of the Kinsey data, including sexual histories on over 17,000 individuals, were collected nearly half a century ago (Kinsey, Pomeroy,

& Martin, 1948; Kinsey, Pomeroy, Martin, & Gebhard, 1953). Nevertheless, their relevance to the present epidemic has been considerable. The other scientific data that do exist to guide AIDS researchers and planners are, in general, scattered, of varied reliability, and known only to limited circles of experts, nearly all of whom were outside the AIDS research domain. Much more is needed to respond to the present urgent needs.

Since the Kinsey data were collected, the impact of many intense social, cultural, and scientific changes on sexual behavior has not been adequately measured. Several wars necessitated nearly universal military training that threw together men from highly varied backgrounds under circumstances of high stress and male bravado. Under these conditions, sexual release was highly competitive and STDs became the leading cause of absence from duty in the armed forces. The Women's Movement and the Gay Movement conducted vigorous media campaigns to change gender role perceptions and behaviors and to open public discussion of the existence of varied sexual roles and practices. Widespread access to more effective contraceptive measures and "miracle" drugs to cure or prevent some of the worst STDs decreased the fear of unwanted pregnancy and venereal disease. Recently, former Surgeon General C. Everett Koop successfully campaigned to place condoms into polite conversation as well as into the national media.

Thus, although an undeniably important sexual data base exists, the need for more current information generated from well-designed research is quite evident. We simply possess considerably less information on sexuality than is necessary for responding to the AIDS crisis as well as the other sex-related public health problems that must be solved.

Political pressures still affect the conduct of sex research and interfere with the building of a current scientific data base on the sexual behavior of people in the United States and around the world. For example, the respected British medical publication *The Lancet* reported in September 1989 that

> A £750,000 study of the sexual behaviour of 20,000 people in Britain has been blocked by the Prime Minister on grounds that it would be an invasion of privacy. The study . . . was expected to provide valuable information about the spread of HIV infection. 1000 people have already taken part in a pilot study. ("Sexual Behavior," 1989, p. 696).

Need for Interdisciplinary Communication

Much of the information we do have about human sexuality is unknown to most AIDS researchers. Moreover, the credentials of biomedical researchers rarely include training in the biobehavioral aspects of sexuality, in taking sex histories free of the interviewer's own ethnocentrism and inhibitions, or in techniques for encouraging patients or subjects to reveal secrets of this most taboo aspect of private life. Special training is necessary because many aspects of sexuality or drug use (both of exceptional importance to the AIDS crisis) carry public or peer opprobrium, or contravene laws. Many researchers naively assume that an acknowledged intravenous drug user, or homosexual, will answer any question related to that status with candor. This is generally untrue unless the appropriate environment of nonjudgmental interest and safety is provided.

Obliviousness to these gaps in knowledge and training is even more destructive than the gaps themselves. Most researchers will remedy deficiencies once they are perceived. In sexuality the perception is often slow to come. Ignorance of these omissions in training is unwittingly fostered by our educational institutions, which leave researchers and health care providers with the erroneous impression that their training has equipped them to evaluate all the important health-related areas of human activity and understanding.

In fact, there are only a few programs devoted to sexuality training, and those usually focus on clinical practice of sex therapy and sex education. Although the public, legislators, courts, and the media commonly turn to physicians and the clergy for information and counsel about sexuality, medical schools and seminaries do not require and rarely offer courses in human sexuality, other than those relating to urology, obstetrics and gynecology, or religious dogma. Physicians are not even taught to take proper sex histories and, as noted in several chapters in this volume, are commonly embarrassed to inquire into a patient's sexuality or even to answer the patient's questions. Physicians at the AIDS and Sex conference repeatedly noted how ill-prepared their training had left them for handling the sexual issues at the core of the AIDS epidemic.

Many of us who have participated on the federal panels that have set AIDS policy and priorities and allocated the resources available for AIDS research, education, and health have noted both the limitations of sound information about human sexuality that could be instructive and the limited awareness of many of our colleagues of their own meager scientific knowledge of sexual behavior. In fact, although nearly everyone considers himself or herself to some degree an authority on sexuality, few are. Recasting street beliefs into polite language does not constitute a scholarly foundation for discussion or action.

As a participant in a 1985 federal council on AIDS, one of us witnessed a noted medical authority on hemophilia express disdain and shock when it was suggested that some hemophiliacs might have contracted HIV infection through homosexual activity or drug addition rather than, or in addition to, administration of virally contaminated clotting factors. Our presence allowed confirmation that some hemophiliacs are homosexual. Too often, however, conferences dealing with AIDS have tended to draw members from within one or another discipline, depriving the conferees of the broadening experience of hearing important concepts and knowledge from other fields that bear on understanding the transmission of AIDS, its treatment, and the social, ethical, and political components of controlling this epidemic.

Few governmental AIDS committees included or recruited experts in behavior or sex research. Even the Presidential AIDS Commission put little emphasis on information from the field of sex research. It heard testimony only late in its many months of hearings, and then only in response to a letter from Drs. William Masters, Virginia Johnson, June Reinisch, and Bruce Voeller, who suggested input from the field of sex research. Not only in such committees has a lack of training or knowledge been a source of delay or misadventure in conquering AIDS. One group of researchers at a medical school, for example, studying the possible role of semen antibodies in AIDS, recruited homosexual couples who, when interviewed together as a couple, denied sexual contact outside their relationship, a critically important characteristic needed for the study. In fact, a significant number of the couples eventually proved not to be sexually exclusive, but the researchers had already published misleading results that have since been refuted. Had the subjects been separately and confidentially interviewed by trained sex researchers, sound research data might have been obtained (and the researchers saved from embarrassment). In another case, government researchers published data indicating that United States armed forces personnel infected with HIV-1 had caught the virus from prostitutes, triggering calls for increasing campaigns against prostitution. When infected soldiers were interviewed by nonmilitary researchers, whom they trusted, it became clear that nearly all had been infected through intravenous drug use or homosexual contact, acts for which they could be expelled from the armed services, which prevented them from being candid with the original military researchers. In each of these flawed published studies, researchers, journal editors, and peer reviewers failed to correct mistakes that should have been recognized.

Similarly, as discussed in one chapter in this volume, the potential role of heterosexual anal intercourse in the spread of AIDS has been discounted because few physicians are willing to discuss anal intercourse with their patients, regarding the inquiry as offensive and the practice as homosexual. Much evidence indicates that receptive anal intercourse (having a penis placed into the rectum) is the sexual act carrying the highest risk of AIDS infection (in the Western industrialized nations) and that anal intercourse may be practiced by a much larger absolute population of heterosexuals than of homosexuals. Such a viewpoint is discomforting to many researchers. All of us sometimes forget that we ourselves are members of one of the planet's many cultures, and that we, too, are selective in our perceptions and judgmental in our understanding because of this. Yet we do not have the luxury of allowing our embarrassment or biases to blind us in collecting full sexual information or in assessing its implications for the spread of AIDS. In addition, we will chill the rapport with our patients and subjects, and misjudge our results, if we remain oblivious to the prejudice and judgmentalism in words or phrases such as "promiscuous" and "admitted to" when used to describe anal intercourse or oral sex.

Final Comments

In short, we lack precious information for much of our planning and much of our theoretical and research thinking. This is dramatically true even in the United States, where there has been more research on human sexuality than in almost any other quarter of the world. We know precious little about the sexual practices, habits, mores, and ways to approach people in our own culture. We know significantly less about people from other subcultural groups in this country. In fact, what little research-based sex data we have is largely limited to the educated white middle class.

Finally, a word about how this volume was edited. Each author drafted a preliminary chapter for presentation at the conference. Following the several days of interaction with colleagues at the meeting, each author was asked to revise and edit his or her manuscript to reflect the experience and deeper perspectives gained through participating in this interdisciplinary conference. Although we in turn have made extensive editorial suggestions to each of the authors represented in this volume, we have been careful about pressing our own notions on them, permitting them instead to express themselves in their own manner. Because the participants came from diverse parts of the world and diverse cultures, we were particularly anxious not to have our perceptions unduly color or filter the points made by the authors, or the manner in which these views were expressed. As members of one particular culture, we editors also are limited by that culture's organization of the universe and its modes of thinking.

References

Kinsey, A. C., Pomeroy, W. B., & Martin, C. E. (1948). *Sexual behavior in the human male*. Philadelphia: W. B. Saunders Company.
Kinsey, A. C., Pomeroy, W. B., Martin, C. E., & Gebhard, P. H. (1953). *Sexual behavior in the human female*. Philadelphia: W. B. Saunders Company.
Sexual behaviour survey banned. (1989). *The Lancet, 2*, 696.

Part I
AIDS RESEARCH AND THE DEMAND FOR INFORMATION ON HUMAN SEXUALITY

1

Sexual Activity, Condom Use, and HIV-1 Seroconversion

Roger Detels, Barbara R. Visscher, Lisa P. Jacobson, Lawrence A. Kingsley, Joan S. Chmiel, Lois J. Eldred, Patricia English, and Harold Ginzburg

The highest risk factor for transmission of the human immunodeficiency virus (HIV-1) among homosexual men has been shown to be receptive anal-genital intercourse. The degree to which other sexual practices are associated with transmission of HIV-1 has not yet been established. The likelihood of success in controlling AIDS through public health education activities would be greatly enhanced if other sexual activities are not associated with a high risk of transmission of HIV-1.

Former Surgeon General C. Everett Koop has advocated the use of condoms for individuals who cannot or will not abstain from high-risk activities. The Food and Drug Administration, and at least one investigator, however, have reported condom failures in *in vitro* testing and in one small cohort study of hemophiliacs ("Condom Recalls," 1987; Fischl, et al., 1987). The clinical study suggests that the use of condoms may reduce but does not eliminate the risk of transmission of HIV-1 (Fischl et al., 1987).

In 1983, the Multicenter AIDS Cohort Study (MACS) was established with the support of the National Institute of Allergy and Infectious Diseases at the Johns Hopkins University School of Hygiene and Public Health in Baltimore, the Northwestern University Medical School/Howard Brown Memorial Clinic in Chicago, the UCLA Schools of Public Health and Medicine in Los Angeles, and the University of Pittsburgh School of Public Health in Pittsburgh (Kaslow et al., 1987; Kingsley et al., 1987). We report herein the relationship of specific sexual activities and condom use to the HIV-1 seroconversion rate over the first 2 years of follow-up of the 2,915 men who were seronegative at first visit to the MACS.

Methods

Homosexual and bisexual men were recruited into cohorts in the four centers of the MACS in late 1984 and early 1985. The volunteers were recruited through the use of service announcements in the media, the cooperation of gay organizations, public presentations by staff, posters at gay meeting places, and personal recruitment by participants and staff.

The baseline interview schedule, physical examination, and specimen collections were completed over a 12-month period between April 1984 and March 1985. Neither the interviewer nor the participant was aware of the current status of the participant at the time of the interview. Follow-up visits were scheduled at 6-month intervals from the date of the baseline visit. The interview schedule was developed with the assistance of experts in the field of sociology and in administering interviews for sexual activities. Drafts of the interview schedule were pilot tested and appropriate revisions made before the interview schedule was finalized. The final interview schedule included questions on demographic characteristics of the respondents, history of major illnesses and sexually transmitted diseases, history of symptoms characteristic of AIDS, use of recreational drugs, and type and frequency of specific sexual activities. In addition, the following question was asked regarding the use of condoms: "With how many of your partners did you use a condom?" We did not ask questions regarding the frequency of breakage, leakage, or slippage and whether the individual used condoms in every encounter with each partner.

A brief screening examination was performed to identify signs and symptoms of AIDS. Blood specimens were collected for determination of antibodies to HIV-1 and other organisms and for the levels of both CD-4 and CD-8 cells.

The presence of antibody to HIV-1 was determined by enzyme-linked immunosorbent assay (ELISA) either by the E. I. Du Pont de Nemours Company or by the individual center's laboratory using the Genetic Systems kits. All ELISA-positive antibody tests were confirmed by use of an immunoblot technique that was performed on sera from the first ELISA-positive visit and the immediately preceding visit. A score of 0 was assigned for a negative band, 1 for a weakly reactive band, 2 for a moderately reactive band, and 3 for a band strongly reactive to proteins p15, p24, p31, p41, p45, p53, p55, p64, or p120. The scores of all bands were summed. A value of greater than or equal to 3 was defined as a positive test.

The interview schedule, physical examination, specimen collection, and antibody determination were repeated at approximately 6-month intervals. The interval between two successive visits by one participant was designated a "person interval." A participant was assigned to a specific exposure category for each person interval acording to his specific sex practices or condom use in the preceding two visit intervals. The mean duration of the intervals ranged from 199 days between visits 1 and 2 to 173 days between visits 4 and 5. Details of the methodology have been previously reported (Chmiel, Detels, Kaslow, et al., 1987; Kaslow, et al., 1987; Kingsley et al., 1987).

Results

Eight percent (232) of the 2,915 men who had no evidence of antibody at baseline developed antibodies within the subsequent 24 months. The annualized incidence of HIV-1 seroconversion fell from 9.2% per year between visits 1 and 2 to 1.8% per year between visits 4 and 5.

The proportion of men who seroconverted differed greatly according to the specific sexual activities in which they reported engaging during the 12 months prior to the first antibody-positive visit. The rate of seroconversion among men practicing specific sexual activities was calculated by determining the proportion of men seronegative at the beginning of each visit interval who developed antibody by the end of the 6-month visit interval, according to the specific sexual activities in which they had engaged for the two previous visit intervals.

The sexual activities were ranked according to a specific hierarchy. The highest ranking was both receptive and insertive anal-genital intercourse, second was receptive but not insertive anal-genital intercourse, third insertive but not receptive anal-genital intercourse, and the lowest no anal-genital intercourse. In this hierarchical system, individuals who are listed in the categories of both receptive and insertive, insertive only, or receptive only may also have been engaging in various other sexual activities.

Of the 232 seronegative men who seroconverted, 191 practiced both receptive and insertive anal-genital intercourse within the previous two intervals. Only 1 of the 232 seroconverters had not engaged in anal-genital intercourse in the 6 to 12 months prior to the appearance of antibody. This gentleman had reported receptive anal-genital intercourse and swallowing of semen 24 months before the appearance of antibodies.

If we assume that the risk associated with no anal-genital intercourse is 1, then the increase in risk associated with each of the other anal-genital practices can be expressed as the risk ratio, as indicated in Table 1-1. The risk ratio is 10 for men who practiced only insertive anal-genital

Table 1-1
Risk Ratio for HIV-1 Seroconversion by Reported Anal-Genital Activity in Previous 12 Months

Anal-Genital Activity	Number Seroconverting	Estimated Incidence[a]	Risk Ratio
No anal-genital intercourse	1	.0005	1
Insertive anal-genital *only*	20	.005	10
Receptive anal-genital *only*	14	.018	36
Both receptive and insertive	191	.038	76

[a] Per six-month interval.

intercourse, 36 for men who practiced only receptive anal-genital intercourse, and 76 for men who practiced both insertive and receptive anal-genital intercourse. The seroconversion ratio for men who practiced receptive only compared to insertive only intercourse was 3.6. The insertive only group, however, had the highest mean number of different partners in the previous two visit intervals, even though it had the lowest seroconversion rate among those practicing some form of anal intercourse.

The low rate of seroconversion among the men *not* engaging in anal-genital intercourse was not due to abstinence from sexual activities. Most of the men in this group reported participating in one or more other sexual activities. Among the men not seroconverting who did not practice anal-genital intercourse, there was a total of 170 person intervals of exposure to receptive oral-genital intercourse with the introduction of ejaculate in the oral cavity.

The HIV-1 seroconversion rate was calculated among men practicing anal-genital intercourse by their reported use of condoms (Table 1-2). Seven of the men using condoms with no partners, 111 of the men using condoms with some partners, and 97 of the men using condoms with no partners seroconverted during the four 6-month follow-up visit intervals. For those men practicing insertive anal-genital intercourse, this referred to their personal use of the condom. For those men engaging in receptive anal-genital intercourse, the condom use referred to their insertive partner. If we consider the risk among men using condoms with all their partners as 1, the risk ratio among men using condoms with only some of their partners was 6 and for men using condoms with none of their partners was 3. This comparison is confounded to some extent by the fact that many of the men using condoms with none of their partners were monogamous and a greater proportion of men using condoms with only some of their partners had multiple partners.

The risk ratio rose with increasing number of different partners in the

Table 1-2
**Risk Ratio for HIV-1 Seroconversion by Reported Condom Use
in Previous 12 Months**

Condom Use[a]	Number Seroconverting	Estimated Incidence[b]	Risk Ratio
All partners used condoms			
Total	7	.007	1[c]
>9 partners	1	.004	1[d]
Some partners used condoms			
Total	111	.043	6
>9 partners	79	.05	13
No partners used condoms			
Total	97	.024	3
>9 partners	65	.038	10

[a] Among those practicing anal-genital intercourse.
[b] Per 6-month person interval.
[c] Reference category for total group.
[d] Reference category for >9 partners.

men using condoms with some or none of their partners. The risk ratio
for men using condoms with none of their partners who reported more
than eight partners was 10.

Discussion

This study reconfirms the association of receptive anal-genital intercourse with the highest risk of HIV-1 infection among homosexual/bi-
sexual men. (See also Chapter 2.) The study, however, also demonstrates a risk of transmission to the insertive partner in anal-genital
intercourse as well. It is very difficult to evaluate the significance of the
observation that only 1 of the 234 men who seroconverted had not practiced anal-genital intercourse within more than 24 months of the appearance of antibody to HIV-1. Documentation of abstinence from anal-
genital intercourse is, of course, dependent on the veracity of the participant. In this instance, the man was reinterviewed twice by a senior
member of our staff who believes that he is, in fact, telling the truth.
Interviewing for sexual activities is very difficult since many individuals
are reluctant to admit to unusual sexual practices. This man, however,
had admitted to both insertive and receptive anal-genital intercourse at
the baseline visit and, therefore, was not likely to have been reluctant
about discussing specific sexual activities. Futhermore, he readily admitted to practicing oral-genital intercourse during the entire period of
follow-up. If we assume that the participant is telling the truth, then

we must either assume that some transmission occurs via the oral-genital route or that the induction period between exposure to the virus and development of antibodies was approximately 2 years. This is beyond the usual interval between exposure and development of antibodies, although three recent reports suggested that in some individuals the interval between time of exposure and development of antibodies may occasionally exceed 12 months (Imagawa et al., 1989; Ranki et al., 1987; Wolinsky et al., 1989).

Because we did not ask about condom failure, slippage, and leakage or whether the men used condoms 100% of the time with each of their partners, a bias was present in favor of showing a low efficacy of condoms in prevention of HIV-1 infection. The observation of a significant difference in the risk ratio among men who reported using condoms with all their partners as compared to those reporting use either with none of their partners or with only some of their partners suggests that condoms as used during the period of follow-up by the MACS participants did provide significant, although *not* complete, protection against infection with HIV-1. We were not able to determine whether the seven failures among the men reporting condom use with all their partners was due to condom failure, improper usage, occasional nonuse, or other types of sexual activity.

This paper underscores the fact that transmission of HIV-1 is associated with specific sexual activities, not with sexual orientation. Anal-genital intercourse is not an exclusively homosexual practice (Bolling & Voeller, 1987; Melbye et al., 1985; Chapter 3, Chapter 19). Thus, the current classification scheme for reporting AIDS cases by sexual orientation does not actually identify the activities that are associated with transmission. Reporting of cases by presumed activity causing infection with HIV-1 (e.g., anal-genital intercourse, intravenous drug use, transfusion) may be helpful in guiding the public to avoid those specific activities that are associated with transmission of HIV-1.

Acknowledgments

This work was conducted by the Multicenter AIDS Cohort Study, including Johns Hopkins School of Hygiene and Public Health, Baltimore, MD (Data Coordinating Center); Howard Brown Memorial Clinic/Northwestern University Medical School, and Northwestern University Cancer Center, Chicago, IL; UCLA Schools of Medicine and Public Health and the Jonsson Comprehensive Cancer Center, Los Angeles, CA; University of Pittsburgh School of Hygiene and Public Health, Pittsburgh, PA; and the National Institutes of Health, National Institute of Allergy and Infectious Diseases, and the National Cancer Institute, Bethesda, MD. It was supported by grants from the National Institutes of Health (AI-72631).

References

Bolling, D. R., & Voeller, B. (1987). Anal intercourse among heterosexuals. *Journal of the American Medical Association, 258*, 474.

Chmiel, J., Detels, R., Kaslow, R., Van Raden, M., Kingsley, L. A., & Brookmeyer, R. (1987). Factors associated with prevalent Human Immunodeficiency Virus (HIV) infection in the Multicenter AIDS Cohort Study. *American Journal of Epidemiology, 126*(4), 568–577.

Condom recalls reported after inspections by FDA. (1987, June 19). *Wall Street Journal.*

Fischl, M. A., Dickinson, G. M., Scott, G. B., Klimas, N., Fletcher, M. A., & Parks, W. (1987). Evaluation of heterosexual partners, children, and household contacts of adults with AIDS. *Journal of the American Medical Association, 257*, 640–644.

Imagawa, D. T., Lee, M. H., Wolinsky, S. M., Sano, K., Morales, F., Kwok, S., Sninsky, J., Nishanian, P., Giorgi, J., Fahey, J., Dudley, J., Visscher, B., & Detels, R. (1989). Long latency of human immunodeficiency virus-1 in seronegative high risk homosexual men determined by prospective virus isolation and DNA amplification studies. *New England Journal of Medicine, 321*, 1458–1462.

Kaslow, R., Ostrow, D., Detels, R., Phair, J. P., Polk, B. F., & Rinaldo, C. R. (1987). The Multicenter AIDS Cohort Study: Rationale, organization and selected characteristics of the participants. *American Journal of Epidemiology, 126*(2), 310–318.

Kingsley, L., Detels, R., Kaslow, R., Polk, B. F., Rinaldo, C. R., Chmiel, J., Detre, K., Kelsey, S. F., Odaka, N., & Ostrow, D. (1987). Risk factors for seroconversion to Human Immunodeficiency Virus among male homosexuals. *Lancet, i*, 345–348.

Melbye, M., Ingerslev, J., Biggar, R. J., Alexander, S., Sarin, P., Goedert, J., Zachariae, E., Ebbsen, P., & Stebjerg, S. (1985). Anal intercourse as a possible risk factor in heterosexual transmission of HTLV-III to spouses of hemophiliacs. *New England Journal of Medicine, 312*, 857.

Ranki, A., Valle, S.-L., Krohn, M., Antonen, J., Allain, J. P., Leuther, M., Franchini, G., & Krohn, K. (1987). Long latency precedes overt seroconversion in sexually transmitted Human Immunodeficiency Virus infection. *Lancet, ii*, 589–593.

Wolinsky, S. M., Rinaldo, C. R., Kwok, S., Sninsky, J. J., Gupta, P., Imagawa, D., Farzadegan, H., Jacobson, L. P., Grovit, K. S., Lee, M. H., Chmiel, J. S., Ginzburg, H., Kaslow, R. A., & Phair, J. P. (1989). Human immunodeficiency virus type 1 (HIV-1) infection a median of 18 months before a diagnostic Western blot. Evidence from a cohort of homosexual men. *Annals of Internal Medicine 111*, 961–972.

2

Sexual Activity and Human Immunodeficiency Virus Type 1 Infection in a Cohort of Homosexual Men in New York City

Cladd E. Stevens, Patricia E. Taylor, Santiago Rodriguez de Cordoba, Edith A. Zang, and Pablo Rubinstein

Homosexually active men have been the group most affected in the United States by AIDS and by infection with the etiological agent of AIDS, human immunodeficiency virus type 1 (HIV-1) ("Follow-up On," 1981; Friedman-Kein, 1981; Gottlieb et al., 1981; Hymes et al., 1981; "Kaposi's Sarcoma," 1981; Masur et al., 1981; "Persistent, Generalized," 1982; "*Pneumocystis* pneumonia," 1981; Siegal et al., 1981). New York City has been a major focus of AIDS since the epidemic began, with 25% of cases reported to the Centers for Disease Control by November 1987 ("Human Immunodeficiency," 1987). Early in 1984 we began a prospective study of AIDS in a group of homosexually active men in New York City, most of whom had participated in the late 1970s in studies of hepatitis B virus (HBV) infection. The advantage of using this population was the availability of information and stored serum samples from this earlier period. We report herein data from this cohort on sexual activity reported at the time of entry into the project in 1984 and during subsequent follow-up and the relationship of this activity to serological evidence of HIV-1 infection.

Methods

Men from the New York City area were recruited into our prospective study of AIDS if they were homosexually active and had not previously been diagnosed as having an AIDS-related disease. There were no other eligibility criteria. The presence of nonspecific symptoms, such as lymphadenopathy, did not exclude a volunteer from participating. Partic-

Table 2-1
Characteristics of Homosexually Active Men in 1978 and 1979[a]

	1984 AIDS Study		
Characteristics in 1978–1979	Did Not Join (N = 3,679)	Joined (N = 714)	p Value
White	79.6	89.1	<.0001
≥30 years old	45.6	59.7	<.0001
≥10 years homosexual activity	49.8	61.1	<.0001
Syphilis/gonorrhea history	57.3	63.4	<.01
≥10 male sex partners/6 months	54.0	64.2	<.0001

[a] Includes men who joined or did not join the prospective study of AIDS in 1984.

ipants were recruited predominantly from among men who had participated in studies begun in the late 1970s of the epidemiology of HBV infection and a hepatitis B vaccine efficacy trial (Szmuness et al., 1975, 1980, 1981). All such men for whom we had names and addresses were mailed an announcement about the study and invited to join. Those who volunteered and were eligible were enrolled over a 4-month period from mid-February through June 1984. Table 2-1 compares data obtained during the studies in 1978 and 1979 on characteristics of men who were invited to participate but did not join the project in 1984 with 714 men who did join. Volunteers were more likely to come from "white" ethnic groups and to be older and had a longer duration of homosexual activity than nonparticipants. Volunteers were also more likely to have had a history of syphilis or gonorrhea and 10 or more male sex partners in the preceding 6 months.

At the time of enrollment, each participant completed a self-administered questionnaire. The questionnaire was reviewed immediately by project staff for completeness and consistency. Questions were asked regarding the participant's medical history, recreational drug use, and sexual activity. Quantitative data on specific sexual practices were obtained (Table 2-2). A steady sex partner was defined as a man with whom the participant had had sex at least four times in a 2-month period. By the time of enrollment in 1984, the AIDS epidemic had been in existence for at least 5 years and public health authorities had already begun to advise homosexual men on safer sex practices. In anticipation that men may have been exposed to the etiological agent of AIDS at various points in time during the epidemic and that sexual practices had probably changed for many men, we asked questions about sexual activity for three time periods: the 6-month period prior to entry into the study, the period just before the participant became aware of AIDS, and the period of peak activity. While there was an overlap in these periods for some

Table 2-2
History of Sexual Activity Obtained on
Enrollment

In 6 months
   Number of steady male partners
   Number of nonsteady male partners
Frequency in a typical month of
   Kissing
   Masturbation
   Oral-genital sex
     Insertive
     Receptive ⎱ Condom use
             ⎰ Ejaculation
             ⎰ Swallowing semen

   Rectal intercourse
     Insertive
     Receptive ⎱ Condom use
             ⎰ Ejaculation
             ⎰ Rectal bleeding
             ⎰ Douching
             ⎰ Lubricants
   Oral-rectal sex
     Insertive
     Receptive

men, the period of peak activity for most men was prior to 1980 and the time when most became aware of AIDS was in 1981 or 1982. Physical examinations were done to detect evidence of AIDS or AIDS-associated signs. Blood specimens were drawn for a variety of serological and immunological studies. Residual serum and peripheral blood mononuclear cells were stored frozen for future testing as needed.

After enrollment, participants were followed at 4½-month intervals. At each follow-up visit, they completed self-administered questionnaires on interim events, including sexual activity, and had physical examinations and repeat blood sampling. At the end of the third year of the study, questions related to sexual activity were revised to obtain separate information on practices for sex with steady male partners and for sex with nonsteady male partners.

When the etiological agent of AIDS was identified and serological methods developed to detect antibody to the virus (anti-HIV-1), we obtained specific consent for anti-HIV-1 testing. Of the original participants, 781 (91.9%) consented for testing. Most of those who did not consent had already been lost to follow-up. A few who remained in the study refused permission for testing. Anti-HIV-1 testing was initially performed in collaboration with Dr. Robert Gallo at the National Cancer Institute (Stevens et al., 1986). Subsequently, screening for anti-HIV-1 was accomplished by enzyme-linked immunosorbent assay (ELISA) using commercially available test kits with seropositives defined according to kit instructions (E. I. du Pont de Nemours, Wilmington, DE). Western blot assay was used for confirmation of anti-HIV-1 positivity (E. I. du Pont de Nemours, DE). An individual was considered to be confirmed as positive if reactivity was detected to HIV-1 core protein, specifically p24, and to envelope glycoproteins (gp41, gp120, or gp160) on two or more blood specimens. For seroconverters, the first positive blood specimen was taken as the one that first had detectable antibody, whether this was for antibody to p24 or to envelope glycoproteins detected only on Western blot or for antibody detected by both ELISA and Western blot. Antibody to hepatitis A virus (HAV) was detected by a radioimmunoassay inhibition procedure (HAVAB; Abbott Laboratories, Chicago, IL).

In this report, analyses of sexual activity prior to and after enrollment were limited to men who have been tested for anti-HIV-1. For data that were skewed to the right, means were calculated from log-transformed values giving geometric mean values. Comparisons of differences between means were made by Student's t test. Comparisons of categorical variables or continuous variables grouped categorically were made using contingency tables and the χ^2 statistic. The strength of associations for combinations of variables was determined by stepwise, multiple logistic regression (Cox, 1970). The annual incidences of infection with HIV-1 or HAV prior to enrollment among men from the hepatitis B vaccine efficacy trial were calculated as the percentage seroconverters per year of the trial among men who were susceptible at the start of the year. For 1979, the rate was adjusted for the numbers of months the men were under study, since not all were followed from the first of the year. The incidence of HIV-1 infections during follow-up among men who were seronegative at enrollment was calculated by life-table methods, with comparisons between life-table attack rates made by Cox regression analysis (Cox, 1970).

Results

Among the 781 men tested for anti-HIV-1, 339 (43.4%) were positive at the time of enrollment in 1984 (Table 2-3). Anti-HIV-1 prevalence was significantly lower among men who had participated in the hepatitis B vaccine efficacy trial than among participants recruited from other sources, probably reflecting the lower levels of reported sexual activity among trial participants than among the other men (Szmuness et al., 1980). Among hepatitis B vaccine trial participants, anti-HIV-1 preva-

Table 2-3
Anti-HIV-1 Prevalence at Enrollment into 1984 Study by Study Subgroup

| | | | Anti-HIV-1 Positive | |
Subgroup	Total Number	Number Tested	Number	Percentage Tested
Baseline survey	389	346	172	49.7
Hepatitis B (HB) vaccine trial	325	307	102	33.2[a]
Special HB vaccine study	28	26	12	46.2
New recruits	108	102	53	52.0
Total	850	781	339	43.4

[a] HB vaccine trial versus survey, $\chi^2 = 18.2$, $p < .0001$.

lence in 1984 did not differ significantly between men who had originally been randomly assigned to the vaccine group and those assigned to the placebo group (52/151 or 34.4% and 50/156 or 32.1%, respectively).

Anti-HIV-1 prevalence at enrollment decreased with increasing age, was more common among black and Hispanic men than among white men, and was associated with a history of use of one or more of some 21 recreational and intravenous drugs. (Table 2-4 includes data on the three most frequently used drugs.) When multivariate analysis was done on these variables with variables on sexual activity, only intravenous drug use was predictive of anti-HIV-1 positivity independent of reported sexual activity.

Analyses of sexual activity focused primarily on practices that might relate to the risk of HIV-1 infection. Anti-HIV-1 prevalence in 1984 correlated with the reported frequency of each sexual activity during any of the three time periods queried about at enrollment. Since frequencies of different sexual practices correlated highly with each other, multivariate analysis was done on each time period to identify those that were independently predictive. Frequency of receptive rectal intercourse at the periods prior to awareness of AIDS and at entry was the only variable that was consistently predictive of anti-HIV-1 positivity. For the period just prior to entry, numbers of sex partners also correlated with seropositivity. Table 2-5 shows the relationship between anti-HIV-1 prevalence at enrollment and receptive rectal intercourse reported for the three periods. Men who reported having receptive rectal intercourse in all three periods had the highest anti-HIV-1 prevalence. Table 2-5 also illustrates the fact that a relatively large proportion of men who had been rectally receptive during their peak period of activity subsequently stopped this practice, especially after they became aware of AIDS. Anti-HIV-1 prevalence increased with increasing numbers of sex partners

Table 2-4
**Anti-HIV-1 Prevalence at Enrollment in 1984 by Age, Ethnicity, and
Recreational Drug Use**

Participant Characteristic	Number Tested	Percentage Anti-HIV-1 Positive	p Value
Age			
<35	163	54.6	
35–39	242	45.9	<.002
≥40	376	37.0	
Ethnic group			
White	704	41.8	<.05
Black or Hispanic	66	59.1	
Recreational drug use			
Marijuana			
No	34	5.9	<.0001
Yes	745	45.1	
Cocaine			
No	210	21.9	<.0001
Yes	562	51.2	
Nitrites			
No	70	14.3	<.0001
Yes	708	46.5	
Intravenous drug use			
No	696	41.4	<.0001
Yes	55	69.1	

Table 2-5
**Correlation between Receptive Rectal Intercourse during Prior Periods of
Sexual Activity and Anti-HIV-1 Prevalence at Enrollment in 1984**

Receptive Rectal Intercourse by Period of Sexual Activity[a]			Number Tested	Percentage Anti-HIV-1 Positive
Peak	Preawareness of AIDS	6 Months Preenrollment		
No	No	No	120	11.8
Yes	No	No	57	21.9
Yes	Yes	No	217	45.9
Yes	Yes	Yes	312	59.3

[a] Forty-two men had other combinations.

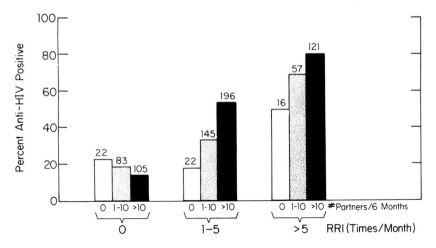

Figure 2-1. Relationship of anti-HIV prevalence on entry with receptive rectal intercourse (RRI) and numbers of nonsteady male sex partners before awareness of AIDS. #Partners indicates the number of nonsteady male sex partners.

only among men who also practiced receptive rectal intercourse (Fig. 2-1).

As predicted, the frequency of various sexual practices reported decreased from the time of peak activity to the time of enrollment, especially after men became aware of AIDS (Tables 2-5 and 2-6). Table 2-6 shows these changes for the frequency of receptive rectal intercourse and numbers of sex partners. Not only did the proportions of men in these categories decrease over time, but the frequency of receptive rectal intercourse decreased, as did the numbers of steady and nonsteady sex partners.

Like the entire study population, men who had participated in the hepatitis B vaccine efficacy trial also reported changes in sexual activity prior to enrollment into the prospective study of AIDS in 1984. The proportion of men who reported having receptive rectal intercourse, for example, decreased from 73.4% at the peak period to only 41.2% at enrollment. Testing stored sequential sera collected from these men since the start of the vaccine trial in 1978 and 1979 for anti-HIV-1 revealed a fairly constant annual infection rate ranging between 4.5% and 8.4% (Fig. 2-2). No significant decrease in incidence was detected that paralleled the reported decrease in sexual activity. In contrast with this pattern, the incidence of HAV infections increased from 1979 through 1982 but subsequently decreased dramatically to nearly negligible levels. Over this 5-year period, numbers of hepatitis A cases reported in New York City actually increased from less than 400 per year in 1979 and

Table 2-6
Reported Frequency of Receptive Rectal Intercourse and Numbers of Male
Sex Partners Prior to Enrollment for All Participants

	Period of Reported Sexual Activity		
	Peak	Preawareness of AIDS	6 Months Preenrollment
Receptive rectal intercourse (RRI)			
Percentage who had RRI	77.8	71.5	45.6
Geometric mean[a] frequency/month	4.7	3.7	2.8
Steady partners			
Percentage who had steady partners	83.4	79.8	74.8
Geometric mean[a] number/6 months	2.4	1.9	1.4
Nonsteady partners			
Percentage who had nonsteady partners	96.7	91.8	82.7
Geometric mean[a] number/6 months	27.4	16.2	7.4

[a] Of those who reported the activity.

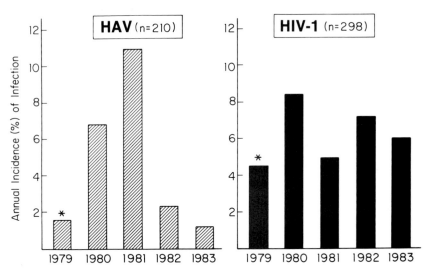

Figure 2-2. Annual incidence of hepatitis A virus (HAV) and human immuno-deficiency virus type 1 (HIV-1) among homosexual men in hepatitis B vaccine efficacy trial. *Incidence adjusted for length of follow-up in participants who entered the study during 1979. (Reprinted with permission from Szmuness et al., 1980.)

Table 2-7

Incidence of HIV-1 Infection (Anti-HIV- Seroconversion) since 1984 among Susceptible Men by Practice of Receptive Rectal Intercourse and Sex with Nonsteady Male Partners at Enrollment

Sexual Activity at Enrollment		Number Susceptible Men	Subsequent HIV-1 Incidence[a]
Receptive Rectal Intercourse	Nonsteady Male Sex Partners		
None	None	60	0.0
None	Yes	223	5.2
Yes	None	29	10.7
Yes	Yes	126	20.3

[a] Life-table attack rate ($p < .0001$).

1980 to 689 in 1982 and 507 in 1983 (information courtesy of Eleanor Bell, New York City Department of Health).

Since enrollment, HIV-1–susceptible men have had their sera tested at each follow-up visit for anti-HIV-1. Among 442 men who were initially seronegative, 38 have become seropositive, giving a 3-year incidence by life-table methods of 9.2%. The incidence of seroconversion was highest among men who were still rectally receptive at the time of enrollment, especially those who also had nonsteady male sex partners (Table 2-7). In contrast with the data obtained before 1984 among hepatitis B vaccine efficacy trial participants, the incidence of HIV-1 infection during the past 3 years has declined steadily (Fig. 2-3). Men from the hepatitis B vaccine efficacy trial who were still seronegative for antibody to HIV-1 in 1984 have shown the same pattern of decreasing incidence, from 4.5% in the first year to none at all in the third. Very few HIV-1–susceptible men had a history of intravenous drug use. Moreover, none of the incident infections were attributable to this practice. Men who had a history of intravenous drug use at the time of enrollment were therefore not excluded from analyses of incidence as related to sexual activity.

Since enrollment in 1984, our population has continued to modify their sexual activity. Among men who were susceptible to HIV-1 (i.e., anti-HIV-1 negative), 35.1% reported having receptive rectal intercourse with either steady or nonsteady sex partners just prior to enrollment. By July through November of 1987, after 3 years of study, only 15% remained in this category. Moreover, the frequency of receptive rectal intercourse and the numbers of nonsteady partners decreased steadily. Data on sexual activity for the most recent follow-up interval are now collected separately on sexual practices with steady and nonsteady partners. Table 2-8 shows sexual practices related to receptive rectal intercourse at the most recent follow-up visit for men who are still susceptible to HIV-1.

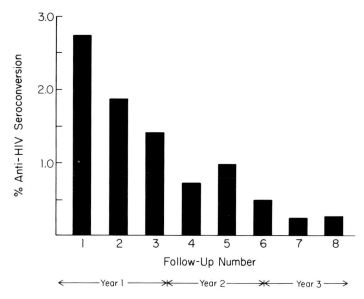

Figure 2-3. Incidence (life-table attack rate) of anti-HIV seroconversion during follow-up by 4-month intervals between routine visits.

At present, only 6% of susceptible men in our study still practice receptive rectal intercourse with nonsteady male sex partners and, in more than half of those who did, their partners used condoms. A larger proportion had receptive rectal sex with their steady partners, but condoms were frequently used, and for many of the others the partner is known to be anti-HIV-1 negative.

In view of the further changes in sexual practices reported during follow-up, it is likely that practices reported at the time of enrollment

Table 2-8
Most Recent Sexual Practices Related to Receptive Rectal Intercourse Reported by Anti-HIV-1–Negative Men with Steady and Nonsteady Male Sex Partners[a]

Sexual Practices	With Steady Male Partners	With Nonsteady Male Partners
Percentage practicing receptive rectal intercourse	29.9	11.1
Geometric mean frequency/4 months	3.5	1.6
Percentage whose partners ever used condoms	66.1	71.8

[a] Follow-up number 9, July through November, 1987.

Table 2-9
Prior Sexual Activity among Men Who Became Infected with HIV-1 during Follow-up and in Seronegative Controls[a]

	Seronegative Controls (N = 70)		Seroconverters (N = 35)	
Sexual Activity with Men	Number	Percentage of Total	Number	Percentage of Total
None	3	4.3	1	2.9
Steady only—no RRI[b]	6	8.6	0	0.0
Steady only—RRI	5	7.1	1	2.9
Nonsteady—no RRI	42	60.0	6	17.1
Nonsteady—RRI	14	20.0	27	77.1

Note: Three seroconverters and their controls were excluded because data on sexual activity were incomplete.
[a] Reported in the 4 to 8 months before anti-HIV-1 was detected.
[b] RRI, receptive rectal intercourse.

may not reflect those engaged in just prior to seroconversion. Limited data suggest that the lag time between infection and seroconversion is usually only a few months (Fincher, de Silva, Lobel, & Spencer, 1985; Neisson-Vernant, Arti, Mathez, Leibowitch, & Monplaisir, 1986; Vittecoq, Autran, Bourstyn, & Chermann, 1986). We therefore examined reported sexual activity for each seroconverter for the two 4-month intervals prior to first anti-HIV-1 detection. Assuming that antibody actually appeared just after the last negative serum sample or just at the time of the first positive sample, these data would represent a minimum of 4 months and a maximum of 8 months between the reported sexual activity and seroconversion. For comparison, we examined sexual activity reported for the same time intervals by two seronegative men for each case seen on the same day as anti-HIV-1 was first detected in the seroconverter. Recent receptive rectal intercourse, especially when the participant also had nonsteady sex partners, was the only sexual activity reported more frequently by seroconverters than by negative controls (Table 2-9). Moreover, among the men who reported having both the nonsteady sex partners and receptive rectal intercourse, the frequency of rectal intercouse was more than twofold higher among seroconverters than among negative controls (geometric means of 8.7 and 4.1 times higher during the two intervals, respectively). The numbers of steady and nonsteady male sex partners did not differ significantly, however.

Seven seroconverters reported having had no receptive rectal intercourse during the two preceding follow-up intervals. However, five of these men had engaged in this practice within an interval that could have been as short as 6 months prior to seroconversion. One of the remaining seroconverters denied having receptive rectal intercourse for a minimum 9-month interval. The other denied ever engaging in this practice. These latter two men did engage in other sexual activities, but their practices did not differ from those reported by seronegative controls. No possible source of HIV-1 other than sexual activity could be identified for these two men. They each denied any possible exposure through accidental inoculation, intravenous drug use, dental work, or blood transfusion. No p24 antigen was detected in sera from these men collected prior to antibody appearance using Abbott's ELISA (Allain et al., 1987).

Discussion

This study, as have several others before it and those reported at the AIDS and Sex Conference by Detels and his colleagues, suggests that receptive rectal intercourse is the primary route of HIV-1 transmission during male homosexual activity (Detels, et al., 1983; Darrow, Jaffe, & Curran, 1983; Goedert et al., 1984; Jaffe et al., 1983; Kingsley et al., 1987; Stevens et al., 1986; Winkelstein et al., 1987). (See also Chapter 1.) Receptive rectal intercourse was the only sexual practice consistently associated with HIV-1 infection at a statistically significant level in this study, both from an analysis of long past practices among prevalent infections at enrollment and from analysis of recent practices among seroconverters. While no other sexual practice was independently associated with HIV-1 infection, transmission by other practices could not be excluded. Two men, in fact, who denied a recent history of receptive rectal intercourse became anti-HIV-1 positive during follow-up. One man had not had receptive rectal intercourse for at least 9 months prior to the appearance of antibody and the other denied ever engaging in this practice.

The explanation for infections in these cases is not clear. If the sexual practice history is accurate, either of two explanations is possible. On occasion, the period between exposure and antibody development may be longer than the few months suggested from previous data. The detection of p24 antigen for 14 months before antibody became detectable in an HIV-1 infected man from Finland suggests this possibility (Ranki et al., 1987). No p24 antigen was detected prior to antibody in our two cases. Alternatively, the virus might be transmitted by some route other than receptive rectal intercourse. No nonsexual source was identified in either case and no other specific sexual practice was implicated. However, transmission at a relatively low efficiency might not be detectable as statistically significant in view of the small number of seroconverters available for analysis. In any event, these two cases raise questions about the absolute safety of so-called safer sex practices.

We are now more than 10 years into the epidemic of HIV-1 infection among homosexually active men in the United States. During this time the virus has spread extensively and the sexual practices of homosexual men have changed dramatically. Data from our earlier study suggested that the initial changes in sexual activity did little to reduce the incidence of HIV-1 infection, although one would suspect that the incidence might have been even higher if some changes had not taken place. The dramatic decrease in the incidence of HAV infection, a virus readily transmitted through male homosexual activity, corroborates the decrease in sexual activity reported by our participants. That the risk of HAV infection was more sensitive to changes in behavior than was HIV-1 is not surprising. While transmission of HAV involves contact with an infected individual, the period of infectivity with HAV lasts only days or weeks and these infections do not become chronic. Therefore, relatively minor changes in behavior could significantly reduce the chance of exposure to a case during the period of infectivity. With HIV-1, however, all infections are persistent and infectivity presumably lasts for years. One would expect, therefore, that more drastic changes in risk behavior would be required to reduce transmission rates. Such changes have apparently taken place in our study population, as in others (Kingsley et al., 1987; Winkelstein et al., 1987). Over the past 3 years, the incidence of new HIV-1 infections has steadily declined, reaching less than 1% in the most recent year. Very few HIV-1 susceptible or infected men now engage in high-risk sexual behavior. Thus homosexual men have become the first risk group, through voluntary intervention on their own, to affect the course of the epidemic of HIV-1 infection.

Homosexual men in our study now have relatively few sex partners and a higher frequency of long-term monogamous relationships, and are avoiding unprotected rectal intercourse when their partners' HIV-1 status is uncertain. Most susceptible men who engage in receptive rectal intercourse now either have a single partner who is known to be uninfected or have their partners use condoms. These kinds of behavior patterns, along with the serological testing needed to allow adoption of the most appropriate behavior, should, in effect, stop the HIV-1 epidemic in homosexual men. Despite the dramatic and apparently effective changes in sexual behavior seen in our study population, however, it is of concern that any recent HIV-1 infections occurred. Two new infections were detected in this past year. While it is encouraging that so few men in our study now practice risky sex, it is potentially tragic that any put themselves or others at risk. The questions remaining for us to answer, therefore, are why risk behavior persists in some men, would they change, and, if so how could we reach them more effectively?

Acknowledgments

This investigation was supported in part by grant HL-09011 from the National Heart, Lung and Blood Institute, a grant from the Calder Foundation, New York, and contract NO1-CM-37609 from the National Cancer Institute.

References

Allain, J-P., Laurian, Y., Paul, D. A., Verroust, F., Leuther, M., Gazengel, C., Senn, D., Larrieu, M. J., & Bosser, C. (1987). Long-term evaluation of HIV antigen and antibodies to p24 and gp41 in patients with hemophilia: Potential clinical importance. *New England Journal of Medicine, 317*, 1114–1121.

Cox, D. R. (1970). *Analysis of binary data*. New York: Methuen Inc.

Darrow, W. W., Jaffe, H. W., & Curran, J. W. (1983). Passive anal intercourse as a risk factor for AIDS in homosexual men. *Lancet, 2*, 160.

Detels, R., Fahey, J. L., Schwartz, D., Greene, R. S., Visscher, B. R., & Gottlieb, M. S. (1983). Relation between sexual practices and T-cell subsets in homosexually active men. *Lancet, 1*, 609–611.

Fincher, R-M. E., de Silva, M., Lobel, S., & Spencer, M. (1985). AIDS-related complex in a heterosexual man seven weeks after a transfusion. *New England Journal of Medicine, 313*, 1226–1227.

Follow-up on Kaposi's sarcoma and *Pneumocystis* pneumonia. (1981). *Morbidity and Mortality Weekly Report, 30*, 409–410.

Friedman-Kien, A. E. (1981). Disseminated Kaposi's sarcoma syndrome in young homosexual men. *Journal of the American Academy of Dermatology, 5*, 468–471.

Goedert, J. J., Sarngadharan, M. G., Biggar, R. J., Weiss, S. H., Winn, D. M., Grossman, R. J., Greene, M. H., Bodner, A. J., Mann, D. L., Strong, D. M., Gallo, R. C., & Blattner, W. A. (1984). Determinants of retrovirus (HTLV-III) antibody and immunodeficiency conditions in homosexual men. *Lancet, 2*, 711–716.

Gottlieb, M. S., Schroff, R., Schanker, H. M., Weisman, J. D., Fan, P. T., Wolf, R. A., & Saxon, A. (1981). *Pneumocystis carinii* pneumonia and mucosal candidiasis in previously healthy homosexual men: Evidence of a new acquired cellular immunodeficiency. *New England Journal of Medicine, 305*, 1425–1431.

Human immunodeficiency virus infection in the United States. (1987). *Morbidity and Mortality Weekly Report, 36*, 801–804.

Hymes, K. B., Cheung, T., Greene, J. B., Prose, N. S., Marcus, A., Ballard, H., William, D. C., & Laubenstein, L. J. (1981). Kaposi's sarcoma in homosexual men: A report of eight cases. *Lancet, 2*, 598–600.

Jaffe, H. W., Choi, K., Thomas, P. A., Haverkos, H. W., Averbach, D. M., Guinan, M. E., Rogers, M. F., Spira, T. J., Darrow, M. W., Kramer, M. A., Friedman, S. M., Monroe, J. M., Friedman-Kien, A. E., Laubenstien, L. J., Marmor, M., Safai, B., Dritz, S. K., Crispi, S. J., Fannin, S. L., Orkwis, J. P., Kelter, A., Rushing, W. R., Thacker, S. B., & Curran, J. W. (1983). National case-control study of Kaposi's sarcoma and *Pneumocystis carinii* pneumonia in homosexual men: I. Epidemiologic results. *Annals of Internal Medicine, 99*, 145–151.

Kaposi's sarcoma and *Pneumocystis pneumonia* among homosexual men—New York City and California. (1981). *Morbidity and Mortality Weekly Report, 30*, 305–308.

Kingsley, L. A., Detels, R., Kaslow, R., Polk, B. F., Rinaldo, C. R., Chmiel, J., Detre, K., Kelsey, S. F., Odaka, N., & Ostrow D. (1987). Risk factors for seroconversion to human immunodeficiency virus among male homosexuals. *Lancet, 1,* 345–349.

Masur, H., Michelis, M. A., Greene, J. B., Onorato, I., VandeStouwe, R. A., Holzman, R. S., Wormser, G., Brettman, L., Lange, M., Murray, H. W., & Cunningham-Rundles, S. (1981). An outbreak of community-acquired *Pneumocystis carinii* pneumonia: Initial manifestation of cellular immune dysfunction. *New England Journal of Medicine, 305,* 1431–1438.

Neisson-Vernant, C., Arfi, S., Mathez, D., Leibowitch, J., & Monplaisir, N. (1986). Needlestick HIV seroconversion in a nurse. *Lancet, 2,* 814.

Persistent, generalized lymphadenopathy among homosexual males. (1982). *Morbidity and Mortality Weekly Report, 31,* 249–251.

Pneumocystis pneumonia—Los Angeles. (1981). *Morbidity and Mortality Weekly Report, 30,* 250–252.

Ranki, A., Valle, S-L., Krohn, M., Antonen, J., Allain, J. P., Leuther, M., Franchini, G., & Krohn, K. (1987). Long latency precedes overt seroconversion in sexually transmitted human-immunodeficiency-virus infection. *Lancet, 2,* 589–593.

Siegal, F. P., Lopez, C., Hammer, S. G., Brown, A. E., Kornfeld, S. J., Gold, J., Hassett, J., Hirshman, S. Z., Cunningham-Rundles, C., Adelsberg, B. R., Parham, D. M., Siegal, M., Cunningham-Rundles, S., & Armstrong, D. (1981). Severe acquired immunodeficiency in male homosexuals, manifested by chronic perianal ulcerative herpes simplex lesions. *New England Journal of Medicine, 305,* 1439–1444.

Stevens, C. E., Taylor, P. E., Zang, E. A., et al. (1986). Human T-cell lymphotropic virus type III infection in a cohort of homosexual men in New York City. *Journal of the American Medical Association, 255,* 2167–2172.

Szmuness, W., Much, M. I., Prince, A. M., Hoofnagle, J. H., Cherubin, C. E., Harley, E. J., & Block, G. H. (1975). On the role of sexual behavior in the spread of hepatitis B infection. *Annals of Internal Medicine, 83,* 489–495.

Szmuness, W., Stevens, C. E., Harley, E. J., Zang, E. A., Oleszko, W. R., William, D. C., Sadovsky, R., Morrison, J. M., & Kellner, A. (1980). Hepatitis B vaccine: Demonstration of efficacy in a controlled clinical trial in a high risk population in the United States. *New England Journal of Medicine, 303,* 833–841.

Szmuness, W., Stevens, C. E., Zang, E. A., Harley, E. J., & Kellner, A. (1981). A controlled clinical trial of the efficacy of the hepatitis B vaccine (Heptavax B): A final report. *Hepatology, 1,* 377–385.

Vittecoq, D., Autran, B., Bourstyn, E., & Chermann, J. C. (1986). Lymphadenopathy syndrome and seroconversion two months after single use of needle shared with an AIDS patient. *Lancet, 1,* 1280.

Winkelstein, Jr., W., Lyman, D. M., Padian, N., Grant, R., Samuel, M., Wiley, J. A., Anderson, R. E., Lang, W., Riggs, J., & Levy, J. A. (1987). Sexual practices and risk of infection by the human immunodeficiency virus: The San Francisco Men's Health Study. *Journal of the American Medical Association, 257,* 321–325.

II
CONTRIBUTIONS FROM RESEARCH ON HUMAN SEXUALITY

3

Sexual Behavior and AIDS: Lessons from Art and Sex Research

June Machover Reinisch, Mary Ziemba-Davis, and
Stephanie A. Sanders

The rapid emergence and worldwide spread of human immunodeficiency virus (HIV) during the last decade has given rise to a wide and confusing array of prescriptions for "safe" sexual behavior and prompted grave concerns, both rational and irrational, about human sexual behavior. In spite of the recent proliferation of educational campaigns in the United States and other countries, scientists, clinicians, and educators are facing a situation in which not only the spread of HIV, but also many of the erroneous beliefs that are associated with the disease, continue to increase. Ignorance about the history of human sexual behavior from ancient to modern times and the demography and epidemiology of sexually transmitted diseases contribute to the belief that AIDS is a unique phenomenon and "the result of the evil machinations of some group distinct from the majority" (Boswell, 1980, p. 38). A grave lack of basic and specific information about the sexual behavior, knowledge, and attitudes of people in our heterogeneous society makes it possible for our strongly held, deeply rooted, and culturally cherished notions about sexuality to go unchallenged. For example, in spite of evidence to the contrary (Bolling, 1977; Bolling & Voeller, 1987; Voeller, 1983, 1988), particular risk behaviors (such as anal intercourse) are believed to be characteristic of only some individuals (i.e., gay men) and rarely are regarded as characteristic of others (such as heterosexual men and women). The commonly held belief that sexual orientation self-labels (i.e., "heterosexual" and "homosexual") accurately reflect partner choice as well as the types of sexual behaviors in which one has engaged may be the most dangerous notion of all. Recent data from our laboratory

(Reinisch, Sanders, & Ziemba-Davis, in press), supported by data from other studies (A.P. Bell & Weinberg, 1978; Kinsey, Pomeroy, & Martin, 1948; Kinsey, Pomeroy, Martin, & Gebhard, 1953), indicate that, although sexual orientation labels may accurately reflect one aspect of sexual identity, they do not necessarily predict current sexual behavior or behavioral patterns across the life span.

In the United States, AIDS has been concentrated primarily among gay men, intravenous (IV) drug users, and urban minorities. Depending upon the unknown frequency of sexual interaction across socioeconomic and racial groups (see Nichols, Chapter 26), however, it is generally believed that AIDS will take an increasing toll on other segments of our society, including the white, middle-class, "heterosexual" population. One source of this spread will undoubtedly be the sexual behavior of nominally heterosexual or homosexual but behaviorally bisexual men. In fact, the "hidden" bisexual behavior of men whose female partners are typically uninformed and unsuspecting, and therefore not likely to protect themselves from possible infection with HIV, has been implicated as a major conduit of HIV transmission into the heterosexual population (see Nichols, chapter 26) (Padian, et al., 1987; Winkelstein, Wiley, Padian, & Levy, 1986). As a typical case in point, Nichols described a 29-year-old Caucasian woman who was diagnosed as having AIDS-related complex (ARC) in the third month of pregnancy. Only after her diagnosis did the father of her child, her steady partner of 4 years, acknowledge that he had engaged in sexual behavior with men: "He told Susan that he had been afraid to tell her . . . because . . . she would leave him. He . . . never considered himself to be at risk for AIDS because [after all] 'AIDS happens to homosexuals, not bisexuals'." (Chapter 26, p. 376). In spite of similar accounts reported in popular magazines and on prime-time television, many find the suggestion that people, including ostensibly "heterosexual" men and women, may occasionally behave bisexually not only perplexing, but incomprehensible.

Within this kaleidoscope of human attitudes, beliefs, and behaviors, researchers, educators, public health officials, government leaders, and representatives of the media must collaborate in their efforts to build and communicate a solid foundation for AIDS education and realistic programs for effective behavior change. Educational efforts have been thwarted, in part, by the widespread belief that sexual behaviors associated with the transmission of HIV are rare, confined to the latter half of the 20th century, and practiced only by a minority of individuals. Ignorance about the actual patterns of sexual behavior among men and women in America's ethnically, socially, and culturally diverse society has contributed to a substantial amount of confusion regarding the sexual transmission of HIV. Even more serious is the appalling lack of basic, accurate, and representative data on human sexual behavior and attitudes with which to estimate the spread of HIV and design educational programs. This chapter attempts to respond to some of these concerns by (a) illuminating the universal nature of various sexual behaviors, including those implicated in the transmission of HIV, by reviewing their depiction in art, artifacts, and ephemera throughout history; (b) reviewing data from studies of human sexual behavior conducted in the United States during the last 40 years; and (c) briefly outlining a set of guidelines for research on human sexual behavior and attitudes.

A Cross-Cultural Perspective on Selected Sexual Behaviors as Depicted in Art Throughout History

> From the dawn of human history . . . men have left a record of their sexual activities . . . poetic, pornographic, literary, philosophic, traditional, and moral. All of them give evidence of what people think and do sexually, and that is sufficient to make them scientifically significant. (Kinsey et al., 1948, pp. 21–22)

Despite the misconceptions of some that behaviors associated with AIDS are the product of Western civilization in the second half of the 20th century, even the most cursory examination of the archeological and ethnographic record reveals art, artifacts, literature, and ephemera that reflect the ubiquitous, pan-historic, and omnicultural nature of sexual behaviors, both procreative and nonprocreative, some of which have been implicated in the transmission of HIV. In order to make the point that these sexual behaviors are neither new nor limited to the culture of post–sexual revolution America, we will briefly discuss their representation in art and artifact from cultures around the world dating from 400 years before the Christian era to the present. Drawing primarily on the archives of The Kinsey Institute, our review begins with the classical art of Greece and Rome, where some of the earliest realistic representations of sexual behavior have been found (Brendel, 1970), and continues with examples of erotic materials from the cultures of prehistoric Peru, China, Japan, Persia, Africa, Latin America, Western Europe, and the United States. Because only a small sample of these materials could be included in the present discussion, when available, additional sources are cited to facilitate access to a more comprehensive account.

Erotic Art in Antiquity

Even among the most casual observers of art, the depiction of both heterosexual and homosexual activities is a well-known motif on erotic vase paintings of ancient Greece. For example, the portrayal of heterosexual intercourse, either vaginal or anal, in the dorsal-ventral or "rear-entry" position is a relatively common theme (Dover, 1978; Johns, 1982).

In a Kinsey Institute publication, Brendel (1970, Fig. 17) discussed a red-figure cup attributed to the Greek Pedieus Painter in which a woman lying on her side performs fellatio while engaging in dorsal-ventral coitus with a second male partner. Eliciting controversy among some historians of art, Dover (1978) argued that the majority of rear-entry coital depictions in art from the Classical period, such as those rendered by the Pedieus Painter, portray anal, rather than vaginal, penetration. To resolve this controversy, we sought additional information by examining the sexual iconography of ancient Rome and found that the literary record also establishes a role for anal intercourse. In his famous epigrams, the Roman poet Marcus Valerius Martialis (40–104 A.D.) wrote:

> You refuse anal intercourse; but the wives of Gracchus, Pompey and Brutus had no such objections, and everyone knows that before young Ganymede arrived to provide the service, Juno stood in for him with Zeus. (Mountfield, 1982, p. 27)

Additional data were gathered by examining erotic representations from the Hellenistic and Roman period, during which the frequency of erotic expressions in art began to rise after a brief decline between 450 and 300 B.C. (Wilson, 1973). As Wilson (1973) noted, the representation of sexual behavior on household utensils and other common, everyday artifacts from the Roman Empire is indicative of the "popular character" (p. 12) of erotic art during this period. In addition to portraying the act of fellatio, a terracotta oil lamp, a typical vehicle for erotic themes, in the collections of The Kinsey Institute (ISR/KI 801) is adorned with a scene in which a second male partner approaches the woman from behind. Although it has not definitively been established that this and the many other depictions of dorsal-ventral coitus on artifacts from the Classical period portray anal, rather than vaginal, intercourse, taken together, these historical records can be conservatively interpreted as indicating that anal intercourse was at least occasionally, if not commonly, practiced by men and women as early as the first to the third century A.D. Minimally, the representation of "unconventional" sexual positions and the realistic depiction of sexual behavior on prosaic objects used by common citizens indicate that sexual themes and pleasures were regarded as an acceptable part of everyday life in these societies. Although the influence of the Christian church with its teachings on the sinfulness of sexual pleasure greatly changed the face of erotic art in subsequent years, art and artifacts from both Greece and Rome attest to the fact that cultures other than our own have celebrated the pleasurable functions of sexual behavior, both procreative and nonprocreative.

Erotic Art and Prehistoric Peru

The northern coast of prehistoric Peru has been referred to as "the heart-land of sexual ceramics" (Gebhard, 1970, p. 139). Here, the representation of sexual behavior is preserved on ceramic works produced primarily by artisans of the north coastal Mochican and Chimu cultures (Gebhard, 1970). Drawing on the work of Gebhard, who, with Alfred C. Kinsey in 1954, documented, photographed, and collected ceramic erotica from Peru, a few highlights from this culturally rich period of human history are presented below.

In spite of an economic and artistic depression during the Peruvian Fusional period (circa 600–1100 A.D.), there was a surprising amount of artistic continuity in the production of ceramics with sexual motifs throughout the Pre-Columbian era. This continuity is demonstrated by comparing ceramics made by artistis of the Mochican (Classic period; 200 B.C. to 600 A.D.), Lambayeque (Fusional period; 600–1100 A.D.), and Chimu (City-Builder period; 1100–1400 A.D.) cultures.

THE MOCHICAN CULTURE

Heterosexual coitus emerged as the most popular theme in Gebhard's (1970) classification of sexual behaviors depicted on Mochican erotic ceramics: 39% of a sample of 456 vessels included scenes of vaginal or anal intercourse. Counting only those instances in which the penis could definitively be seen entering the anus, 21% of all coital scenes in this sample depicted anal intercourse. The archives of The Kinsey Institute house many examples and photographs of Mochican wares on which anal intercourse, an apparently prevalent activity perhaps used as a method of contraception, is portrayed (e.g., ISR/KI 357).

THE LAMBAYEQUE CULTURE

In spite of a marked decline in sexual art during the Fusional period (600–1100 A.D.), artists of the Lambayeque culture produced a number of ceramic vessels with sexual motifs, thus paving the way for a full-scale revival of erotic art by the later Chimu culture, which thrived on the northern coast of Peru. Of 37 Lambayeque wares studied by The Kinsey Institute (Gebhard, 1970), 13 include depictions of penile-vaginal intercourse and 5 portray heterosexual anal intercourse (e.g., ISR/KI 958).

THE CHIMU CULTURE

Following the Fusional period, there was a renaissance of sexual art during the City-Builder period of 1100–1400 A.D. (Gebhard, 1970). Although erotic art in Peru would never again flourish as it had during the Classical period in which the Mochican potters worked, artists of the Chimu culture revived many of the erotic styles indigenous to the northern coast of Peru. Again, a large proportion (approximately one half) of Chimu vessels with sexual themes portray heterosexual coitus. Although anal intercourse appears less often than on Mochican ceramics, it is by no means a rare theme on artifacts produced during this period just prior to the Incan conquest of Peru.

Although there is not space here to include an exhaustive listing, it is quite clear from Gebhard's (1970) documentation, that virtually all Peruvian cultures produced artifacts with erotic themes during the 650 years that mark the rise and fall of the Pre-Columbian period. Collectively, these artifacts confer considerable insight into the history of human sexual behavior. The realistic representation of human sexual behavior, including anal intercourse, on artifacts from this period makes it unequivocally clear that these behaviors were not rare or confined to Western industrial societies of the 20th century.

Erotic Art in China

The majority of surviving Chinese erotic art derives from the later, rather than ancient, eras, such as the late Ming and Ch'ing dynasties (Rawson, 1968). Drawing heavily on literary and cultural symbols and allusions, its primary purpose, as noted by Rawson (1968), was for stimulation and instruction in the art of lovemaking. Unlike Western artists, who traditionally attempted to adhere to mathematical formulas for correct proportions, Chinese artists strove to "express the spirit of forms, not their mere outward likenesses" (p. 254) by focusing their artistic energy on the spiritual or "cosmic" motions of the Tao (Rawson, 1968). The resulting fluid and sometimes ephemeral character of Chinese erotic art is captured in two pieces selected for our review.

Dated to the 20th century, a Chinese book illustration in the collections of The Kinsey Institute depicts two women and a man engaged in heterosexual oral sex. Another work housed at The Institute dated to 18th or 19th century China, depicts male actors engaged in homosexual anal intercourse. The participants are readily identified as men by their unbound feet. Although homosexual behavior was thought to waste the vital energies Yang and Yin, there is little question that homosexual male prostitutes thrived in many of the larger Chinese cities (Rawson, 1968). Similarly, female homosexual behavior was accepted and encouraged in China's traditionally polygamous society, where men typically engaged in sexual behavior with many women simultaneously.

Erotic Art in Japan

Although Japan has a rich tradition of erotic painting and sculpture, in which sexual behavior is celebrated and even glorified (Rawson, 1968), Japanese erotic art did not begin to flourish until the 17th century (Bowie

& Christianson, 1970). The Kinsey Institute collections house many Japanese depictions of a variety of sexual behaviors in addition to heterosexual intercourse. For the present paper, we review two 19th century color woodcuts, one of which has been attributed to Utagawa Kunisada, a prolific producer of Japanese erotic art in the modern period. Kunisada's elegant portfolio includes a striking and skillfully executed color woodcut (A250Q, K965.3a), depicting mutual heterosexual oral-genital behavior. This woodcut is one of the illustrations in an early 19th century Japanese book titled *Oyagari no Koe* (*O Great Cry of Woman*). Kunisada's erotic illustration is in keeping with the theme of the book, whose title refers to a sexual cry.

Another 19th century Japanese color woodcut in The Kinsey Institute collections (A250Q, 014) depicts lesbian sexual activity. Although it has not been attributed to Kunisada, consonant with his style, this work also reflects two characteristic Japanese artistic conventions for erotic art, in which the act of lovemaking is enhanced by the use of fashionable and colorful costumes and the exaggerated size of genitalia. Whereas the former convention provided a multiplicity of outlines, edges, and folds through which the Japanese artist's characteristic interest in contour was expressed, the latter convention was used to highlight the fact that genitals are the primary focus of sensation and interest during sex (Rawson, 1968). The exaggeration of genitalia also provided the Japanese artists an opportunity to express their interest in linearity and the extravagant representation of contour.

Erotic Art in Persia

Two gouache paintings from Persia (circa 1160–1720) are also part of The Kinsey Institute's collections. One depicts a women engaging in both vaginal and anal intercourse with two men wearing the handsomely rendered robes characteristic of Persian art from this period (A172P, A001.1). In another anal intercourse is again depicted, only this time the artist portrays anal sexual activity between two men (A172P, A001.3). In the tradition established by artists who created illuminated manuscripts and miniature paintings, signatures of Persia's distinctive artistic tradition throughout the world, the painter of these works capitalized on the use of fine detail and exquisite color to create a decorative representation of human sexual behavior.

Erotic Art in Africa

More often than not, traditional African artifacts are appreciated for what they do not express, or what is expressed in the abstract, rather than for the extent to which they mirror the natural world through the manipulation of precise detail (Bamert, 1980). Inspired by traditional mysteries of magic and ritual, African artists have continued to produce vessels, masks, jewelry, and figures of great harmony and proportion, such as the cast brass lovers attributed to a 20th century artist and housed at The Kinsey Institute. An anonymous artist of the Fon tribe, descendants of the Yoruba and natives of Dahomey, situated between the Ivory Coast and Nigeria, is credited with creating two elegantly sculpted portraits of a man and a woman engaged in dorsal-ventral coitus (ISR/KI 1277) and a man and a women engaged in mutual oral-genital behavior (ISR/KI 1276).

Erotic Art in Latin America

One example of art from modern Latin America, unlike some of the other pieces covered in our review, is more overtly erotic. Most easily classified as kitsch art, The Kinsey Institute collections contain a Mexican pipe holder (ISR/KI 272) on which finely detailed, realistic male and female wax figures are presented engaging in anal coitus. Although relatively little is known about the artistic heritage of such figures, this piece, deriving from the modern period (circa 1950), provides further evidence for our belief that interest in a variety of sexual behaviors, as depicted in art and artifacts, is not limited to particular cultures or historic periods.

Erotic Art in Western Europe

Although religious and mythical themes, such as celebrations of the wine god Bacchus, and Leda's seduction by Jupiter disguised as a swan (Wilson, 1973), were the source of much early Western European erotic art, it is important to note that a large proportion of ancient artifacts with sexual motifs, especially those deriving from Greece and Rome after about 530 B.C., depict the behavior of *real*, not mythical, people (Brendel, 1970; Wilson, 1973). After the triumph of Christianity in the fourth century, however, the acceptance of sexual pleasure, not to mention the presentation of erotic motifs, was severely curtailed (Bentley, 1984). For several centuries subsequent to the advent of Christianity, when erotic themes did appear in Western European art, they typically addressed man's knowledge of sexual shame acquired in the Garden of Eden, or the tortures of those damned to hell or cruel and often grotesque punishment for a lack of "purity" in sexual conduct and intention (Wilson, 1973). Although license to depict sexual or erotic themes underwent many transformations during the 14th, 15th, and 16th centuries, even during the Gothic and "enlightened" Renaissance periods, women and sexual minorities were still often depicted as "the personification of lust . . . frequently [with] Death as a companion" (Wilson, 1973, p. 15). Thus, the belief held by some today that AIDS is a "punishment" for those they consider to be engaging in illicit or immoral sexual behavior is reminiscent of this older view of sexuality.

Nonetheless, artists throughout history, including masters such as Rembrandt Van Rijn (1606–1669), provided frank and nonjudgmental depictions of human sexuality and procreation. Indeed, as shown in his famous etching titled "Ledikant" or "the Great Bed" (reproduction held by The Kinsey Institute; A468N R3857.1), Rembrandt has presented a straightforward look at a contemporary couple making love in an opulent four-poster bed.

Represented by the light, often frivolous, lovers' escapades painted by Jean-Antoine Watteau (1684–1721) and the often blantantly erotic and genitally explicit scenes depicted by Francois Boucher (1703–1770), artists of the French Rococo period (18th century) created fantastically elaborate and celebratory expressions of the erotic in art. In so doing, they largely omitted any suggestion of sin, shame, or anxiety and "unhibitedly raised erotic art to a new level of elegance" (Carr, 1972, p. 78). The archives of The Kinsey Institute house an anonymous and untitled painting dated to late 18th or early 19th century France, in which heterosexual group sex involving two men and two women engaged in anal intercourse and other sexual behaviors is depicted in a style highly reminiscent of the lively Rococo period.

During this period in which much explicit erotica was being produced, it was not at all uncommon to find simple, everyday trinkets such as snuffboxes and lockets adorned with sexual scenes, albeit sometimes covert or hidden. One example of covert erotica is a colorfully enameled French pocket watch dated to approximately the Rococo period, the outside of which contains a simple, nonsexual idyllic scene. Inside, however, there is a separate latched compartment that when opened reveals erotic depictions of two couples engaged in intercourse.

In contrast to this seemingly more liberal atmosphere in 18th century France, described by Lorenzoni (1984b, p. 24) as "the century of voluptuousness," the social climate in early 18th century England was, depending on the royal court, considerably more restrictive (Lorenzoni, 1984a). The title page from a rare English book dated to the early 1700s housed in the Institute's archives gives strong indication of just how restrictive it could be:

> The trial of Mervin Lord Audley, Earl of Castlehaven, for committing a rape and sodomy with two of his servants, viz. Laurence Fitz Patrick and Thomas Brodway, who has try'd and condemn'd by his peers . . . and beheaded on Tower-Hill . . .

This passage attests to the fact that individuals throughout history have participated in a wide variety of sexual behaviors, including those related to AIDS risk, even when the punishment involved imprisonment or death.

Erotic Art in the United States

Although many well-known artists in the United States have produced erotic paintings, sculptures, and drawings, we elected to discuss only one of the American works in the Institute's archives, a watercolor by Howard Chandler Christy (1873–1952; A630R, C5569.1). Dated to the 1920s or 1930s, it depicts heterosexual group sex in which the participants engage in coitus, cunnilingus, and fellatio. In light of the artist's considerable recognition as the creator of the American feminine type known as the "Christy" girl, and his prestigious commission as the painter of "Signing the Constitution" in the Capital Building in Washington, D.C., this watercolor makes the important point that almost all artists, masters and apprentices alike, produce erotic art (Bentley, 1984; Bowie & Christianson, 1970; Johns, 1982; Melville, 1973; Rawson, 1968, 1973), either in response to or as a reflection of the culture in which they live.

Summary

Our purpose in presenting this abbreviated survey of erotic images is to underscore the universality of various sexual activities, including AIDS risk–related behaviors, throughout nearly all of history and in nearly every culture. Although some of these behaviors might have been stigmatized or disapproved of at various points in time or in various cultures, few would deny that their depiction in art indicates that they were at least known, if not widely practiced.

Research on Human Sexual Behavior from 1948 to 1988

> Because all of the technical problems in sex research, social scientists have to regard each study as part of a mosaic. Each fragment adds something to the understanding of whether and how behavior changed over the years, who is affected, and the direction in which we are headed. When the results of many studies start to converge, we can be sure we are on to something. (Tavris & Sadd, 1975, p. 15)

To estimate the prevalence of sexual behaviors associated with the transmission of HIV, we reviewed studies on human sexual behavior conducted in the United States during the last 40 years. All of the currently available data sets on human sexual behavior are characterized by serious limitations such as small sample size, biased samples, insufficient data on subcultural groups, and incomplete or ambiguous information. Despite these limitations, we have drawn from 17 of the most thoughtfully designed scientific investigations and the largest, most comprehensive commercial surveys published between 1948 and 1988 to derive estimates of the prevalence of several AIDS risk–related behaviors for white, middle-class, relatively well-educated Americans between 20 and 45 years of age (see also Reinisch, Sanders, & Ziemba-Davis, 1988). Unfortunately, because even fewer accurate data sets are available, no information on the behavior of other ethnic, racial, social, age, and regional groups that comprise the United States population can be presented. Although commercial magazine surveys typically are not the most scientifically sound studies, they were included in our review because they constitute a major (with regard to number of respondents), albeit limited, source of information in the notable absence of more scientifically designed investigations.

Before summarizing these data, key methodological and demographic characteristics are presented for each study, in the text and in Tables 3-1 and 3-2, respectively. Whenever possible, we report (a) when and how subjects were recruited; (b) how the data were collected (interview, on-site questionnaire, or magazine survey); and (c) the number of questions asked. In addition to describing the number of heterosexual and homosexual male and female participants, information regarding the average age and/or age range of participants, their ethnic/racial and religious background, education level, marital status, and geographic location is presented. In some cases, to facilitate comparisons among the 17 studies, multiple levels of complex demographic variables (such as educational background) were condensed by summing across levels and creating more uniform, simple categories. Presented in order by the years in which they were published, each of the studies included in our review is described below.

Methodological Characteristics of the Seventeen Studies on Human Sexual Behavior

1948 AND 1953: KINSEY'S BASIC SAMPLE

The classic volumes *Sexual Behavior in the Human Male* (Kinsey et al., 1948) and *Sexual Behavior in the Human Female* (Kinsey et al., 1953) are the product of the prodigious efforts of Alfred C. Kinsey and his colleagues, who conducted what is still, after 40 years, regarded as the largest and most comprehensive scientific interview study of human sexual behavior. Beginning with students at Indiana University in 1938, and ultimately including individuals from throughout the United States, a face-to-face structured interview was used to gather as many as 521 items pertaining to the sexual behavior of American men and women. It was not until 25 years later, in 1963, that Kinsey's unparalleled study of American sexual behavior ended after more than 18,000 men and women had been interviewed.

Kinsey's basic approach to sampling was to increase the size and di-

Table 3-1
Demographic Characteristics of the 17 Studies of Human Sexual Behavior

Study	N^a	Age	Race[b] %W	%B	%O	Religion[c] %P	%C	%J	%O	%N	Education Level[d] %HS	%C	%G	Percentage Ever Married
Kinsey data cited in Gebhard & Johnson (1979)	5,460 Males[e]	17.89% [Under 19] 35.55% [20 to 24] 16.59% [25 to 29] 16.01% [30 to 39] 7.97% [40 to 49] 4.34% [50 to 59] 1.65% [60 or over] Mean: 27.7 Median: 24.0 Range: 15–81+	100.0	—	—	66.3	11.0	20.7	0.7	1.3	14.0	45.4	40.6	39.4
	5,386 Females[e]	22.91% [Under 19] 27.43% [20 to 24] 13.24% [25 to 29] 19.51% [30 to 39] 11.40%	100.0	—	—	60.8	9.5	28.7	0.3	0.7	19.1	60.0	20.9	43.5

continued

[40 to 49]
4.14%
[50 to 59]
1.37%
[60 or over]
Mean: 28.4
Median: 24.0
Range: 15–81 +

Study												
Athanasiou et al. (1970) — 9,400 Males, 10,600 Females	—	—	(29.0)	(15.0)	(7.0)	(20.0)	(29.0)	11.0	55.0	34.0	55.0	
Saghir and Robins (1973) — 89 HO males	—	100.0	68.0	27.0	5.0	0.0	0.0	—	—	—	18.0	

Athanasiou et al. (1970)
9,400 Males
10,600 Females

11%
[Under 20]
32%
[20 to 24]
34%
[25 to 34]
14%
[35 to 44]
6%
[45 to 54]
3%
[55 or over]

Saghir and Robins (1973)
89 HO males

2%
[Under 20]
33%
[20 to 29]
40%
[30 to 39]
25%
[40 or over]
Mean: 35.0
Median: 32.7
Range: 19–70

Table 3-1 (*Continued*)
Demographic Characteristics of the 17 Studies of Human Sexual Behavior

Study	N^a	Age	Race^b			Religion^c					Education Level^d			Percentage Ever Married
			%W	%B	%O	%P	%C	%J	%O	%N	%HS	%C	%G	
Saghir and Robins (1973)	35 HT males	0% [Under 20] 66% [20 to 29] 31% [30 to 39] 3% [40 or over] Mean: 30.0 Median: 28.0 Range: 21–50	100.0	—	—	46.0	48.0	6.0	0.0	0.0	—	—	—	14.0
	57 HO females	0% [Under 20] 52% [20 to 29] 33% [30 to 39] 15% [40 or over] Mean: 31.0 Median: — Range: 20–54	100.0	—	—	65.0	22.0	9.0	0.0	4.0	—	—	—	25.0
	43 HT females	0% [Under 20] 69%	100.0	—	—	56.0	35.0	7.0	0.0	2.0	—	—	—	21.0

Study	Sample	Age												
		[20 to 29] 26% [30 to 39] 5% [40 or over] Mean: 29.0 Median: — Range: 21–50	—											
Hunt (1974)[f]	982 Males 1,044 Females	—	90.0	10.0	—	—	—	—	—	—	—	—	—	75.0
R. R. Bell et al. (1975)	2,262 Females	25 to 51+ (34.5)	—	—	—	(68.0)	(18.0)	(9.0)	(1.0)	(4.0)	29.0	48.0	23.0	100.0
Tavris and Sadd (1975)	2,278 Females	25% [Under 25] 52% [25 to 34] 23% [35 or over]	—	—	—	(57.0)	(27.0)	(3.0)	(7.0)	(6.0)	38.0	38.0	13.0	100.0
Bolling (1977)	526 Females	84.8% [Under 30] 15.2% [Over 30]	36.5	8.0	55.5	—	—	—	—	—	—	—	—	—
Pietropinto and Simenauer (1977)	3,797 HT males 52 HO males 25 BI males 92 U Males	6.3% [18 to 19] 31.7% [20 to 29] 22.8%	87.5	5.9	3.4	—	—	—	—	—	33.2	52.0	13.1	74.3

continued

Table 3-1 (Continued)
Demographic Characteristics of the 17 Studies of Human Sexual Behavior

Study	N[a]	Age	Race[b] %W	%B	%O	Religion[c] %P	%C	%J	%O	%N	Education Level[d] %HS	%C	%G	Percentage Ever Married
		[30 to 39] 7.9% [40 to 44] 20.2% [45 to 54] 8.9% [55 to 64] 1.7% [65 or over]												
Reanalysis of data from A. P. Bell and Weinberg (1978)	625 HO males	Mean: 35.83	84.8	15.2	—	—	—	—	—	—	25.3	49.6	25.1	16.8
	315 HT males	Mean: 34.78	83.8	16.2	—	—	—	—	—	—	26.7	47.9	25.4	71.1
	248 HO females	Mean: 34.78	79.8	20.2	—	—	—	—	—	—	24.2	46.4	29.4	33.9
	137 HT females	Mean: 32.91	73.0	27.0	—	—	—	—	—	—	25.5	57.7	16.8	73.0
Yablonsky (1979)[g]	771 Males	Mean: 36.0	—	—	—	—	—	—	—	—	—	80.0	—	100.0
Wolfe (1981)	10,000 Females	2.3% [Under 18] 46.9% [18 to 24] 26.7%	—	—	—	—	—	—	—	—	31.3	51.8	17.0	46.2

	[25 to 29] 13.0%											
	[30 to 34] 11.3%											
	[35 or over]											
Blumstein and Schwartz (1983)	3,638 Married males 1.0% [Under 21] 27.0% [21 to 30] 31.0% [31 to 40] 19.0% [41 to 50] 15.0% [51 to 60] 7.0% [Over 60] Mean: 39.9 Median: 36.4	96.0	1.0	3.0	(38.0)	(14.0)	(13.0)	(6.0)	(29.0)	13.0	54.0	33.0	100.0
	3,634 Married females 1.0% [Under 21] 35.0% [21 to 30] 31.0% [31 to 40] 17.0% [41 to 50] 11.0% [51 to 60] 5.0% [Over 60]	97.0	1.0	2.0	(43.0)	(16.0)	(13.0)	(6.0)	(22.0)	19.0	62.0	19.0	100.0

continued

Table 3-1 (*Continued*)
Demographic Characteristics of the 17 Studies of Human Sexual Behavior

Study	N[a]	Age	Race[b]			Religion[c]					Education Level[d]			Percentage Ever Married
			%W	%B	%O	%P	%C	%J	%O	%N	%HS	%C	%G	
Blumstein and Schwartz (1983)	649 Unmarried males	[Over 60] 2.0% Mean: 37.3 Median: 33.9 [Under 21] 48.0% [21 to 30] 35.0% [31 to 40] 10.0% [41 to 50] 4.0% [51 to 60] 1.0%	94.0	2.0	4.0	(16.0)	(8.0)	(14.0)	(9.0)	(53.0)	14.0	59.0	27.0	50.0
	650 Unmarried females	[Over 60] 6.0% Mean: 32.3 Median: 30.5 [Under 21] 59.0% [21 to 30] 26.0% [31 to 40] 6.0% [41 to 50] 2.0%	96.0	2.0	2.0	(20.0)	(11.0)	(12.0)	(9.0)	(48.0)	15.0	66.0	19.0	45.0

continued

	[51 to 60] 1.0% [Over 60] Mean: 29.7 Median: 28.2											
Brecher (1984)	1,895 Married[h] males	Median: 64.0	—	—	—	—	—	—	—	21.0	79.0	100.0
	1,245 Married[h] females	Median: 60.0	—	—	—	—	—	—	—	21.0	79.0	100.0
	507 Unmarried[h] males	37% [In their 50's] 31% [In their 60's] 32% [Age 70 or Over]	—	—	—	—	—	—	—	—	—	85.0
	599 Unmarried[h] females	36% [In their 50's] 36% [In their 60's] 28% [Age 70 or over]	—	—	—	—	—	—	—	—	—	88.0
McWhirter and Mattison (1984)[i]	312 HO males	5.4% [20 to 25] 21.8% [25 to 30] 15.4% [30 to 35] 21.2% [35 to 40]										

Table 3-1 (*Continued*)
Demographic Characteristics of the 17 Studies of Human Sexual Behavior

Study	N^a	Age	Race[b] %W	%B	%O	Religion[c] %P	%C	%J	%O	%N	Education Level[a] %HS	%C	%G	Percentage Ever Married
McWhirter and Mattison (1984)		15.1% [40 to 45] 8.7% [45 to 50] 5.4% [50 to 55] 4.2% [55 to 60] 1.9% [60 to 65] 0.9% [65 to 70] Mean: 37.5 Median: 36.0 Range: 20–69	—	—	—	55.1	27.9	13.5	3.2	0.3	43.6	43.9	12.5	15.1
Rubenstein and Tavris (1987)	26,000 Females	Mean: 34.0	—	—	—	—	—	—	—	—	—	34.0	—	93.0
Wyatt et al. (1988a)	122 Females	39% [18 to 26] 61% [27 to 36]	100.00	—	—	(25.0)	(13.0)	(11.0)	(6.0)	(45.0)	—	54.0	—	74.0
Reinisch et al. (in press)	252 HO females	Mean: 33.7 Range: 19–70	94.0	3.5	2.5	58.0	24.0	8.0	10.0	3.0	46.0	47.0	20.0	

[a] Whenever the sexual orientation of participants was identified, we present separate N's for heterosexual (HT), homosexual (i.e., gay men or lesbian women; HO), and bisexual males and females (BI). When the sexual orientation of participants could not be identified, these subjects are noted by the letter "U."

[b] Because of the absence of, or inconsistent use of, categories for ethnic/racial backgrounds used across studies, we only present data for the percentage of respondents who could be classified as white (W), black (B), or other (O). The category "Other" includes, for example, Asians or Asian-Americans, Hispanics, and Native Americans.

[c] When available, religion of *rearing* is presented. When religion of rearing was not available, the religious affiliation of participants at the time of the study is presented. Because of the inconsistency of religious categories used across studies, we only present data for the percentage of respondents who could be classified as Protestant (P), Catholic (C), Jewish (J), other than Protestant, Catholic, or Jewish (O), or having no religious background/affiliation (N, none).

[d] To achieve comparability across studies, we present educational level as follows: HS, having graduated from or attended high school or technical school level courses or less; C, having graduated from or attended college; and G, having graduated from or attended graduate/professional school.

[e] Only Caucasian participants from Kinsey's "Basic Sample" (see text) were used. Although none of these individuals met the criteria for being included in the "Homosexual Sample" (50 homosexual contacts or 20 partners of the same sex since puberty), 37% (2,008/5,456) of the men and 14% (757/5,385) of the women had had at least one homosexual experience since puberty.

[f] Although demographics such as average age or religious and educational background were not reported by Hunt, there characteristics were, according to the author, "represented in about the same proportion as in the adult American population" (p. 16).

[g] Although Yablonsky did not present descriptive data on such variables as race and religion, the author reported that "An effort was made to acquire a reasonable proportion of responses from various minority, religious, occupational, income, and educational categories of married men" (p. 11).

[h] Although Brecher's sample was presumably largely heterosexual, 13% of all male respondents and 8% of all female respondents reported having had at least one homosexual experience.

[i] Although McWhirter and Mattison did not present detailed descriptive statistics on the ethnic/racial background of their participants, they reported that their sample was predominantly white with small representations of blacks, Mexican Americans, and Asians.

Table 3-2
Regional Distribution of Participants in the 17 Studies
of Human Sexual Behavior

Study	Subsample[a]	Regions	Percentage[b]
Kinsey data cited in Gebhard and Johnson (1979)	Males	North Central	45.5
		Northeast	35.7
		Pacific	7.5
		South	7.1
		Mountain	0.9
		Non-U.S.A.	2.2
	Females	Northeast	46.2
		North Central	31.1
		Pacific	10.3
		South	8.5
		Mountain	0.7
		Non-U.S.A.	1.8
Athanasiou et al. (1970)	Males & females	New England	32.0
		Midwest	26.0
		West	26.0
		South	8.0
		Southwest and Mountain states	8.0
Saghir and Robins (1973)	HO males HT males HO females HT females		c
Hunt (1974)	Males Females		d
R. R. Bell et al. (1975)	Females	Northeast	24.0
		Midwest	23.0
		States west of the Mississippi (Mountain and Prairie)	23.0
		West Coast	17.0
		South	13.0
Tavris and Sadd (1975)	Females	North Central	25.0
		Mid-Atlantic	19.0
		Pacific	15.0
		South Central	15.0
		South Atlantic	10.0
		New England	9.0
		Mountain	4.0
		Outside continental United States	4.0

Table 3-2 (*Continued*)
**Regional Distribution of Participants in the 17 Studies
of Human Sexual Behavior**

Study	Subsample[a]	Regions	Percentage[b]
Bolling (1977)	Females	c	
Pietropinto and Simenauer (1977)	HT males HO males BI males U males	Northeast North Central West South	32.7 25.2 20.6 20.0
Reanalysis of data from A. P. Bell and Weinberg (1978)	HO males HT males HO females HT females	c	
Yablonsky (1979)	Males	e	
Wolfe (1981)	Females	West South Northeast North Central Outside U.S.	25.8 24.2 23.5 22.6 3.9
Blumstein and Schwartz (1983)	Married males and females	North Central Middle Atlantic California and Hawaii South Atlantic Pacific Northwest South Central Mountain New England U. S. territories, Canada, and other foreign	23.0 18.0 18.0 12.0 10.0 8.0 5.0 5.0 1.0
	Unmarried males and females	California and Hawaii Middle Atlantic Pacific Northwest North Central South Atlantic Mountain New England South Central U.S. territories, Canada, and other foreign	24.0 22.0 18.0 14.0 8.0 5.0 4.0 4.0 1.0

continued

Table 3-2 (*Continued*)
**Regional Distribution of Participants in the 17 Studies
of Human Sexual Behavior**

Study	Subsample[a]	Regions	Percentage[b]
Brecher (1984)	Married males Married females Unmarried males Unmarried females	*f*	
McWhirter and Mattison (1984)	HO males	*c*	
Rubenstein and Tavris (1987)	Females	*f*	
Wyatt et al. (1988a)	Females	*c*	
Reinisch et al. (in press)	HO Females	*g*	

[a] HT, heterosexual; HO, homosexual (i.e., gay men or lesbian women); BI, bisexual; U, unidentified sexual orientation.

[b] Percentages may not add up to 100% because of missing data.

[c] Regional breakdowns were not provided by Saghir and Robins (1973), Bolling (1977), A. P. Bell and Weinberg (1978), McWhirter and Mattison (1984), and Wyatt et al. (1988a), because these studies were all conducted in specific regions of the United States (see text).

[d] Although Hunt (1974) did not report data on geographical location, according to the author, this characteristic was "represented in about the same proportion as in the adult American population" (p. 16) (see text).

[e] Although regional breakdowns were not provided, Yablonsky (1979) reported that, "The survey was randomly administered . . . in several geographic locations including the East Coast, West Coast, Midwest, and South. The sample included married men from Los Angeles, New York City, Syracuse, St. Louis, Baltimore, New Haven, and San Francisco" (p. 11).

[f] Although regional breakdowns were not provided, Brecher's (1984) and Rubenstein and Tavris's (1987) studies were comprised of self-selected readers of two magazines with relatively wide circulation. The latter study was based on a survey published in the same magazine as that used by Tavris and Sadd (1975).

[g] Reinisch et al.'s (in press) sample includes residents of 34 states, the District of Columbia, Puerto Rico, and two countries outside of the United States: Canada and New Zealand.

versity of his sample by establishing quotas for the number of respondents required to meaningfully address a given topic (fewer subjects for relatively common or stable phenomena and more subjects for less common phenomena). In addition, Kinsey targeted particular groups (e.g., schoolteachers, presumably comprised of relatively more unmarried individuals) either as needed or as encountered. A more detailed description of his interview and sampling procedures is provided in Kinsey et al. (1948, 1953) and Gebhard and Johnson (1979).

The "Basic Sample" to which we refer in the present review is comprised of all original subjects who were never convicted of a criminal offense and who did not come from sources with known sexual bias, such as homes for unwed mothers (Gebhard & Johnson, 1979). Only data pertaining to the larger and more representative Caucasian groups (male, $n = 5,460$; female, $n = 5,386$) were used in our review of data on behaviors implicated in the transmission of HIV.

1970: The Psychology Today Survey

In July of 1969, a 100-item survey on sexual attitudes and behavior appeared in the popular professional magazine *Psychology Today*. Data from the convenient sample of 20,000 self-selected readers who completed the survey were reported in a 1970 publication by Robert Athanasiou, Phillip Shaver, and Carol Tavris.

1973: A Study of Male and Female Homosexuality

In their study of the "behavioral manifestations and correlates of homosexuality" (p. 3), Marcel Saghir and Eli Robins (1973) utilized a nonpatient sample of homosexuals (male, $n = 89$; female, $n = 57$) recruited in Chicago and San Francisco by directly requesting their participation during meetings of gay organizations, through friends, or at gay bars. The nonpatient heterosexual comparison sample (male, $n = 35$; female, $n = 43$) was obtained by recruiting occupants living in a 500-unit apartment complex renting mostly to single individuals of relatively high socioeconomic status.

Only those homosexual men and women who met the criteria for inclusion—a self-report of homosexual orientation, a history of overt homosexual behavior after age 18, no previous hospitalization for a psychiatric illness, and no history of confinement in a state penitentiary or federal prison—were accepted as participants in the study. Heterosexuals had to meet the same selection criteria required of the homosexual sample except that their self-report of orientation and sexual behavior since age 18 had to be heterosexual and they had to have been single for at least 2 years before entry into the study. Finally, only those heterosexual subjects who could be matched to homosexual participants on age, marital status, religious, and socioeconomic background were accepted as participants.

After appropriate modifications were made, the same structured, 50-page interview was used with both the homosexual and heterosexual participants.

1974: THE PLAYBOY FOUNDATION DATA

With the financial support of the Playboy Foundation, Morton Hunt (1974) gathered information on the sexual attitudes and behavior of men and women across the United States in 1972. Utilizing four versions of a self-report questionnaire (one each for male, female, married, and unmarried participants), Hunt and his associates collected 1,000 to 2,000 items of information on each of the 2,026 participants. Professional public opinion survey teams located in 24 states randomly selected the study participants by telephone. In addition to this "original" sample, several hundred participants younger than those that could be contacted by telephone were solicited on an individual basis in an attempt to increase the study's generalizability. Data from this additional sample were merged with those gathered on the original sample to form the basic data set.

1975: A STUDY OF EXTRAMARITAL SEX AMONG MARRIED WOMEN

In their study of the extramarital sexual behavior of 2,262 women, Robert R. Bell, Stanley Turner, and Lawrence Rosen (1975) recruited married women through the offices of professionals listed with the American Sociological Association as providers of family services. Seventy-five family practitioners with offices in various parts of the United States helped place questionnaires with married women of various backgrounds. Respondents completed the questionnaires anonymously and returned them to the principal investigators using a self-addressed, stamped envelope provided in the study packet.

1975: THE FIRST *REDBOOK* SURVEY

In 1974, 100,000 women throughout the United States replied to *Redbook* magazine's first survey on the sexual experiences of their readers. The anonymous survey was comprised of 75 questions. The published report, written by Carol Tavris and Susan Sadd (1975), is based only on a random sample of 2,278 married women drawn from the larger pool of 100,000 participants. An additional larger random sample of 18,000 respondents, also drawn from the original pool of 100,000 were used when more subjects were needed for special comparisons.

1977: A STUDY OF ANAL INTERCOURSE AMONG HETEROSEXUAL WOMEN

Between March 1974 and October 1975, David Bolling (1977) interviewed 526 women sequentially presenting in one of four gynecological settings, including a cancer detection clinic, a free clinic for young adults, a free reproductive care clinic, and a private gynecological practice. In addition to conducting a standard sexual history, the investigator administered a five-item standardized interview on anal intercourse in the presence of a female staff member of the clinic. Patients who reported that they had engaged in anal intercourse were asked to participate in a second, more extensive interview on the frequency, goals, and complications of anal sexual activity, the results of which are included in the present review.

1977: A Study of What Women Want to Know About Men's Sexuality

In the late 1970s, Anthony Pietropinto and Jacqueline Simenauer asked women of various backgrounds to rate a series of 96 questions indicating what they most wanted to know about the private lives of the opposite sex. After finding that most of the women sampled were interested in men's *feelings* about women and sex (rather than what men actually do), the investigators constructed a sexual attitudes questionnaire consisting of 40 multiple choice items and 32 essay questions. The 32 essay questions were subsequently distributed among eight different versions of the measure so that each participant answered only 4 essay questions along with the set of 40 objective items. Male participants were recruited through a professional research service organization in New York. The data set included in our review consists of 4,066 responses to 40 multiple choice items and 500 responses to each of the 32 essay questions.

1978: The Kinsey Institute's San Francisco Sample

In the tradition established by Alfred C. Kinsey, Alan Bell and Martin Weinberg (1978) used a 175-page structured interview consisting of 528 questions in their investigation of 5,000 men and women living in the San Francisco Bay area. This large group of individuals who referred to themselves as either homosexual or heterosexual was sampled during a 5-month period in 1969.

The recruitment of homosexual subjects was carried out in several phases. Initially, volunteer sign-up cards were distributed in various locations throughout the San Francisco Bay area. The majority of subject recruitment, however, was accomplished by local residents hired as recruiters, many of whom themselves were gay. With the help of these assistants, the principal field staff utilized a variety of resources (e.g., gay bars and organizations) in their efforts to contact gay men and women who might be interested in participating in their study.

Because many of the sources used to contact the homosexual respondents were not commonly utilized by heterosexual men or women, probability sampling with quotas was used to compile the heterosexual comparison groups. Ultimately, the pool of heterosexual respondents was narrowed to those who provided the best match to the homosexual respondents in terms of age, race, sex, and education.

After identifying hundreds of heterosexual and homosexual men and women, the investigators examined each subject's self-ratings on sexual feelings and behavior in relation to Kinsey's (1948, 1953) 7-point scale (0, exclusively heterosexual, to 6, exclusively homosexual). In our reanalysis of this data base for the current review, we only included those subjects who rated themselves as exclusively heterosexual (Kinsey 0s) or predominantly homosexual (Kinsey 5s and 6s) in terms of actual behavioral patterns. Utilizing these more stringent criteria, we were able to identify a subsample of 315 exclusively heterosexual men, 137 exclusively heterosexual women, 625 predominantly gay men, and 248 predominantly lesbian women.

1979: A STUDY OF EXTRAMARITAL SEX AMONG MARRIED MEN

In 1979, Lewis Yablonsky published the results of his survey on the extramarital sexual behavior of men. In the first phase of his research, Yablonsky interviewed 50 married men involved in extramarital sexual behavior. He also interviewed 16 women regarding their responses to their husband's affairs and their own extramarital involvement with married men. In a second, follow-up phase of the study, utilizing the assistance of more than 100 sociology students and professional colleagues around the country, a 16-item questionnaire was randomly distributed to 771 men.

1981: THE COSMOPOLITAN MAGAZINE SURVEY

In January, 1980, *Cosmopolitan* magazine featured a 79-item questionnaire titled "The Cosmo Sex Survey." Data from a randomly selected subsample of 10,000 of the original 106,000 self-selected respondents form the basis of an extensive report written by Linda Wolfe (1981) and included in our review.

1983: A STUDY OF AMERICAN COUPLES

In 1975, Phillip Blumstein and Pepper Schwartz (1983) initiated a questionnaire study of American couples. They focused on behaviors, attitudes, and beliefs associated with three primary issues: money, work, and sex.

In phase one of their study, nationwide advertising (including television and radio spots and announcements in newspapers and magazines), appeals before social, service, and church groups, and notices posted in supermarkets and laundromats were used to inform people about the study. As a result, 22,000 questionnaires were either mailed upon request or left in places where they could be picked up anonymously. The basic questionnaire was approximately 38 pages in length.

Married and cohabitating heterosexuals received either a male or a female version, whereas members of gay couples completed a version tailored either for gay men or lesbians. In addition to extensive demographic information, a composite version of all four questionnaires contained approximately 175 questions. At the time data analyses were begun, approximately 12,000 usable questionnaires had been returned.

In a second phase of the study, the investigators identified all of the participating couples who lived within an hour of New York, San Francisco, or Seattle or subjects who might be recruited to take part in a more in-depth, face-to-face interview. After grouping these couples into married, cohabiting, gay male, or lesbian groups and subdividing them according to number of years together as a couple and education level, the investigators randomly selected 320 heterosexual, gay male, and lesbian couples to be individually interviewed.

In the third phase of their study, commencing approximately 18 months after the interviews were conducted, the investigators sent a final, follow-up questionnaire to participants who had been interviewed as well as to a randomly selected number of those who had only completed the basic questionnaire in the first phase of the study. The follow-up questionnaire was designed to ascertain if the couples were still together and in what ways, if any, their relationship had changed.

1984: A STUDY OF LOVE, SEX, AND AGING

In November 1977, the Consumers Union announced their forthcoming study on the personal relationships of people age 50 and over in their monthly magazine *Consumer Reports*. In response to their advertisement, 4,246 self-selected men and women completed a 130-item questionnaire. Data from this questionnaire, produced in separate versions for men and women, are included in our review and discussed in more detail in a Consumers Union Report by Edward M. Brecher (1984).

1984: A STUDY OF MALE COUPLES

Between 1974 and 1979, David McWhirter and Andrew Mattison (1984) gathered extensive data on 156 male couples (312 individuals) in a study focusing on relationships among gay men. Members of these male couples were not in psychotherapy nor were they located through gay organizations. Subjects were recruited via the "friendship method" in which couples who had already consented to an interview helped the investigators locate other couples who might be interested in participating. Unlike newspaper or magazine advertisements advertising in public places where gay men gather for social or political reasons, the "friendship network method" facilitated access to "hidden" couples, including older men and less socially active individuals. After identification, the investigators established separate appointments with each individual by telephone providing that the referred couple met their criteria for participation—having lived together in the same home for at least 1 year while identifying themselves as a couple. Both an interview schedule (consisting of 185 questions) and a 63-item questionnaire were used with each of the 312 men.

1987: THE SECOND *REDBOOK* SURVEY

In January 1987, *Redbook* magazine published their second survey on the sexual lives of women. Responses of the 26,000 self-selected women included in our review were analyzed and reported by Carin Rubenstein and Carol Tavris (1987).

1988: A SURVEY OF WOMEN IN LOS ANGELES COUNTY, CALIFORNIA

Gail Wyatt, Stefanie Doyle Peters, and Donald Guthrie (1988a, 1988b) have published two studies in which data on the sexual socialization and experiences of white and African-American women in Los Angeles County, California were compared to data obtained from women interviewed by Kinsey and his colleagues (1953) 33 years ago. Only data from the report on white women (1988a) were included in this review.

A multistage stratified probability sampling technique (for details see Wyatt & Peters, 1986) was used to recruit the 122 white women studied by Wyatt and her colleagues (1988a). The demographic variables used to establish quotas for the number of subjects to be obtained within each stratum were age, education, marital status, and the presence of children. Based on these quotas, simple random sampling techniques were employed to select individuals who met the demographic criteria established for participation. Subjects were recruited by random-digit dialing of telephone prefixes in Los Angeles County, combined with four randomly generated numbers.

A 478-item face-to-face structured interview, the Wyatt Sex History Questionnaire (WSHQ), was used to obtain the data included in our review.

IN PRESS: THE KINSEY INSTITUTE'S FEMALE SAMPLE

As part of an ongoing Kinsey Institute research project focused on the sexual lives and health of women, we studied a volunteer sample of 262 lesbian women derived from a population of approximately 5,000 to 7,000 women attending an annual national meeting held in a Midwestern state (Reinisch et al., in press).

The questionnaire was developed with the assistance of a group of women and men ranging in age from 24 to 51 years who had resided in widely dispersed geographic regions of the United States and represented heterosexual, homosexual, and bisexual orientations as well as diverse ethnic, religious, educational, and socioeconomic backgrounds.

Each of the 262 subjects was given a 42-page questionnaire divided into color-coded sections so that she could complete only those sections that were relevant to her sexual history. Depending on a subject's sexual history, this questionnaire yielded from 300 to 500 data points for each subject.

Questionnaires were distributed by a seven-member trained research team from a centrally located site at the meeting. By establishing regular daily hours at the distribution site, we ensured that a trained researcher was present to provide an introduction to the study, offer clarification when necessary, and preserve subjects' confidentiality.

Findings from the Seventeen Studies on Human Sexual Behavior

In spite of their limitations and significant variation in methodologies, scientific expertise, and sample size, the 17 studies included in our review contain useful descriptions of sexual behavior patterns as they pertain primarily to the specific subpopulation of white, middle-class, relatively well-educated Americans. Taken together, these studies can be used to generate estimates of the prevalence of several AIDS risk–related sexual behaviors in this subpopulation, including (a) heterosexual and homosexual anal intercourse (the highest risk behavior); (b) behavioral bisexuality; (c) extramarital contacts; (d) sex with prostitutes; (e) fellatio; and (f) cunnilingus.

Before presenting our summary of data from the 17 studies, we want to clarify that it is not engaging in these behaviors per se that places an individual at increased risk for contracting AIDS. Rather, it is engaging in unprotected sexual activity with an HIV-seropositive partner that may result in the transmission of the AIDS virus. However, many people are not aware of their own or their sexual partners' HIV antibody status or may be unwilling to divulge information regarding their sexual history or HIV status. In fact, recent data (Cochran & Mays, 1990) indicate that people will lie about their personal history in order to convince a partner to have sex with them. Because behavioral bisexuality, extramarital contacts, and sex with prostitutes are often regarded as illicit or illegal, we included information on these activities in order to highlight their potential implications for the continued spread of HIV throughout the American population. With respect to the inclusion of data on fellatio and cunnilingus, although these behaviors are not considered to carry the same risk as anal or vaginal intercourse, they may present some risk given that they involve an exchange of bodily fluids.

All of our estimates on the prevalence of the six behaviors discussed below are conservative. They represent weighted means derived by multiplying the results from each of the relevant studies by the sample size, summing across studies, and dividing this sum by the total number of subjects across all studies. In general, because many of the available studies are commercial surveys conducted by women's magazines on convenient samples of readers, there is considerably more information on females than males.

HETEROSEXUAL AND HOMOSEXUAL ANAL INTERCOURSE

The identification of anal intercourse as a high-risk behavior in the transmission of HIV arose as a result of investigations into the sexual behavior of gay men—members of the first of several groups whose sexual behavior patterns eventually were associated with an increased prevalence of HIV infection. Although experience with anal intercourse among gay men is relatively prevalent, ranging from 50% (Kinsey data tabulated by Gebhard & Johnson, 1979) to 95% (reanalysis of data from A. P. Bell & Weinberg, 1978) (see also Chapter 19), it is often mistakenly assumed that *only* gay men, and not heterosexual men and women, engage in this behavior. However, data from three of the studies we reviewed, (reanalysis of data from A. P. Bell & Weinberg, 1978; Gebhard & Johnson, 1979; Hunt, 1974) conservatively indicate that at least 18% of heterosexual men (range: 8 to 30%) have engaged in anal intercourse. Data from a sample of older men indicated that 16% had engaged in heterosexual anal intercourse *since* the age of 50 (Brecher, 1984).

Our conservative estimate for the proportion of heterosexual women who have ever engaged in anal intercourse, based on data from seven studies (reanalysis of data from A. P. Bell & Weinberg, 1978; Bolling, 1977; Gebhard & Johnson, 1979; Hunt, 1974; Rubenstein & Tavris, 1987; Tavris & Sadd, 1975; Wyatt et al., 1988a), is 39% (range: 20 to 43%). In addition, one commercial survey (Wolfe, 1981) reported that 13% and one clinical study (Bolling, 1977) reported that 9% of Caucasian women *regularly* participated in anal intercourse. In an investigation of older adults, Brecher (1984) reported that 16% of the women studied had engaged in anal intercourse *since* the age of 50.

The fact that more women than men were estimated to have ever participated in anal intercourse (39% compared to 18%, respectively) suggests that the estimate for men, which was derived from much less abundant data than that for women, should be adjusted upward. Similar discrepancies in the number of women and men reporting incidence data for nonprocreative activities, including premarital fellatio, cunnilingus, and anal intercourse, can be found in Gebhard and Johnson (1979; see also Chapters 18 and 19, Voeller, 1988).

BEHAVIORAL BISEXUALITY

Data gathered during the last 40 years indicate that, even among people who identify an interest only in opposite-sex or same-sex partners and, thus, adopt either the label "heterosexual" or "homosexual," *behavioral* bisexuality is not rare. Discrepancies between self-labeled sexual orientation and actual behavior highlight the importance of obtaining extensive behavioral data rather than simply relying on self-labels in studies concerned with the sexual behaviors implicated in the transmission of HIV.

In 1948, Kinsey and his colleagues reported that, since puberty, 37% of single, married, and previously married Caucasian men between the ages of 16 and 55 had engaged in sexual activity to the point of orgasm with another man. A magazine survey in which male subjects, regardless of sexual orientation, were asked whether they had ever had sex with another man corroborates the data collected by Kinsey et al. (1948): 37% of the men surveyed reported that they had had at least one homosexual experience in adulthood (Athanasiou et al., 1970). A later Kinsey Institute study of people living in the San Francisco Bay area (reanalysis of data from A. P. Bell & Weinberg, 1978) found that 4% of heterosexual men had received and 3% had performed anal intercourse with another man. In this context, it is useful to note, data that provide indirect evidence of homosexual activity in at least some *married*, presumably "heterosexual" men: 70% of the white gay males studied by Kinsey et al. (1948) reported that they had had sex with a married man and 20% of these reported having done so with six or more married men.

Three of the studies under consideration in this paper asked self-identified gay men whether they had ever engaged in heterosexual intercourse (Athanasiou et al., 1970; reanalysis of data from A. P. Bell & Weinberg, 1978; McWhirter & Mattison, 1984). Sixty-two to 79% of gay men reported that they had engaged in coitus with a woman at least once. With regard to other heterosexual activities, the gay men participating in A. P. Bell and Weinberg's (reanalysis of data from 1978) study also reported that they had engaged in anal intercourse (9%), performed cunnilingus (28%), and received fellatio (41%) during sexual activity with women. Furthermore, data from four studies (reanalysis of data from A. P. Bell & Weinberg, 1978; Gebhard & Johnson, 1979; McWhirter & Mattison, 1984; Saghir & Robins, 1973) revealed that from 15 to 26% of self-identified gay men had been married. In McWhirter and Mattison's (1984) study, 26% of these men had children.

Although studies investigating homosexual activity among nominally "heterosexual" women are generally lacking, Kinsey and his associates (1953) found that 28% of single, married, and previously married Caucasian women between the ages of 12 and 45 had engaged in sexual activity since puberty to the point of orgasm with another woman. With regard to the cross-orientation behavior of *lesbian* women (based on two Kinsey Institute studies: reanalysis of data from A. P. Bell & Weinberg, 1978; Reinisch et al., in press), it appears that 81% have engaged in heterosexual intercourse. The more recent study (Reinisch et al., in press) included data that revealed that 45% of a sample of lesbian women had had sex with men since 1980, a year that marks the accelerated spread of HIV throughout our population. Underscoring the degree to which labels may be misleading, 43% of the women who had *always* (since age 18) labeled themselves as lesbian reported that they had had sex with a man since age 18, and 21% had done so since 1980 (Reinisch et al., in press).

Of all the lesbian women who had had sex with men since 1980, one third reported that at least one of their male partners had had sex with other men and, therefore, was behaviorally bisexual. It is especially significant that the women with behaviorally bisexual male partners were more likely to have engaged in anal intercourse than those who believed that all of their male partners were exclusively heterosexual (Reinisch et al., in press). In the earlier Kinsey Institute study, which addressed heterosexual behaviors among homosexual women (reanalysis of data from A. P. Bell & Weinberg, 1978), in addition to having engaged in coitus, 20% of lesbian women reported that they had also engaged in anal intercourse, 52% had performed fellatio, and 60% had received cunnilingus during sexual activity with a man.

With regard to the previous marital history of lesbian women, three studies (reanalysis of data from A. P. Bell & Weinberg, 1978; Reinisch et al., in press; Saghir & Robins, 1973) reported that from 20 to 35% have been married at least once. In addition, 15% of the lesbian women studied by Reinisch et al. (in press) had children.

EXTRAMARITAL CONTACTS

Extramarital sexual interactions are relevant to the AIDS epidemic because of the importance of multiple partners in predicting level of risk, at least among men, and the potential danger of infection to a monogamous partner, especially an unsuspecting and, thus, potentially unprotected one. Based upon data from six studies (Athanasiou et al., 1970; Blumstein & Schwartz, 1983; Gebhard & Johnson, 1979; Hunt, 1974; Pietropinto & Simenauer, 1977; Yablonsky, 1979), we conservatively estimate that at least 37% (range: 26 to 40%) of heterosexual men have engaged in sex outside of a marital relationship. To corroborate these data, we turned to information gathered by Kinsey and his colleagues, who noted that 38% of heterosexually active women report having engaged in coitus with a married man (Gebhard & Johnson, 1979). With regard to the extramarital activity of men after 10 or more years of marriage, 30% of the married men interviewed by Blumstein and Schwartz (1983) reported that they had had at least one affair. In a study of older men, Brecher (1984) reported that 23% of men in the Consumers Union sample had engaged in sex outside of marriage *since* the age of 50.

Data from nine studies (Athanasiou et al., 1970; R. R. Bell et al., 1975; Blumstein & Schwartz, 1983; Gebhard & Johnson, 1979; Hunt, 1974; Rubenstein & Tavris, 1987; Tavris & Sadd, 1975; Wolfe, 1981; Wyatt et al., 1988a) entered into our estimate of the prevalence of extramarital sexual behavior among married women. The conservative estimate for wives, based on these nine studies, is 29% (range: 20 to 54%). Data from men who participated in the original Kinsey studies also support this estimate: 36% of heterosexually active men interviewed by Kinsey and his colleagues reported that they had engaged in coitus with a married woman (Gebhard & Johnson, 1979). Addressing extramarital contacts among married women after 10 or more years of marriage, Blumstein and Schwartz (1983) noted that 22% of these women reported having had at least one extramarital affair. With respect to older women, 8% of the women studied by Brecher (1984) reported that they had engaged in extramarital sexual activity *since* the age of 50.

Sex With Female and Male Prostitutes

Although a large proportion of prostitutes have taken measures to protect themselves and their clients from sexually transmitted diseases and male-to-female transmission of HIV is somewhat more common than the reverse, sex with female prostitutes has been a source of concern with regard to the transmission of HIV. Limited information on the number of men who have paid for sex with a woman are provided by four studies. From these data (reanalysis of data from A. P. Bell & Weinberg, 1978; Brecher, 1984; Gebhard & Johnson, 1979; Hunt, 1974), we conservatively estimate that 33% (range: 30 to 45%) of men have had at least one sexual experience with a female prostitute. In a sample of older men (Brecher, 1984), 34% reported that they had paid a woman for sex prior to age 50, and 7% reported having done so *after* the age of 50. Interestingly, 13% of *gay* men interviewed by A. P. Bell and Weinberg (reanalysis of data from 1978) also reported having paid a woman for sex at least once.

Limited data exist for the number of gay men who have paid, or been paid by, another *man* for sex. Based on three studies that asked gay men if they had ever accepted money for sex with another man (reanalysis of data from A. P. Bell & Weinberg, 1978; Gebhard & Johnson, 1979; Saghir & Robins, 1973), we estimate that approximately 26% (range: 18 to 27%) have done so. Using data from the same three studies, the estimate for the proportion of gay men who themselves have paid another man to engage in sex is 25% (range: 24 to 26%).

Fellatio

Data on the prevalence of oral sex among heterosexual men and women were obtained in several of the studies we reviewed. Approximately 27% of the heterosexual men interviewed by Kinsey and his colleagues

(Gebhard & Johnson, 1979) reported that they had been orally stimulated during *premarital* petting or foreplay, and data from two studies (Gebhard & Johnson, 1979; Hunt, 1974) indicated that approximately 48% (range: 43 to 57%) of men have received oral sex during *marriage*. Three additional studies that asked subjects if they had *ever* participated in fellatio (Athanasiou et al. 1970; reanalysis of data from A. P. Bell & Weinberg, 1978; Blumstein & Schwartz, 1983) indicated that 83% (range: 79 to 90%) of married and single men have been orally stimulated by a female partner. In a study of men age 50 and older, 49% reported oral stimulation during sexual activity *since* the age 50 (Brecher, 1984). (See also Chapter 18.)

Data on the number of women who have performed fellatio corroborate the above findings: 41% of the women studied by Kinsey and his colleagues during the 1940s and 1950s (Gebhard & Johnson, 1979) had performed fellatio during *premarital* sexual activity and, based on three data sets (Gebhard & Johnson, 1979; Hunt, 1974; Tavris & Sadd, 1975), approximately 67% (range: 45 to 91%) of women had done so during *marriage*. Relying on data from the four studies that asked whether subjects had *ever* engaged in fellatio (Athanasiou et al., 1970; reanalysis of data from A. P. Bell & Weinberg, 1978; Blumstein & Schwartz, 1983; Wyatt et al., 1988a), our estimate of the proportion of women who have performed fellatio either *premaritally or maritally* is 84% (range: 59 to 93%), a proportion similar to the number of men who have received oral stimulation either before or after marriage. In addition, 43% of older women in the Consumers Union sample reported engaging in fellatio *since* the age of 50 (Brecher, 1984).

With regard to frequency, 40% of the women participating in *Redbook's* first survey (Tavris & Sadd, 1975) and 50% of those who completed *Redbook's* second survey (Rubenstein & Tavris, 1987) reported that they often perform fellatio during sexual activity with men—for some, at least half of the time. Eighty-four percent of women in a third commercial magazine survey said that they *regularly* performed fellatio during sexual activity with men (Wolfe, 1981).

CUNNILINGUS

Approximately 46% of the women interviewed by Kinsey and his colleagues (Gebhard & Johnson, 1979) reported that they had received oral stimulation from a male partner during *premarital* foreplay. Data from three studies (Gebhard & Johnson, 1979; Hunt, 1974; Tavris & Sadd, 1975) indicate that approximately 70% (range: 50 to 93%) of women have engaged in cunnilingus during *marriage*. These estimates of the prevalence of heterosexual cunnilingus both before and after marriage are supported by data from the five studies that asked if subjects had *ever* participated in cunnilingus (Athanasiou et al., 1970; reanalysis of data from A. P. Bell & Weinberg, 1978; Blumstein & Schwartz, 1983; Rubenstein & Tavris, 1987; Wyatt et al., 1988a) indicating that 94% (Range: 68 to 95%) of women have been orally stimulated by a man at least once. In a study of older women (Brecher, 1984), 49% reported that they had received cunnilingus *since* the age of 50.

Frequency can be established from the following studies. In the commercial survey analyzed by Wolfe (1981) 84% of the women reported *regularly* receiving oral sex. In the first (Tavris & Sadd, 1975) and second (Rubenstein & Tavris, 1987), *Redbook* surveys, respectively, 39% and 45% of this magazine's female respondents reported that they often receive oral stimulation during sex—for some, at least half of the time.

With regard to the proportion of men who have performed cunnilingus, Kinsey and his colleagues (Gebhard & Johnson, 1979) found that 14% of heterosexual men had orally stimulated a female partner during *premarital* sexual activity. The proportion of men who have done so during *marriage*, based on two studies (Gebhard & Johnson, 1979; Hunt, 1974) was 46% and 61%, respectively. Using data from the four studies that asked men whether they had ever participated in cunnilingus (Athanasiou et al., 1970; reanalysis of data from A. P. Bell & Weinberg, 1978; Blumstein & Schwartz, 1983; Pietropinto & Simenauer, 1977), we conservatively estimate that 82% (range: 76 to 93%) of single and married men have performed cunnilingus at least once. In a study of individuals age 50 and older, 56% of the men queried reported that they had engaged in cunnilingus *since* the age of 50. (See also Chapter 18.)

Conclusions

To underscore the point that sexual behaviors that place people at risk for infection with the AIDS virus (HIV) are not new, or limited to the culture of post–sexual revolution America, this chapter included a discussion of the representation of several AIDS risk–related behaviors in art and artifact from around the world dating from 400 years before the Christian era to the present. We also reviewed data from 17 sexual behavior studies conducted in the United States during the last 40 years in order to provide conservative estimates of the prevalence of behaviors associated with the transmission of HIV. Because the available data are limited, these estimates chiefly pertain to white, middle-class, relatively well-educated Americans, primarily between 20 and 45 years of age.

Collectively, these brief reviews of art and data provide evidence that behaviors implicated in the transmission of HIV, including anal intercourse, are not new or rare, either historically or culturally. Many behaviors associated with the sexual transmission of HIV appear to be universally practiced and are not limited to people who identify themselves as gay. In ancient times, sexual behaviors such as heterosexual and homosexual anal intercourse were depicted on common household items such as lamps and bowls and the highly revered funerary offerings that typically portrayed the daily life of people in some traditional cultures. During the centuries that followed, in both oriental and occidental cultures, these behaviors commonly appeared in the erotic renderings produced by fine artists for the nobility and aristocracy. Even as anal intercourse and homosexual relations became taboo and illegal in Western culture, artists still depicted them in their artistic creations. In short, even if the punishment involved imprisonment or death, as people continued to participate in these behaviors, artists continued to produce erotic images of them.

In some cultures, despite centuries of effort, societal regulation or eradication of heterosexual and homosexual anal intercourse, cunnilingus, fellatio, and a number of other socially disapproved behaviors has been unsuccessful.In the face of the AIDS crisis, it is accurate information and increased understanding, not denial and condemnation, that will assist in the development of techniques for helping people to be responsible in their sexual behavior. Although all too scarce and often flawed, research during the last 40 years on American sexual behavior highlights the grave lack of precise information on human sexuality and alerts us to the fact that none of us are really isolated or completely safe from the threat of AIDS. Precise research on sexual behavior—that is, who is doing what with whom, how often, and under what circumstances—as well as information on the attitudes that relate to these behaviors will ultimately provide the essential data base upon which effective programs of education and behavior change, so necessary to stemming the tide of AIDS and other sexually transmitted diseases, must be built. These data will also provide crucial information for conducting research on the efficacy of barrier methods to block infection with HIV and assist in the identification of appropriate subjects for biomedical solutions to AIDS, such as the development of effective vaccines. Because an appreciation of the methods and procedures necessary to derive accurate data on the sexual behavior patterns of the various ethnic, racial, social, regional, age, and sexual orientation groups that comprise our heterogeneous society are essential to the identification of people at risk for AIDS, we present below some key methodological considerations for the conduct of research on human sexual behavior, particularly as it relates to the transmission of HIV.

Some Key Methodological Considerations for Research on Human Sexual Behavior

The concept most essential to the conduct of scientifically sound research on American sexual behavior is the understanding that our society is comprised of many subcultural groups that are very likely to differ, both behaviorally and attitudinally, when it comes to the highly personal and sensitive matter of sex. In their attempt to assemble representative samples, researchers must be aware that subcultural groups such as blacks, Asians, and Hispanics are not homogeneous. For example, American blacks are minimally comprised of African-Americans, Caribbean blacks, and Haitians; American Hispanics include individuals of Cuban, Mexican, Puerto Rican, and other Central and South American cultural heritages; and American Asians are from Chinese, Indian, Japanese, Korean, Philippine, and Vietnamese cultural descent. Once representative samples have been secured, specific techniques must be employed to ensure the collection of meaningful and valid data. Several strategies that may be used to develop representative samples and collect valid and reliable data are briefly discussed below and addressed in greater detail in an earlier publication (Reinisch et al., 1988).

SAMPLING AND SUBJECT RECRUITMENT

Most sex research is conducted using convenience rather than probability samples. Although such studies provide useful insights into the range and complexity of human sexuality, such as the discrepancies that may exist in sexual orientation labels and sexual behavior (e.g., Reinisch et al., in press), they are inadequate for developing valid epidemiological estimates of sexual behaviors related to AIDS risk in our population. A number of factors should be considered in designing sex research:

1. The advantages of various sampling techniques that can increase representativeness, including stratified probability sampling, obtaining 100% participation from targeted groups, and use of multiple appeals to obtain participation from initially hesitant subjects.
2. The need to "oversample" minority groups, that is, obtain more subjects than required by simple probability or random sampling in order to capture the true diversity of behaviors and attitudes within such groups (see Wyatt, Chapter 4).
3. The importance of obtaining complete demographic information on all subjects so that comparability to other studies and limits on the generalizability of findings can be determined.

Subject recruitment can be assisted by (a) advertising the study with a title that will seem relevant and be inoffensive to all potential subjects, (b) assuring confidentiality, (c) convincing subjects that the investigators are not biased or judgmental, and (d) appealing to altruism, such as explaining to subjects that their data are important to public health concerns.

DATA COLLECTION AND INSTRUMENT DESIGN

There are also a number of considerations in the selection of methods and design of the data collection instruments. Generally speaking, interviews are preferable to questionnaires for several reasons, including the avoidance of validity problems deriving from a subject's failure to understand the meaning of questions and/or subject illiteracy. There are a number of techniques that can be used to enhance subjects' willingness to divulge highly personal data and to increase accuracy. In general, questions should be sequenced from those that are less sensitive to those that are more sensitive. Given the importance of obtaining accurate sexual histories, however, behavior questions should precede those on attitudes or knowledge. Sex histories should be obtained from at least 1980, a year that can be used to mark the accelerated spread of HIV in the United States.

Most important is the use of techniques that can help assure that the data collected will be meaningful. Survey instruments used in the collection of data on human sexual behavior must reflect a keen awareness of the different, sometimes multiple, meanings associated with various sexual behaviors as well as the subculturally specific terms used to refer to these behaviors. For example, in northwestern Mexico not all men who engage in anal intercourse with other men are regarded as homosexual (Carrier, 1985). Although the receptive male partner may be regarded as homosexual or gay, the insertive partner retains both his masculine and heterosexual status. Consequently, if the insertive partner were asked if he had ever engaged in "homosexual" activity, it is likely that he would answer no. In addition to their ambiguity, euphemisms such as "homosexual activity" reflect culturally specific assumptions about the relationship between sexual orientation labels and behavior and should always be replaced with behaviorally specific questions that identify the directionality of behaviors that may be either performed or received.

When it comes to terminology, there is not a universal nomenclature for sexual behavior. Textbook terms such as "coitus," "fellatio," and "cunnilingus" and "street" or slang terms such as "rimming," "fisting," and "going down" are usually educationally and subculturally specific. In addition, AIDS-related medical terms such as "seropositive" are most certainly not universally understood. A case in point was printed by *The Palm Beach Post* in December, 1987. During an educational program on AIDS presented before a Hispanic audience in which the term "seropositives" was frequently used to refer to persons who had developed antibodies to HIV, a woman who thought that the speakers were referring to her blood type (O+) stood up and said "I've been zero positive all my life. I haven't done anything wrong, and now you're telling me I'm going to die of AIDS" ("Cultural Barriers," 1987).

Investigators can enhance their confidence in sexual behavior data immeasurably by asking clear and straightforward questions, avoiding euphemisms, and using subculturally appropriate vernacular rather than sanitized, often incomprehensible words. In addition to incorporating the multidisciplinary input of experts in the fields of psychology, anthropology, sociology, demography, epidemiology, and medicine, as needed, any study that involves selecting and questioning human subjects about sexual behaviors should be designed and conducted with assistance from members of the community to be studied, including men and women of various ages, sexual orientations, ethnic, religious, educational, and professional backgrounds. Collectively, these individuals can contribute both expert and practical advice regarding the construction and administration of questionnaires and interviews, thus enabling researchers to avoid the collection of contaminated or invalid data through the use of inappropriate, imprecise, or ill-defined questions.

Although limited space does not permit us to include them here, additional considerations regarding the content of questions, use of recall data, selection of interviewers, and the limitations that face certain types of institutions when they attempt to conduct sex research are presented in Reinisch et al. (1988).

Summary

Comprehensive and accurate data on human sexual behavior and attitudes are requisite to coping with the AIDS crisis. They are important not only for identifying behavioral risk factors in nominally low-risk groups and designing and evaluating education and behavioral change programs, but also for developing effective communication about the sensitive matter of human sexuality. Although these data would greatly facilitate our efforts to control the AIDS epidemic, the task of obtaining valid and reliable data on human sexual behavior poses an especially difficult and complex task in a society in which sexuality is regarded as intimate, private, often embarrassing, perhaps socially disapproved, and even illicit or illegal. By illuminating the universal nature of those sexual behaviors implicated in the transmission of HIV in our review of art, artifact, and ephemera from around the world and providing estimates of the prevalence of these behaviors among those members of our society for whom data exist, we hope to illustrate that, in spite of our varying beliefs, at-risk sexual behaviors are not new, nor do they simply reflect the behavioral patterns of particular minority groups. Evidence attesting to the cross-orientation behaviors of both heterosexual and homosexual men and women questions the assumption that sexual orientation labels and other nominal indices of an individual's sexual "preferences" (such as whether or nor he or she is married) predict actual behavioral patterns. Although many individuals may behave in ways consistent with their orientation label (i.e., heterosexual or homosexual), in light of the serious threat of AIDS, researchers cannot afford to assume that current labels accurately reflect an individual's behavior over the life span or, in some cases, at the present time. Rather, they must demonstrate an awareness of the fact that in spite of our cultural and sexual diversity, all members of our society potentially live in overlapping communities of risk.

Acknowledgments

This work was supported in part by Public Health Service grants HD20263, HD17655, and DA05056 (to J. M. Reinisch) and The Kinsey Institute and partial faculty support of S. A. Sanders by MATEC grant HHS BRT 000033-02-0 (to J. Johnson-Deutsch).

We gratefully acknowledge the library and research assistance of C. Kaufman, E. Roberge, and N. Alfonso; the editorial comments of C. A. Hill; and the general support of S. Stewart Ham and The Kinsey Institute staff.

References

Athanasiou, R., Shaver, P., & Tavris, C. (1970, July). Sex. *Psychology Today, 4*(2), 39–42.

Bamert, A. (1980). *Africa: Tribal art of forest and savanna*. New York: Thames and Hudson.

Bell, A. P., & Weinberg, M. S. (1978). *Homosexualities: A study of diversity among men and women*. New York: Simon and Schuster.

Bell, R. R., Turner, S., & Rosen, L. (1975). A multivariate analysis of female extramarital coitus. *Journal of Marriage and the Family, 37*(2), 375–384.

Bentley, R. (1984). *Erotic Art*. New York: Gallery Books.

Blumstein, P., & Schwartz, P. (1983). *American couples*. New York: William Morrow and Company, Inc.

Bolling, D. R. (1977). Prevalence, goals and complications of heterosexual anal intercourse in a gynecologic population. *Journal of Reproductive Medicine, 19*, 120–124.

Bolling, D. R., & Voeller, B. (1987). AIDS and heterosexual anal intercourse. *Journal of the American Medical Association, 258*(4), 474.

Boswell, J. (1980). *Christianity, social tolerance and homosexuality*. Chicago: The University of Chicago Press.

Bowie T., & Christianson, C. V. (1970). *Studies in erotic art*. New York: Basic Books, Inc.

Brecher, E. M. (1984). *Love, sex, and aging: A Consumers Union report*. Boston: Little, Brown and Company.

Brendel, O. (1970). The scope and temperament of erotic art in the Greco-Roman world. In T. Bowie & C. V. Christiansen (Eds.), *Studies in erotic art* (pp. 3–69). New York: Basic Books, Inc.

Carr, F. (1972). *European erotic art*. London: Luxor Press Ltd.

Carrier, J. M. (1985). Mexican male bisexuality. *Journal of Homosexuality, 11*(1/2), 75–85.

Cochran, S. D., & Mays, V. M. (1990). Sex, lies, and HIV. *New England Journal of Medicine, 322*(11), 774–775.

Cultural barriers increase AIDS risk for Hispanics. (1987, December 29). *The Palm Beach Post.*

Dover, K. S. (1978). *Greek homosexuality.* London: Duckworth.

Gebhard, P. H. (1970). Sexual motifs in prehistoric Peruvian ceramics. In T. Bowie & C. V. Christiansen (Eds.), *Studies in erotic art* (pp. 109–144). New York: Basic Books, Inc.

Gebhard, P. H., & Johnson, A. B. (1979). *The Kinsey data: Marginal tabulations of the 1938–1963 interviews conducted by the Institute for Sex Research.* Philadelphia: W. B. Saunders Company.

Hunt, M. (1974). *Sexual behavior in the 1970's.* New York: Dell Publishing Co., Inc.

Johns, C. (1982). *Sex or symbol: Erotic images of Greece and Rome.* Austin: University of Texas Press.

Kinsey, A. C., Pomeroy, W. B., & Martin, C. E. (1948). *Sexual behavior in the human male.* Philadelphia: W. B. Saunders Company.

Kinsey, A. C., Pomeroy, W. B., Martin, C. E., & Gebhard, P. H. (1953). *Sexual behavior in the human female.* Philadelphia: W. B. Saunders Company.

Lorenzoni, P. (1984a). *English eroticism.* New York: Crescent Books.

Lorenzoni, P. (1984b). *French eroticism: The joy of life.* New York: Crescent Books.

McWhirter, D. P., & Mattison, A. M. (1984). *The male couple: How relationships develop.* Englewood Cliffs, NJ: Prentice Hall, Inc.

Melville, R. (1973). *Erotic art of the West.* New York: G. P. Putnam's Sons.

Mountfield, D. (1982). *Greek and Roman erotica.* New York: Crescent Books.

Padian, N., Marquis, L., Francis, D. P., Anderson, R. E., Rutherford, G. W., O'Malley, P. M., & Winkelstein, W. (1987). Male-to-female transmission of Human Immunodeficiency Virus. *Journal of the American Medical Association, 258*(6), 788–790.

Pietropinto, A., & Simenauer, J. (1977). *Beyond the male myth: What women want to know about men's sexuality.* New York: New York Times Books.

Rawson, P. (1968). *Erotic art of the East: The sexual theme in oriental painting and sculpture.* New York: G. P. Putnam's Sons.

Rawson, P. (1973). *Primitive erotic art.* New York: G. P. Putnam's Sons.

Reinisch, J. M., Sanders, S. A., & Ziemba-Davis, M. (1988). The study of sexual behavior in relation to the transmission of human immunodeficiency virus: Caveats and recommendations. *American Psychologist, 43*(11), 921–927.

Reinisch, J. M., Sanders, S. A., & Ziemba-Davis, M. (in press). Self-labeled sexual orientation, sexual behavior, and knowledge about AIDS: Implications for biomedical research and education programs. In S. J. Blumenthal, A. Eichler, & G. Weissman (Eds.), Proceedings of NIMH/NIDA Workshop, *Women and AIDS: Promoting Healthy Behaviors.* Washington, D. C.: American Psychiatric Press.

Rubenstein, C., & Tavris, C. (1987, September). Special survey results: 26,000 women reveal the secrets of intimacy. *Redbook,* pp. 147–149 and 214–215.

Saghir, M. T., & Robins, E. (1973). *Male and female homosexuality: A comprehensive investigation.* Baltimore: The Williams & Wilkins Company.

Tavris, C., & Sadd, S. (1975). *The Redbook report on female sexuality: 100,000 married women disclose the good news about sex.* New York: Delacorte Press.

Voeller, B. (1983). Heterosexual anal intercourse. *Mariposa Occasional Paper #1B,* 1–8.

Voeller, B. (1988, December). Heterosexual anorectal intercourse: An AIDS risk factor. *Mariposa Occasional Paper #10*, 1–19.

Wilson, S. (1973). Short history of Western erotic art. In R. Melville (Ed.), *Erotic art of the West* (pp. 11–31). New York: G. P. Putnam's Sons.

Winkelstein, W., Wiley, J. A., Padian, N., & Levy, J. (1986). Potential for transmission of AIDS-associated retrovirus from bisexual men in San Francisco to their female sexual contacts. *Journal of the American Medical Association, 255* (7), 901.

Wolfe, L. (1981). *The Cosmo report.* New York: Arbor House.

Wyatt, G. E., & Peters, S. D. (1986). Issues in the definition of child sexual abuse in prevalence research. *Child Abuse and Neglect, 10,* 231–240.

Wyatt, G. E., Peters, S. D., & Guthrie, D. (1988a). Kinsey revisited, Part I: Comparisons of the sexual socialization and sexual behavior of white women over 33 years. *Archives of Sexual Behavior, 17,* 201–239.

Wyatt, G. E., Peters, S. D., & Guthrie, D. (1988b). Kinsey revisited, Part II: Comparisons of the sexual behavior of black women over 33 years. *Archives of Sexual Behavior, 17,* 289–332.

Yablonsky, L. (1979). *The extra-sex factor: Why over half of America's married men play around.* New York: New York Times Books.

4

Maximizing Appropriate Populations and Responses for Sex Research

Gail Elizabeth Wyatt

Few would argue that the sexual patterns and preferences of men and women are one of the most difficult and sensitive areas of research (Martin & Vance, 1984). Sex research is controversial for a variety of reasons. Potential respondents see themselves as having an insatiable interest in other people's sexual behavior, but are often reluctant to discuss theirs, fearing that they will reveal a conflict within themselves about what they said they would not do but in reality have already done.

From the researcher's perspective, it is difficult to obtain funds to support such highly intimate research for two reasons: because it is so often fraught with methodological and sampling problems, and because social science research has not been a priority in federal funding for over 10 years (Tucker, 1984). Consequently, many scholars have sought funding through private or commercial sources, such as popular magazines and organizations whose primary motives may be to exploit the data for monetary gains rather than to broaden our understanding of human sexual patterns (Wyatt, in press).

From the reader's perspective, two assumptions are often made about what they read: the results are accurate, and the findings apply to them. Unfortunately, considering the methodological and funding problems in the field, those assumptions are rarely met.

We have learned so little about human sexuality and the factors that influence sexual patterns. With the reality, however, that diseases like AIDS are reaching and affecting a substantial proportion of people regardless of sexual orientation, ethnicity, income, or gender, we are faced with overcoming the obstacles that have limited what we know and generating research that (a) is as accurate as possible, (b) identifies "at-risk" sexual behaviors for large populations, and (c) is relevant to various ethnic, cultural, and religious groups and life-styles.

This chapter will review techniques to maximize responses of male and female multiethnic groups in sex research by examining sampling and other methodological issues that may help to reveal at-risk sexual patterns for diseases such as AIDS.

Obtaining Diverse Samples

Much of what we know about sexuality concerns either white, middle-class females or lower socioeconomic status black adolescents (Catania, Gibson, Chitwood & Coates, in press). Regardless of economic strata, age, or ethnicity, males have been less the focus of epidemiological studies in sex research (e.g., Zelnik & Kantner, 1972), because they tend not to volunteer as often, due to the expectation that their responses should reflect a notion of male sexual prowess rather than their own sexual experience (Kinsey, Pomeroy, & Martin, 1948). Consequently, research questions designed for females have sometimes been administered to males, and the findings have often not addressed gender-specific aspects of the socialization and sexual experience of men (Risen & Koss, 1987).

The pluralistic society in which sexual and health problems thrive requires that researchers attempt to obtain samples that represent the ethnic and cultural diversity of larger communities. If the goal of research is to recruit a diverse, multiethnic sample of males and females ranging across other critical demographic variables, then issues regarding sampling need to be reviewed first.

Probability Random Sampling

Probability sampling is a method of increasing the likelihood of selecting elements from a population in such a way that these elements are accurately described (Babbie, 1973). Examining the population parameters and estimates of error can help to determine if those aims were accomplished. A random selection ensures that each element has an equal opportunity of being selected in the sample. Most sex research has not used probability or random sampling techniques. When it has (Zelnik & Kantner, 1972, 1979), ethnic minority groups tend to be underrepresented, from the inner city, and poor. "Oversampling" of areas where ethnic minorities live or conduct other activities (attend religious services or attend school) ensures that samples will be larger and, it is hoped, representative of a more diverse cross-section of the desired group. It is also important, however, to compare the demographic characteristics of oversampled groups with those of the minority group sample generated through random techniques in order to avoid assumptions that both the samples are representative of the desired population.

Random samples are often stratified on such variables as age, gender, or sexual orientation in AIDS research. The results from studies such as these should be discussed with the sampling restrictions in mind, since stratification often requires weighting of variables in order to achieve generalizability.

Nonprobability Sample

This most commonly used sampling technique in sex research usually involves surveying specific residential areas (e.g., the Haight-Ashbury district), clients using mental health centers, homosexuals who frequent bars, or college students. While studies of sexual attitudes and behaviors usually use the latter group, studies of gay and lesbian sexual patterns often use persons from the former groups (Bieber et al., 1962; Curran & Parr, 1957; Haynes & Oziel, 1976; Hooker, 1957, 1958; LoSciuto, 1980; Saghir, 1980; Saghir & Robins, 1973; Weinberg & Williams, 1974). The limitation of purposive or nonprobability sampling is that it tends not to generate a broad range of elements from a population: samples may be void of ethnic minorities, or exclusively ethnic or middle class. Most of these may be clinical samples, persons recruited through friendship networks, or self-selected persons. Those who are not literate or who have another primary language, are mistrustful of well-intended professionals, or consider sex to be a private matter are unlikely to volunteer for sex-related research.

Consequently, purposive or nonprobability samples, although rich in generating responses to behaviors that others are reluctant to discuss (Joseph et al., 1984), are often biased on gender, age, ethnicity, income, language, and sexual preference. Oversampling techniques used with randomized methods appear to be the best approach to obtaining a broad sample, with potential for generalizability.

Question Format

The next important task is to use measures that will obtain information from respondents within the context of their culture, life-style, literacy, and sexual preference. Four methods will be reviewed: telephone surveys, paper-pencil tests, face-to-face interviewing, and the randomized response technique. The manner in which the format has been used, its pros and cons, and its sociocultural relevance will be discussed.

Telephone Survey

Use

This method has been used to assess attitudes regarding consumer product research, political issues, and natural catastrophes, to name just a few topics. Recently, the *Los Angeles Times* conducted a national survey on the prevalence of sexual abuse (Lewis, 1985) and sex-related surveys are increasingly being conducted over the telephone (Catania et al., in press). For example, there are several studies of sexual knowledge, attitudes, and behavior that are being conducted in the state of California on which this author and other professionals are consulting regarding the feasibility of telephone surveys with ethnic groups and adolescents.

PROS

Because the respondent is being interviewed on the telephone, the interview is usually time limited and relatively inexpensive (Bradburn, Rips, & Shevell, 1987). It is most useful in metropolitan areas where the majority of the population has single-line phones. Respondents: (a) are anonymous to the interviewer and consequently may be more likely to reveal socially unacceptable behaviors; (b) are usually not threatened even when the topic is highly intimate (Locander, Sudman, & Bradburn, 1976; Quine, 1985); and (c) tend to answer each question (Locander et al., 1976; Quine, 1985). This format does not encourage respondents to overstate their actual response, as might be the case if heterosexual men were asked about the number of female sexual partners with whom they had sex.

CONS

The length of the interview can negatively influence the reliability of the response (Bradburn et al., 1987). If too many questions are asked within a short period of time, the strategy that a respondent uses in recalling information can be compromised. Additionally, responses given over the telephone tend to be less precise than those obtained in pencil-paper or face-to-face formats.

In sex research, it is important to define and clarify terms used by both the respondent and the interviewer. This can prove to be a time-consuming process over the telephone, depending upon the educational level of the respondent. Consequently, response biases can be just as great as with face-to-face interviews. Finally, if rural or extremely indigent, homeless, or transient samples are sought, telephone surveys may greatly restrict the representativeness of the group selected to be interviewed, because of inconsistent availability of the telephone to these groups.

SOCIOCULTURAL RELEVANCE

Telephone surveys may be useful with third-world populations, if the language of the interviewer matches that of the respondent. Undocumented persons should not be threatened by an anonymous phone contact. However, the social acceptability of discussing sex or related health problems with an unknown person is a major issue among ethnic minorities and highly religious groups. Furthermore, there is less likelihood that people living in densely populated circumstances will disclose intimate and socially unacceptable information in the presence of other adults or children.

Paper-Pencil Measures (Self-Report)

Use

Self-report measures have been used to assess attitudes and behaviors, including sexual patterns (Joseph et al., 1984). They can be administered individually, in groups, or through the mail.

Pros

These measures: (a) are relatively inexpensive, completed in a short period of time, and easily readied for computer input; (b) reduce response error, specifically overstatements of socially desirable responses (Locander et al., 1976); and (c) allow the respondent privacy (LoSciuto, 1980) and self-pacing (Bishop, Hippler, Schwartz, & Strack, 1988). Also, because question construction is usually closed ended, internal consistency, test/retest reliability, or validity can easily be obtained (Quine, 1985).

Cons

In these measures: (a) the language and terms used may not be familiar to the respondent; (b) a certain level of literacy in the English language is required; (c) socially desirable behaviors are often underreported (Catania et al., in press), perhaps because respondents may not appreciate the importance of the accuracy of their response; (d) respondents can experience anxiety regarding the questions, depending upon the level of threat involved; and (e) respondents do not tend to be any more truthful than in a personal interview (Benson & Holmberg, 1985). Research assessing refusal rates on these measures indicates that respondents tend to give more incomplete answers (Johnson & Delameter, 1976). These studies have included the frequency of vaginal intercourse, masturbation, and number of sexual partners in the past year (Bradburn, Sudman, Blair, & Stocking, 1978; Catania, McDermott, & Pollack, 1986; Johnson & Delameter, 1976; Michael, Laumann, Gagnon, & Smith, 1988).

Sociocultural Relevance

Paper-pencil measures are highly biased toward literate, well-educated persons who are comfortable with discussing sex or are at least familiar with the appropriate anatomical labels and terms for body parts, problems, and diseases. People with reading and writing or learning difficulties, a right-to-left reading orientation, or who are not knowledgeable or comfortable with sex-related research may misinterpret the intent of questions and leave them unanswered. It is also important to remember that persons such as the undereducated who are not accustomed to or are unsuccessful at taking tests may not comply with requests to complete these kinds of measures.

Face-to-Face (Structured) Interview

USE

This format is used to assess various aspects of sexual experiences in research (Kinsey, Pomeroy, Martin, & Gebhard, 1953; Kinsey et al., 1948).

PROS

In the structured interview: (a) the use of terms can be modified within the context of the dialogue between interviewer and interviewee; (b) the responses tend to be no less truthful than in self-report measures (Benson & Holmberg, 1985); (c) there tend to be fewer nonresponses to questions (Locander et al., 1976; Quine, 1985); and (d) rapport can be established and the respondent's level of discomfort monitored during the interview. There was no difference found in internal reliability as indicated by consistency of responses repeated within the interview (Quine, 1985), and this format allows topics such as AIDS to be discussed openly in professional and confidential settings, where the respondents questions can also be answered.

CONS

This is an expensive and time-consuming method of data collection (Quine, 1985). Interviewers need to be well trained to conduct face-to-face interviews and to answer questions that respondents may have. Their gender, ethnicity, and language also needs to reflect that of the population being studied (Benson & Holmberg, 1985; Quine, 1985). Respondents may experience anxiety in discussing intimate details of their lives in the presence of a stranger, and the truthfulness of their responses may be influenced (Benson & Holmberg, 1985). They may also assess the social desirability of their responses and attempt to gauge the expectations of the interviewer (Bradburn et al., 1987; Miller, 1986).

SOCIOCULTURAL RELEVANCE

Regardless of its shortcomings, this is the best approach to establishing rapport, screening for the literacy and vocabulary of the respondent, clarifying the intent of questions, using anatomical charts, and allowing respondents to answer questions in more of an open-ended format. It is especially useful in epidemiological research that deals with sexuality, a topic that most people do not openly discuss.

Randomized Response Technique

This is an indirect approach to answering sensitive questions in a face-to-face format. It has been used in survey research (Locander et al., 1976), in telephone surveys of the prevalence of child sexual molestation (Lewis, 1985), and in this author's work on women's sexual experiences (Wyatt, Lawrence, & Vodounon, 1990). A good description of the technique is as follows:

> Suppose you want to ask a man whether he had sex with a prostitute this month. You would ask the question and then ask him to flip a coin. Then you would instruct him to answer "no" if the coin comes up tails and he has not had sex with a prostitute this month. Otherwise, he should answer "yes." Only he knows whether his answer reflects the toss of the coin or his true experience.
>
> Next, you would look at all the responses in your population. You know that half the people—or half the questionnaire population—who have not had sex with a prostitute are expected to get tails and the other half are expected to get heads when they flip the coin. For that reason, half of those who have not had sex with a prostitute will answer "yes" even though they have not done it. So whatever proportion of your group said "no," the true number who did not have sex with a prostitute is double that. For example, if 20% of the population you surveyed said "no," then you can conclude that the true fraction that did not have sex with a prostitute is 40%. (Cohen, 1986, p. 236)

PROS

This technique diminishes the threat that sensitive questions often elicit. Respondents tend to provide an answer to all questions, terms can be defined as they are in the face-to-face format, and the likelihood of over-reporting socially desirable acts is reduced. Randomized responses can be obtained in a "game"-like format, using playing cards, dice, or some other customized method that will allow for random numbers to appear. Consequently, it can be an "ice breaker" in an interview that also includes questions of a very serious nature.

CONS

This method is least effective in reducing overreporting of socially desirable behaviors, when compared to face-to-face, self-report, and telephone survey techniques. Since the method of generating responses is indirect, the respondent may be unaware that the interviewer can determine the accuracy of the response and may inflate his or her answers. The randomized response technique has been found to be equally threatening and anxiety provoking as asking questions directly (Locander et al., 1976). Indeed, responding in an undesirable way, according to the instructions of this method, can result in respondents directly telling the interviewer the accurate response. For example, in a recent study conducted to assess the utility of this technique in sex research (Wyatt et al., 1990), when asked whether or not they had ever had an abortion, respondents who were instructed to randomize their response to this question were often unwilling to do so, because of their objections to abortion.

SOCIOCULTURAL RELEVANCE

Requiring respondents to answer controversial questions in a manner that they perceive suggests behavioral patterns contrary to their religion, culture, or ethics can pose problems in research of this sensitive nature.

Summary

No method of data collection is superior to all others: each one has serious limitations. The goals of the research can best determine which of these is less objectionable.

Framework of Questions

The manner in which questions are worded and ordered can facilitate or limit the information obtained (Sudman & Bradburn, 1982). Questions are most frequently open or closed ended. The latter allow the respondents to respond in their own words, while the former limit the response to alternatives that are selected, based upon previous work or expectations of the researcher (Schuman & Scott, 1987). For example, this author developed the Wyatt Sex History Questionnaire (WSHQ), a 478-item structured interview, from two pilot studies with focus groups identifying critical issues regarding their sexual socialization and sexual experience (Wyatt, 1982, 1985). Many of the questions were open ended because: (a) this was the first structured interview designed for two ethnic groups, and ethnic and cultural sensitivity was being monitored in women's responses; and (b) the items were exploratory in nature and many had never been asked in previous sex research. After the WSHQ had been administered to 248 women, their responses were used to make decisions about closing many previously open-ended questions.

Both types of questions have their limitations. Respondents may choose among alternatives offered them in closed questions, but these may not be the most important issues on their minds (Schuman & Scott,

1987). Furthermore, the alternatives in closed questions may suggest that only normative responses are included (Sudman & Bradburn, 1982). Open-ended questions may be the best index of an individual's concerns, but coding the responses can be time consuming. Additionally, open-ended responses can limit data analyses, in that it may be difficult to order categorical responses.

The latter problems notwithstanding, open-ended questions are currently being piloted in this author's research regarding women's attitudes about the effect of AIDS on their sexual activity (Wyatt et al., 1990). Based upon several open-ended questions, one 25-year-old woman, married for 3 years, responded that she was not afraid of AIDS, nor had it influenced her sexual behavior because she was in a "strictly monogamous" relationship. Yet earlier in the interview, she reported having three extramarital relationships. If she were single, she said, she would "be more careful" in selecting sexual partners. It is doubtful that she considered herself at risk because of her brief marriage or because of her lovers.

The above-noted example demonstrates that in exploratory research such as this, open-ended questions are very useful in describing some of the conflicts that individuals face in their sexual relationships because of AIDS. Some trends and later hypotheses may be generated from initial findings such as these.

Other Issues

Ethnic Matching

Most recently, sex researchers are considering that the ethnicity of the interviewer can influence the respondent's level of comfort and can suggest that people like the respondent do discuss sex-related topics. Research has evolved quite a bit from the days when gender was not an issue (Kinsey et al., 1948, 1953), but ethnic matching is often compromised because reseachers complain about the expense of training additional personnel or the problems of finding well-suited ethnic candidates for interviewer training. All of these reasons do not obviate the importance of ethnic matching in sex research. To overlook its importance to an ethnic group member is to suggest that the research does not really intend to examine issues that can severely limit the data obtained. It is just that important. Likewise, when there is more than one language spoken in the desired sample, ethnic and language matching is also necessary. However, sometimes bilingual interviewers cannot understand the cultural nuances of a question or the response. Consequently, bilingual-bicultural interviewers are optimal for research with multicultural populations. Of course, when the optimum is not attainable, researchers should attempt to approximate it, but should also discuss the limitations of the research design in relation to their findings, if the match between interviewer and interviewee cannot be fully achieved. Other research suggests that both males and females are more likely to offer accurate sex-related responses to female rather than male interviewers (Catania et al., in press). In a recent pilot study of 12 gay and heterosexual Latins and African-American men's sexual practices, this author found that they most consistently preferred a female interviewer and not always someone of the same ethnicity. In Judith Becker's research (in press), adolescent respondents are allowed to select either an ethnic person of the same sex or a female interviewer. Perhaps this is one compromise to demonstrating sensitivity to both ethnic and gender matching.

Location of the Interview

As a part of the effort to ensure the comfort of the respondent, the location of the interview could facilitate or limit responses. In contrast to the more experimental model of research, wherein all stimuli are controlled in the respondent's environment, interviews about sex are rarely void of contextual and environmental elements. Although the setting can vary from individual to individual, this author allows respondents to choose the setting in which they will be most comfortable, rather than attempting to control it. If the home is chosen, the only prerequisite is that there be a room with a door that can be closed, to prevent other family members from interrupting or overhearing the conversation.

Although interviewing respondents at their chosen site increases both transportation and interviewer time and costs, it also suggests that the researchers acknowledge that sex is a highly personal topic, not easily discussed in a strange setting. From a practical perspective, since we found that ethnic minority group women most often declined to participate in a face-to-face interview, when we offered to inconvenience ourselves by traveling to respondents, rather than inconveniencing them, some of the reluctant respondents eventually agreed to be interviewed. It is also important to note that some ethnic minority women refused to come to the university, but agreed to meet at a site that was convenient to them. They were reluctant to be interviewed at home, because family members would not sanction the discussion of their sex lives or any other topic with a stranger, even one matching their ethnicity and language. Consequently, a physician's office that was conveniently located was used as a community site.

These are important issues to consider that can affect the refusal rate of a study and limit its generalizability.

Memory Error

In retrospective research, memory problems can adversely influence the accuracy and consistency of response (Bradburn et al., 1987). The following are common problems identified in research requiring the recollection of events.

The two most common problems are *omissions*, or forgetting, and *commissions*, where information is added or enhanced. In a study of the recollection of personal events, 20% of information was irretrievable after 1 year and 60% of information was lost after 5 years (Bradburn et al., 1987). However, studies of this nature rarely include sex-related information to be recalled. Long-term memory can often depend upon the events being recalled and their salience to the individual (Cash & Moss, 1972; Catania et al., in press).

When events are added or enhanced, they are often misplaced temporally, as if they occured earlier or later (Garobalo & Hindelang, 1977). This is called *telescoping*. *Retrospective interference* refers to more recent events clouding those in the past and *similarity interference* refers to similar events becoming indistinguishable from one another (Bradburn et al., 1987; Hunter, 1957; Klalzky, 1975; Linton, 1982; Loftus, 1980; Murdock, 1974). Sexual encounters with anonymous partners in a dimly lit setting frequented by the interviewee are an example of a case in which similarity interference may occur. Questioning that gradually separates each occasion may help respondents to recall specific aspects of similar events (Loftus & Fathi, 1985; Wagenaar, 1986). Finally, those life experiences that are frequently recalled often seem to be more recent than they really are (Bradburn et al., 1987). In order to avoid the *availability bias*, questions need to be introduced requiring the respondent to recall events that may have occurred long ago (Hastie & Park, 1986). Some of the techniques described below can facilitate the recollection of events that are not necessarily those most often recalled.

These sources of memory error can be remediated with the proper cues, some of which are described below.

Minimizing Memory Performance Error

A number of techniques are used to minimize memory performance errors. For example, *bounding techniques* help the respondent to use a particular time frame (e.g., the period of adolescence, a social occasion, or a certain location) to encompass the recollection of events (Huizinga & Elliott, 1984; Martin & Vance, 1984).

Researchers can also use *anchor points* to help respondents focus on a particular time period (Jenkins, Hurst, & Rose, 1979). Interviewers can briefly discuss personal events that occurred in close proximity to the desired behavior to be recalled, in order to help the respondent to focus on that time frame and enhance the memory of events that occurred during that time. The use of anchor points can also reduce telescoping errors. (Loftus & Marburger, 1983).

When critical information is given during an extensive interview, such as the respondent's age, religion, or age of onset of a particular sexual activity, it is sometimes useful for the interviewer to use a *tab* or a sheet easily in view with this information on it, so that when subsequent questions are asked that refer to the respondent's prior responses, they are more readily available. For example, if the respondent is 18 years old and became sexually active during that year but reports three pregnancies, the interviewer can recheck the responses to all of these questions with the subject to identify where the discrepancy may be found.

When respondents are asked about factual knowledge about a topic such as AIDS, they might use *interpolation* if they are unsure of their response (Bradburn et al., 1987). This technique involves guessing the lowest and highest values and giving a response in the medium range. For example, if a respondent is asked the number of safer sex techniques, he or she might recall between 1 and 10 and give the answer "7." *Relational reasoning* can also be used when the respondent is unsure of how to answer a knowledge-based question but has similar related knowledge. For example, if a respondent is asked the number of gay bars frequented in a certain city during a specific time period, he or she may recall the number visited in another city and answer on the basis of that information, given that the frequency of visitation was similar.

Recall and count involves respondents' remembering an event and being able to count all of their behavior. If asked the number of sexual partners in the last 6 months, some can recall better by naming each partner or by recalling specific events that facilitate the counting procedure. This technique is only useful when events or people are distinguishable from one another. Finally, respondents *decompose* events by breaking them down into subparts (Bradburn, et al., 1987). If respondents are asked the frequency of a behavior per month, they may first need to identify weekly frequency and multiply to obtain the monthly total. These techniques can improve the accuracy of survey responses.

Obtaining and Assessing the Consistency of Responses

When responses are prone to distortion, it is best to build in techniques to monitor the accuracy and consistency of the responses. Kinsey and colleagues (1948, 1953) repeated key questions during a structured interview and, for married couples, asked the spouse a set of questions to verify some of the respondent's responses. Wyatt (1985) asked both demographic and sexual behavior questions on several occasions over time, in different settings, and with different interview formats (Wyatt, et al., 1990). Zelnik and Kantner (1972) asked similar questions in different formats, some within an interview and others at the completion of a data-gathering session. It is also important to establish and maintain interrater reliability when face-to-face interviews are given. All of these and other more standard procedures can help to ensure that, in spite of memory error, information obtained is as reliable as possible.

Conclusions

The tasks of obtaining multiethnic samples, questionnaire construction, and minimizing potential response biases in sex-related research present many challenges to researchers who are examining the social and psychological aspects of AIDS upon male and female populations. This is obviously not an area that should attract those who are anxious to collect data quickly and not carefully. Likewise, those whose expertise is in identifying and treating AIDS patients should not assume that asking about sexual patterns and preferences in a sexual history is a straightforward endeavor, similar to taking a medical history.

Perhaps the tragedy of this disease will facilitate the cooperation of sex researchers and AIDS researchers in projects that will yield the most information that identifies high-risk sexual behaviors in relation to the transmission of the virus. These collaborative efforts cannot be initiated too soon. Enough lives have been lost because so many victims have been overlooked in research. There will not be an easy, inexpensive, or expedient solution to preventing the spread of AIDS until we recognize and become knowledgeable about the various modes of transmission with the ethnicity, income, marital status, sexual preference, gender, and life experience of those most at risk in mind.

Acknowledgments

The author wishes to thank The Women's Project Staff, Don Guthrie, Ph.D., Ray Mickey, Ph.D., and Gwen Gordon for data analyses and programming, and Jennifer Lawrence, M.P.H., for her invaluable assistance in manuscript preparation.

This research was funded by the Center for Prevention and Control of Rape, NIMH Grant RO1 MH33603 and through a Research Scientist Career Development Award, KO1 MH00269.

References

Babbie, E. R. (1973). *Survey research methods*. Belmont, Ca: Wadsworth Publishing Company.
Becker, J. (in press). The effects of child sexual abuse on adolescence sexual offenders. In G. E. Wyatt & G. J. Powell (Eds.), *The lasting effects of child sexual abuse*. Newbury Park, CA: Sage Publisher.

Benson, G., & Holmberg, M. B. (1985). Validity of questionnaires in population studies of drug use. *Acta Psychiatrica Scandinavia, 71,* 919.

Bieber, I., Dain, J. H., Dinco, R. P., Drellich, G. M., Grand, J. H., Gundlach, A. R., Kremer, W. M., Rifkin, H. P., Wilbur, B. G., & Bieber, B. T. (1962). *Homosexuality: A psychoanalytic study of homosexuals.* New York: Basic Books.

Bishop, G., Hippler, H. J. Schwartz, N., & Strack, F. (1988). A comparison of response effects of self-administered and telephone surveys. In R. Groves & P. Biemar (Eds.), *Telephone survey methodology* (pp 321–340). New York: John Wiley & Son.

Bradburn, N. M., Rips, L. J., & Shevell, S. K. (1987). Answering autobiographical questions: The impact of memory and inference on surveys. *Science, 236,* 157–161.

Bradburn, N., Sudman, S., Blair, E., & Stocking, C. (1978). Question threat and response bias. *Public Opinion Quarterly, 42,* 221–234.

Cash, W. A., & Moss, A. J. (1972). Optimum recall period for reporting persons injured in motor vehicle accidents. *Vital and Health Statistic, 50,* 133.

Catania, J. A., Gibson, D. R., Chitwood, D. D., & Coates, T. J. (in press). Methodological problems in AIDS behavioral research: Influences on measurement error and participation bias in studies of sexual behavior. *Psychological Bulletin.*

Catania, J., McDermott, L., & Pollack, L. (1986). Questionnaire response bias and face-to-face interview sample bias in sexuality research. *The Journal of Sex Research, 22,* 52–72.

Curran, D., & Parr, D. (1957). Homosexuality: An analysis of male cases seen in private practice. *British Medical Journal, 1,* 797–801.

Garobalo, J., & Hindelang, J. (1977). *An introduction to the National Crime Survey.* Washington, D. C.: U.S. Department of Justice.

Hastie, R., & Park, B. (1986). The relationship between memory and judgment depends on whether the judgment task is memory-based or on-line. *Psychological Review, 95,* 258.

Haynes, S. N., & Oziel, L. J. (1976). Homosexuality: Behaviors and attitudes. *Archives of Sexual Behavior, 5,* 283–289.

Hooker, E. (1957). The adjustment of the male overt homosexual. *Journal of Projective Techniques and Personality Assessment, 21,* 18–31.

Hooker, E. (1958). Male homosexuality in the Rorschach. *Journal of Projective Techniques and Personality Assessment, 22,* 33–54.

Hunter, I. M. L. (1957). *Memory.* Middlesex, England: Penguin Books.

Huizinga, D., & Elliott, D. S. (1984). *Self-reporting measures of delinquency and crime: Methodological issues and comparative findings* (pp. 58–61). Denver, CO: Behavioral Research Institute.

Jenkins, C. D., Hurst, M. W., & Rose, R. M. (1979). Life changes: Do people really remember? *Archives of General Psychiatry, 36,* 379–384.

Johnson, W., & Delameter, J. (1976, Summer). Response effects in sex surveys. *Public Opinion Quarterly,* pp. 165–181.

Joseph, J. G., Emmons, C., Kessler, R. C., Wortman, C. B., O'Brien, K., Hocker, W. T., & Schaefer, C. (1984). Coping with the threat of AIDS. *American Psychologist, 39,* 1297–1302.

Kinsey, A. C., Pomeroy, W. B., & Martin, C. E. (1948). *Sexual behavior in the human male.* Philadelphia: W. B. Saunders Company.

Kinsey A., Pomeroy, W., Martin, C., & Gebhard, P. (1953). *Sexual behavior in the human female.* Philadelphia: W. B. Saunders Company.

Klalzky, R. L. (1975). *Human memory: Structures and processes.* San Francisco: W. H. Freeman.

Lewis, I. A. (1985). Los Angeles Times Poll #98. Unpublished raw data.

Linton, M. (1982). Explorations in cognition. In U. Neisser (Ed.), *Memory observed* (p. 77). San Francisco: W. H. Freeman.

Locander, W., Sudman, S., & Bradburn, N. (1976). An investigation of interview method, threat and response distortion. *Journal of the American Statistical Association, 71,* 269–274.

Loftus, E. F. (1980). *Memory.* Reading, MA: Addison-Wesley.

Loftus, E. F., & Fathi, D. C. (1985). Retrieving multiple autobiographical memories. *Social Cognition, 3,* 280–295.

Loftus, E. F., & Marburger, W. (1983). Since the eruption of Mt. St. Helens, has anybody beaten you up? Improving the accuracy of retrospective reports with landmark events. *Memory and Cognition, 11,* 114–120.

LoSciuto, L. A. (1980). Discussion: Research on heterosexual relationships. In R. Green & J. Wiener (Eds.), *Methodology in sex research* (pp. 62–66). Rockville, MD: U.S. Department of Health & Human Services.

Martin, J. L., & Vance, C. S. (1984). Behavioral and psychosocial factors in AIDS. *American Psychologist, 30,* 1303–1308.

Michael, R., Laumann, E., Gagnon, J., & Smith, T. (1988, September 23). Number of sex partners and potential risk of sexual exposure to HIV. *Morbidity and Mortality Weekly Report,* No. 37, 565–568.

Miller, P. V. (1986). *Interviewing behavior as response context.* Paper presented at the National Opinion Research Center Conference on Context Effects in Surveys, Chicago, IL.

Murdock, B. B. (1974). *Human memory: Theory and data.* Hillsdale, NJ: Lawrence Erlbaum Associates.

Quine, S. (1985). Does the mode matter: A comparison of three modes of questionnaire completion. *Community Health Studies, 2,* 151–155.

Risen, L. I., & Koss, M. P. (1987). The sexual abuse of boys: Prevalence and descriptive characteristics of childhood victimizations. *Journal of Interpersonal Violence, 2,* 309–323.

Saghir, M. (1980). Homosexuality. In R. Green & J. Wiener (Eds.), *Methodology in sex research* (pp. 280–292). Rockville, MD: U.S. Department of Health & Human Services.

Saghir, M. T., & Robins, E. (1973). *Male & female homosexuality: A comprehensive investigation.* Baltimore: Williams & Wilkins.

Schuman, H., & Scott, J. (1987). Problems in the use of survey questions to measure public opinion. *Science, 236,* 957–959.

Sudman, S., & Bradburn, N. (1982). *Asking questions: A practical guide to questionnaire design.* San Francisco: Jossey-Bass.

Tucker, M. B. (1984). Is Afro-American studies research in jeopardy? A review of recent trends in federal research support. *CAAS Newsletter, 1,* 812.

Wagenaar, W. A. (1986). My memory: A study of autobiographical memory over six years. *Cognitive Psychology, 18,* 225–252.

Weinberg, M. S., & Williams, C. J. (1974). *Male homosexuals: Their problems and adaptations.* New York: Oxford University Press.

Wyatt, G. E. (1982). The sexual experience of Afro-American women: A middle income sample. In M. Kirkpatrick (Ed.), *Women's sexual experience: Explorations of the dark continent.* Plenum Press, New York.

Wyatt, G. E. (1990). Why we don't know more about Afro-American sexuality. In R. L. Jones (Ed.), *Black adult development and aging.* Berkeley, CA: Cobbs and Henry Publishing Co.

Wyatt, G. E., Lawrence, J., & Vodounon, A. (1990). *Using the randomized response technique in sex research*. Unpublished manuscript.

Zelnik, M., & Kantner, J. (1972). Sexuality, contraception, and pregnancy among young unwed females in the United States. U.S. Commission on Population Growth and the American Future. Demographic and Social Aspects of Population Growth. In F. Westoff & Parke, R. (Eds.), *Commission Research Reports, 1*. Washington, D. C.: U.S. Government Printing Office.

Zelnik, M., & Kantner, J. (1979). Sexual and contraceptive experience of young unmarried women in the United States, 1976 & 1971. In C. S. Chilman (Ed.), *Adolescent pregnancy and childbearing: Findings from research* (Publ. No. 79-4381). Washington, D. C.: U.S. Government Printing Office.

5

Methodological Issues in the Assessment and Prediction of AIDS Risk–Related Sexual Behaviors Among Black Americans

Vickie M. Mays and Susan D. Cochran

AIDS poses a grave threat to Americans, particularly for blacks. Blacks, depending upon their gender, region of residence, and history of drug use, may be from 3 to 21 times more likely than whites to be infected with human immunodeficiency virus (HIV), the infectious agent responsible for AIDS (Selik, Castro, & Pappaioanou, 1988). Within the black community, both sexual behavior (heterosexual and male homosexual) and the sharing of drug paraphernalia associated with intravenous (IV) drug use are the primary infection vectors.

To date, research on AIDS-related sexual behaviors among blacks is quite limited. Direct generalizations from studies of nonblack samples to understanding the behaviors of black Americans need to be greatly tempered by an understanding of the influence of cultural differences on the choices, risk assessments, and behaviors of this community. This chapter will focus on methodological issues associated with studying AIDS-related sexual behavior among blacks. These issues involve understanding cultural context as a determinant of behavior; cultural biases inherent in standard methodological, sampling, and measurement procedures; and cultural biases in our attempts as social scientists to model and predict individual risk reduction behaviors within the various subpopulations of the black community.

Understanding Cultural Influences on Behavior Related to AIDS Risk

In general, there is little scientific literature available to provide us with well-grounded information about the sexual lives of black Americans. Clearly, however, cultural differences between blacks and other ethnic groups, including whites, do exist for both heterosexuals (Wyatt, Peters, & Guthrie, 1988) and homosexual men (Bell & Weinberg, 1978; Wyatt et al., 1988).

In assessing sexual behavior, cultural and sex role norms as well as the interpersonal context of behavior are important mediating factors in interpreting the data we collect. That is, we must not forget that in measuring behavior our interests go far beyond the counting of discrete behavioral units, even assuming that the definition of these units is not subtly culture bound. A particular sexual practice, such as anal intercourse, performed by two Euro-American men could have very different social meanings than the same behavior performed by two Mexican-American males (Carrier, 1988). The same sexual practice performed by a heterosexual or a homosexual couple may lead to differing interpretations about the nature of each relationship and the probability of the occurrence of the behavior again. Equally true, among heterosexuals, women's perceptions of a sexual act and its meaning may differ dramatically from those of men. Thus, although behaviors may be equivalent, the psychological precursors of those behaviors may differ, affecting our ability as researchers to predict or modify the risk-related activities of individuals or to generalize findings beyond the specific group of study. This underscores the fact that in examining cultural influences on sexual behavior, we need to understand not only the proximal sexual behaviors themselves, but the behaviors in their cultural and interpersonal contexts.

Cultural Influences on Sexual Behavior

Weinstein (1987) delineated three possible functions of sexual behavior: procreation, recreation, and an expression of emotional connectedness. Clearly these functions are not immune to cultural influences (Cochran, 1989). The relative importance of the procreative aspects of sexual behavior can vary significantly depending upon one's religion (Furstenberg, 1972; Gregersen, 1986; Spilka, Hood, & Gorusch, 1985); sexual orientation (Blumstein & Schwartz, 1983); age as related to cohort differences in fertility (Byrne & Fisher, 1986); or the value that an ethnic group, family, or society places on children (Day & Mackey, 1988; Mays & Cochran, 1988b). Sex, both as a recreational activity and as an expression of emotional connection, can also be influenced in similar ways.

There is yet a fourth component to sexuality: sex as power, or sex as an expression of one's self in the world, an assertion of self, or self-procreation (Cochran, 1989; Mays & Cochran, 1988b). This notion may be particularly salient in the context of being relatively powerless. In the black community, sex sometimes has less of the private meanings implied in the three definitions offered by Weinstein and can take on this fourth, more political, meaning. For example, in the early 1970s, blacks viewed the use of contraceptives as a form of genocide promulgated by white Americans. The ability to reproduce was seen as a powerful tool in the struggle for liberation.

Sexuality, historically, has been regarded quite seriously in the black community. The term "swinging singles" is foreign to the black experience, although black adults are more likely to be unmarried than white adults (Mays & Cochran, 1988b). Swinging has for the most part been viewed by black Americans as a phenomenon characteristic of the loose sexual morals of whites. Yet, blacks are as likely as whites to be sexually active. While there is little recognition by black Americans of the terms "swinging singles" or "wife-swapping" in reference to sexual experiences (Staples, 1973), there are contexts in which similar behaviors occur. Black adults are more likely to be unmarried and experience greater instability in their love and sexual relationships than white adults (Mays & Cochran, 1988b). Black adolescents are more likely to begin sexual intercourse at an earlier age than whites (Brooks-Gunn, Boyer, & Hein, 1988). Overall, in the black community there is a greater likelihood of sexual activities for survival or in exchange/barter for needed resources (Mays & Cochran, 1988b, 1990a).

Sociocultural factors have a definite influence on sexual behaviors and activities (Gibbs, 1986). Underemployed or unemployed teenagers and adults who are coping with the inequalities of society while seeking a sense of belonging, creativity, or achievement may find sexuality a ready means to demonstrate manhood or womanhood through having children and being sexually active (Wilson, 1986).

AIDS researchers are not ignorant of this notion of sexual behavior as a statement of self-identity. Many AIDS-related risk reduction interventions directed at the gay white male community seek to meld individuals' needs for self-expression with the practice of safer sex. This has not always been so. Early in AIDS work, monogamy was advocated for gay men, of course, without any effort aimed at legislative changes in the marriage laws that would encourage this behavior further. This advice ignored the political reality of gay male life and the cultural differences between homosexual and heterosexual sexual behavior and relational attachments. The valuing of nonmonogamy and the acceptance of a different life-style for sexual relating was a key fight in the gay liberation movement (Shilts, 1987). Current risk reduction efforts have tempered early advice out of sensitivity to these realities.

There are similar issues for black Americans. Sexual behavior can sometimes be seen as making a statement in response to the larger political reality of society. This may be forgotten when we experience confusion, rather than understanding, if a poor black woman continues to have children in the face of economic poverty, endangering her health and the health of the child (Mays & Cochran, 1988b). Yet it is fully understood when a white family shows great pride in the birth of a child who is destined to inherit a family fortune or a family tradition. Researchers understand desires for immortality among themselves as they write papers and conduct research on what they might hope will become an oft-cited theory. However, we may fail to understand this same hunger for immortality in others if we lose sight of the fact that, for many people, having a child is the most creative thing they will ever do. This is especially so for poor women, whose only hope for a change in a family's social or economic status may be to bring a child in the world who will achieve in ways she and her family of origin never had the opportunity of doing (Mays & Cochran, 1988b). The dream is that this child, in contrast to any previous offspring, will be the one to begin a family legacy, a legacy in which she, the mother, will be remembered as having equipped that child to overcome all the odds stacked against herself, her family, and her community.

Sex *is* a serious endeavor in the black community, but a lack of cultural sensitivity to this may be translated into inappropriate AIDS risk reduction methods. Many of our risk reduction messages for the practice of safer sex market these activities within a context of fun. Some of our risk reduction materials suggest safer sexual substitutes such as massage, hugging, holding, cuddling, or showers together. The most obvious examples of this theme of fun come in the large printed message of "PLAY IT SAFE" or "PLAY SAFE." The thrust of these messages is sex as a leisure activity or sex as play (Mays & Cochran, 1988b).

The marketing of this message has many embedded assumptions, some of which clash with cultural, religious, and class-related behaviors of black Americans. First, "sex as fun" and "sex as play" stand in direct conflict with many fundamental religious teachings about sex. In both the black and Hispanic communities, traditional religion serves as a cornerstone, organizing not only personal values, but community life (Mays, 1989). Promoting risk reduction within the context of "sex as fun" may lose the attention of many individuals for whom religion is a guiding force and the mainstay of their support system. A message of "sex as play" also creates a barrier to the utilization of churches as a network for the delivery of AIDS education.

Second, sex as fun and leisure paints a vacation-like framework for viewing sexual activity. It presumes that individuals have the time and privacy to fully enjoy their sexuality. Parents with children at home or individuals with demanding work or travel schedules can clearly understand that "sex as play" is a pleasant diversion experienced occasionally, but not necessarily routinely. Yet routine sex must also be safer sex. A poor, single woman, responsible for child care of her young children, having worked either inside or outside of the home full-time, and currently involved with a man who lives elsewhere with perhaps his own commitments, may have a different context for viewing sexual behavior. It may not necessarily be one of play. She may need whatever support he can give her, whether economic or emotional. Her choices of sexual behaviors are complex, determined not only by her fear of AIDS, but the other more pressing realities of her life.

Sexual Attractiveness/Activity as a Resource or Commodity

In an economically impoverished community, sexual attractiveness is an important resource for black women, 65% of whom over the age of 15 years are not married (compared to 43% of white women) (Bureau of the Census, 1983). Sexual involvement may significantly improve a woman's economic position (Mays & Cochran, 1988b). For a woman to insist that her sexual partner use condoms when other readily available partners may not could destroy her tenuous hold on a developing relationship. Indeed, an insistence on premarital celibacy may seriously undermine a poor woman's ability to enter into a serious, committed relationship in the first place, when sexuality is a prime means of generating attachment to her partner.

Sexual Orientation from a Black Perspective

Typical views of sexual orientation postulate that individuals' sexual feelings and behaviors lie somewhere on a bipolar dimension where one extreme is heterosexuality, the other homosexuality, and in between is bisexuality (Bell & Weinberg, 1978). However, ethnicity or culture can function as an interactive factor influencing the expression of sexual object choice and sexual orientation identity (Cochran & Mays, 1988b).

Homosexuality within the black community is not necessarily consistent with the white gay life-style. Many blacks believe that homosexuality is a white phenomenon (Mays & Cochran, 1987). Among black gay and bisexual men, the issue of primary identification along ethnic lines or the dimension of sexual orientation is quite salient, reflecting the multiple social identities that accrue to individuals who are minorities within minorities. Recent research on black lesbians suggests that for these women ethnic identification is primary, while gay identification is secondary (Cochran & Mays, 1986; Mays & Cochran, 1986). This hints that within the black community, there may be greater separation of sexual behavior from labeling of sexual orientation. That is, individuals may engage in homosexual behavior, but not perceive themselves as homosexual. Community ties may also encourage greater levels of heterosexual behavior in those who do label themselves as homosexual (Bell & Weinberg, 1978; Cochran & Mays, 1988a, 1988b).

Researchers not familiar with this issue of multiple identities for black gay men may misunderstand this separation that can occur between sex as a behavior and sex as a statement of sexual orientation. For whites, the hidden assumption is that the act of sexuality has certain predetermined meanings—for example, "I am gay," "I am in love," or "We are getting married." All of these are individual reactions dependent upon one's ability to translate a behavior into a social reality or life direction. For black gay men, economic and emotional commitments to the black community, which, because of its fundamentalist religious ties, may be particularly homophobic, may result in extensive integration into heterosexual life-styles. This, of course, does not preclude homosexual sexual behavior, but may affect one's self-identification as being a gay man.

Therefore, it is important when asking about sexual behaviors that preconceived judgments about the meanings of those behaviors not be made. Determination of meaning is highly dependent on cultural influences. However, we cannot ignore meanings. Comprehensive understanding of sexual behavior within its psychological context is important because that is the way in which behavior can be altered. Risk reduction interventions can be developed that produce change by allowing the core motivations to have other expression while reducing the behavior that is risky.

Differences in Intimate Relationships

Male-Female Relationships

Although it appears that blacks hold the same ideals for marriage and family relationships as whites, Tucker, and Mitchell-Kernan (in press) reported that in practice the sex ratio imbalance among black heterosexuals has influenced the development of alternative forms of male-female relationships. Sex ratio imbalance refers to differences between the genders in number of eligible partners with whom to establish a relationship. Among blacks, there are far fewer eligible males than available females. The researchers observe that the effect of this is greater instability of relationships and higher rates of sexual activity outside of marriage. At a psychological level, the overrepresented gender, in this case women, experience their options in choosing mates as limited and may be more likely to tolerate objectionable behavior. The underrepresented gender, men, may view their options as limitless, resulting in less pressure to develop commitments, greater power within relationships, and fewer behavioral controls. The effects of sex ratio imbalance, coupled with the economic realities of poverty that encourage individuals to seek relationship partners who can provide financial support (Scott, 1980), create an environment in which friendship-based models of close relationships are inappropriate (Mays & Cochran, 1988b).

This has important implications for development of preventive intervention models in the black community. Interventions that emphasize egalitarian negotiation, as among best friends, will fail to address the oftentimes divergent goals of relationship partners.

Male-Male Relationships

The specific experiences of black gay men have not received much attention from researchers (Bell, Weinberg, & Hammersmith, 1981; Cochran & Mays, 1988b). With the appearance of AIDS and the higher than expected infection rates among black gay men, researchers have begun to seek reasons for these differences through comparing data gathered from black and white gay men. However, questionnaires, sampling procedures, and topics of focus usually do not emerge from the concerns of the black gay community (Mays & Jackson, in press). Instead, there has been a tendency to use white gay male patterns of behavior as a template for comparing black and white gay men.

Given the black-white differences in family structure and sexual patterns between black and white heterosexuals (Guttentag & Secord, 1983; Spanier & Glick, 1980; Staples, 1981a, 1981b), there does not appear to be a good empirical basis upon which to assume that black gay men's experiences of homosexuality perfectly conform to those of whites (Bell et al., 1981). Instead, one might predict that many of the same sociological factors that influence black heterosexual relationships would also have an impact on black gay men. Thus, racial discrimination, less availability of same ethnic group partners, fewer social and financial resources, residential immobility, and hampered employment opportunities might result in differential patterns of socializing, stability of relationships, and choices of relationship partners (Beame, 1983; Cochran & Mays, 1988b; Soares, 1979).

Female-Female Relationships

We have not discussed same-sex female relationships because of the extremely low incidence of female homosexual sexually transmitted HIV infection (Marmor et al., 1986). Of the first 2,200 U.S. AIDS cases in women diagnosed since 1981, 46 had reported sexual contact with women (Kahn, 1987). Thirty-six of these women were IV drug users, four had had sexual contact with men in high-risk groups, two had received contaminated blood transfusions, and two came from countries where heterosexual transmission is more common (i.e., Central Africa or Haiti). The remaining two women had unknown risk factors. Thus, as a group, lesbians are not at significant risk for HIV infection. Nevertheless, there are some individuals within this population who, through their behaviors, are at higher risk (Mays & Cochran, 1988a).

Although very little is known about the relationship experiences of black lesbians (Bell & Weinberg, 1978; Mays & Cochran, 1988a; Peplau, Cochran, & Mays, 1986), there is reason to be concerned about the risk to black lesbians of HIV infection relative to white lesbians (Cochran & Mays, 1988a). It is not being a lesbian that inherently removes risk; rather it is the behaviors that are most typical of lesbians that result in the low incidence of HIV infection. Lesbians tend to have far fewer sexual partners than gay men (Blumstein & Schwartz, 1983) and it is not clear how efficient viral transmission is with normative manual-genital or oral-genital sexual behaviors in lesbians (Friedland & Klein, 1987). Yet black lesbians, as a result of activities that occur in higher incidence in the black population (drug use, less separation between black gays and lesbians, poverty, poor health care), may be more at risk for HIV infection not from lesbian-related behaviors but from their drug and sexual involvement in the broader black community (Cochran & Mays, 1988a).

Issues in the Perception of AIDS Risk

In assessing AIDS-related risk reduction behaviors, it is important that we examine perceived risk as a context for understanding behavior change (Mays & Cochran, 1988b). For blacks, this is a particularly important issue. Without the perception of risk the individual may not be motivated to alter sexual practices, and even with the perception of risk, should this risk perception be inaccurate, the individual may change his or her behavior, but not effectively. Unsafe sexual behavior will not be perceived as risky if: (a) the individual is unaware of the relation between behavior and level of risk; (b) the individual is aware but devalues the risk to the group (e.g., the black community); or (c) the individual is aware of risk to the group but devalues the extent of personal risk (Weinstein, 1987).

Most blacks, particularly when their lives have involved poverty, drug abuse, or street prostitution, have lived with risks of some kind (Mays & Cochran, 1988b). AIDS simply joins the list of threats with which one needs to be concerned. These individuals have long coped with both higher levels of omnipresent danger and lower levels of resources with which to combat them (Mason, Ogden, Berrett, & Martin, 1986). Understanding poor ethnic individuals' response to AIDS involves knowledge of both their perception of its relative riskiness in comparison to more proximal threats and the existence of resources available to behave differently.

There are other reasons why blacks who are most at risk for acquiring an HIV infection may be less than optimally concerned. Some blacks even today still consider AIDS to be a white gay disease (Mays & Cochran, 1987). In a sample of black college students surveyed in 1986, we found that almost 50% worried very little or not at all about getting AIDS (Mays & Cochran, 1990c). Approximately 30% had done nothing to reduce their chances of getting a sexually transmitted disease. Also, contrary to the truth, they viewed blacks as significantly less likely than whites to get AIDS.

Cultural Factors Affecting HIV Transmission

As shown in Tables 5-1 through 5-3, blacks are more affected by AIDS and have a greater likelihood of becoming infected. Table 5-1 presents the cumulative incidence rate by ethnic group; Table 5-2 the relative risk, as compared to whites, for each minority group by risk categories as of 1987; and Table 5-3 the number of reported cases through July,

Table 5-1
Number of AIDS Cases and Cumulative Incidence Rates by Ethnic/Race
Group, United States as of March 1, 1990[a]

	White	Black	Hispanic	Asian	Native American
Total cases	69,743	34,431	19,565	772	163
Cumulative incidence[b]	37.2	114.0	98.6	11.8	9.6

[a] AIDS cases from the Centers for Disease Control (1990).
[b] Per 100,000 using July 1, 1988 government estimates of resident U.S. population, except for Asian and Native American ethnic groups, for which total U.S. population, including overseas, is utilized (U.S. Bureau of the Census, 1990).

Table 5-2
Relative Risk of AIDS in Minority Risk Groups Versus White Risk Groups,
United States, 1981–1987[a]

Risk Group	Black	Hispanic	Other
Adults	3.1	3.0	0.4
Adult males	2.8	2.7	0.4
Adult females	13.2	8.6	0.9
Homosexual males	1.4	1.7	0.3
Bisexual males	3.8	2.7	0.5
Heterosexual IV drug abusers	19.9	19.3	0.3
Children	12.1	6.8	0.8

[a] Adapted from Curran et al. (1988).

Table 5-3
Total Number of Reported AIDS Cases by Risk Group for Each Ethnic/Race Group, United States, through July, 1990[a]

	White		Black		Hispanic		Asian/ Pacific Islander		Native American		Total[b]	
Males												
Homosexual/bisexual	60,322	(80%)	14,107	(44%)	8,873	(46%)	643	(80%)	109	(63%)	84,241	(65%)
Intravenous drug user	4,687	(6%)	11,163	(34%)	7,416	(38%)	23	(3%)	18	(10%)	23,379	(18%)
Homosexual/bisexual IV drug user	5,618	(7%)	2,560	(8%)	1,375	(7%)	16	(2%)	26	(15%)	9,609	(7%)
Heterosexual contact	474	(<1%)	2,164	(7%)	253	(1%)	6	(<1%)	2	(1%)	2,904	(2%)
Hemophilia/coagulation disorder	1,027	(1%)	78	(<1%)	96	(<1%)	14	(2%)	8	(4%)	1,227	(<1%)
Receipt of blood products/tissue	1,537	(2%)	299	(<1%)	169	(<1%)	43	(5%)	1	(<1%)	2,055	(2%)
Other/unknown	1,615	(2%)	1,396	(4%)	911	(5%)	46	(6%)	8	(4%)	4,012	(3%)
Children (under age 13)	329	(<1%)	649	(2%)	333	(2%)	8	(1%)	2	(1%)	1,326	(1%)
Total male cases	75,609	(100%)	32,416	(100%)	19,426	(100%)	799	(100%)	174	(100%)	128,753	(100%)
Females												
Intravenous (IV) drug user	1,452	(38%)	3,999	(52%)	1,386	(47%)	12	(16%)	15	(50%)	6,877	(47%)
Heterosexual contact	1,026	(27%)	2,234	(29%)	966	(33%)	24	(32%)	7	(23%)	4,271	(29%)
Hemophilia/coagulation disorder	24	(<1%)	6	(<1%)	1	(<1%)	0	(0%)	0	(0%)	31	(<1%)
Receipt of blood products/tissue	851	(22%)	262	(3%)	155	(5%)	25	(34%)	2	(7%)	1,296	(9%)
Other/Unknown	245	(6%)	505	(7%)	151	(5%)	9	(12%)	3	(10%)	920	(6%)
Children (under age 13)	206	(5%)	623	(8%)	301	(10%)	4	(5%)	3	(10%)	1,138	(8%)
Total female cases	3,804	(100%)	7,629	(100%)	2,960	(100%)	74	(100%)	30	(100%)	14,533	(100%)
Total cases	79,413		40,045		22,386		873		204		143,286	

[a] Data from the Centers for Disease Control (1990). Percentages are for each risk group calculated within gender and ethnic group.
[b] Total includes individuals for whom ethnic background is unknown.

1990 by risk factor and ethnic group (Centers for Disease Control, 1990). In total, whites account for 55% of cases, blacks for 28%, and Hispanics for 16%. In the U.S. population, however, whites represent 76% of all Americans, blacks 12%, and Hispanics 8% (U.S. Bureau of the Census, 1990).

The data in the tables indicate that both blacks and Hispanics are at higher risk of HIV infection than whites. Similarities between blacks and Hispanics in patterns of drug use, poverty, and other cultural factors may account for the higher than expected rate for ethnic groups. However, the implication also is that cultural factors specific to blacks may place them at a higher risk for HIV infection than whites and other ethnic groups.

Friedland and Klein (1987) have made the point that HIV is not an efficient virus. In most instances, it takes frequent and sufficient contact with the virus in order for infection to occur. In certain subpopulations, such as hemophiliacs, the virus was transmitted quickly and efficiently because of repeated and/or substantial contact with the virus through receipt of contaminated blood products.

In the black community, there are behaviors that also facilitate a relatively more efficient transmission of the virus. Some of these are quite proximal to transmission (e.g., high levels of IV drug use with sharing of drug paraphernalia). Some are more distal and socioculturally based. These include instability of relationships, little or no prenatal care, and conditions of poverty. Given the inefficient nature of the virus, it is important to examine these sociocultural factors because they mediate the efficiency of the transmission, much more so than if the virus did not depend on intimate behavioral choices of individuals in order to infect.

Intravenous Drug Use

Over 50% of AIDS cases in blacks are either primarily or secondarily related to IV drug use (Centers for Disease Control, August, 1990). For whites, this applies to 16% of cases. The reason for this difference stems from the fact that IV drug use is more common in the black community (Gary & Berry, 1985). In the urbanized Northeast, HIV infection is endemic among IV drug users, who are most likely to be black (Ginzburg, MacDonald, & Glass, 1987). Questions concerning IV drug use, and social and intimate contact with IV drug users, are therefore more critical when researching AIDS issues in black populations.

Yet, there are few large-scale investigations involving IV drug users, particularly female IV drug users. Gay men have been viewed as ideal subjects for participation in biomedical and psychosocial studies. Their language skills, access to health care, and willingness to participate have resulted in several very large studies, such as the Multicenter AIDS cohort studies described by Detels (Chapter 1) and Stevens et al. (Chapter 2). The result is large bodies of data on white gay men's immunological, serological, and psychosocial profiles. The mode for much of this collection of data has been university based, in which subjects come directly to the university hospital or a satellite clinic. Many universities do not have good access to IV drug users or drug abuse treatment centers. In university-based models of research, IV drug users are often perceived as poor subjects because of their higher risk for attrition, other drug-related infections, and unwillingness to volunteer. Many of these are accurate and valid concerns. Nonetheless, the research designs are frequently experienced as culturally insensitive and are not based on cultivated relationships benefiting *both* the university researcher and the population of IV drug users he or she seeks to study.

This does not have to be so. An alternate research model involves equal collaborations between university researchers and those in community-based substance abuse and health care organizations that traditionally service IV drug users. Investigators worried about attrition will find that in methadone maintenance clinics many participants come to the clinic as often as 4 days a week. Collection of both biomedical and psychosocial data from these individuals is quite feasible.

Heterosexual Behavior

Community activities and norms regarding premarital and extramarital sexuality and condom use play a role in the transmission of HIV infection. Structural demographics such as the ratio of available black men to women and sex as a survival strategy and/or an economic resource affect sexual activities outside of marriage and committed relationships. This complexity of sexual activities in the black community can neither be solved nor researched through mere frequency counts of behavior occurrences. Rather, exploration of the context or behavioral ecology of sexual activities is key (Barker, 1968; Flora & Thorensen, 1988). To best determine influencing factors in perception of risk, negotiation of safer sex, use of condoms, or other AIDS prevention behaviors of black heterosexuals, the questions must incorporate the psychosocial forces that influence sexual and drug use behavior. Instead of focusing exclusively on the use of condoms, additional research questions might address the motivations for sexual activities. Do these activities occur out of a context of psychological needs, economic factors, or gender or cultural roles?

Male Homosexual Behavior

Because of somewhat negative attitudes within the black heterosexual community surrounding homosexuality, the extent to which anal intercourse between men is the actual infection vector of HIV transmission may be difficult to determine accurately (Cochran & Mays, 1988b). Apart from the reports of black gay men, consensual or nonconsensual sexual activities between men in such environments as prison or the armed services, where there are extended periods of male-only isolation, may easily be forgotten when recounting sexual histories. While little empirical information on the frequency and type of same-sex behaviors in prison is available, accounts by prisoners indicate anal intercourse and oral-genital sex occur even between men who view themselves as heterosexual (Harding, 1987). The occurrence of such incidents is likely to be higher in blacks than whites since black men as a group are more likely to be incarcerated or to serve in the armed forces. Some have questioned the veracity of the military recruit study results in which sex with prostitutes was identified as the primary risk in contracting an HIV infection (Potterat, Phillips, & Muth, 1987; Voeller, Chapter 19).

Homophobia may also be one factor that accounts for the bisexual behavior in the black community at large. Some black males attracted to other males find it to difficult to maintain a primary gay life-style and still stay a part of the black community (Cochran & Mays, 1988b). For some the answer is maintenance of a heterosexual life-style with homosexual behavior in the background.

An early study of homosexual sexual behavior in San Francisco suggested that the practice of anal intercourse was higher in black gay/ bisexual men than in white gay men (Bell & Weinberg, 1978). In a more recent study of a cohort of HIV-infected black gay and bisexual men, the proportion of receptive anal-genital contact with ejaculation was not significantly different from a comparative group of white gay and bisexual men (Samuel & Winkelstein, 1987). This latter study suggests that sexual risk factors as currently established may not explain well the differential rates of seropositivity or seroconversion between blacks and whites (see also Voeller, Chapter 19).

Poverty

Economics constitute an important backdrop for the day-to-day behavioral choices of many inner-city black Americans. In poor, urban black communities, selling of drugs can be a vital economic base, thus encouraging IV drug use. Poverty undermines stability of heterosexual relationships by making stable, economically successful individuals more attractive to others and by discouraging commitments, such as by a young female to a young, unemployable male, that can cause potential financial hardship (Tucker & Mitchell-Kernan, in press). Poverty also results in poorer health care, lower levels of education, and higher rates of morbidity and mortality from diseases (U.S. Department of Health and Human Services, 1986).

Summary

When all of these factors are taken into account it may explain the higher numbers of blacks at risk for AIDS. As discussed below, current data used to project future cases are based on models that do not necessarily figure these sociocultural elements into their predictions.

Research Issues in Sampling and Measurement

Recruitment

The difficulty of maintaining a desired response rate by black Americans is a major problem for large-scale survey researchers. Nonresponse is affected by a variety of factors, some more obvious than others, including concerns about confidentiality, lack of interest, time factors, or distrust of researchers (Berk, Wilensky, & Cohen, 1984; Mays & Jackson, in press).

Often overlooked, but critical, is the political context of AIDS-related data collection. Perceptions that the federal government might blame blacks for AIDS and its spread influence participant cooperation. Stories in community newspapers or community forums suggesting that AIDS results from government germ warfare only serve to raise the refusal rate for those AIDS-related studies requiring blood drawing. Fears or beliefs that blacks are perceived as expendable by the federal government and will be injected with an experimental virus influence choices to participate in research studies. Poor response rates are then attributed to a lack of concern on the part of the black community.

Researchers may find it necessary to adapt the introduction of their study to the community's perceptions of AIDS. Given the ongoing nature of the politicalization of AIDS, vigilance to these perceptions is necessary.

Sampling

Traditional sampling procedures that rely on Census data or exclude institutionalized populations result in sampling frames that undercount black Americans, particularly black males (Mays & Jackson, in press). Sampling frames employed to simultaneously recruit both blacks and whites from the same geographic area often result in small numbers of blacks who are nonrepresentative of the black population. This often occurs as a function of segregated housing practices and household income differences. Sampling procedures must be employed that take into account structural and demographic differences present in the black population (Jackson & Hatchett, 1985).

For example, in attempting to recruit a diverse sample of black gay men, the use of outreach efforts channeled through gay community networks will reach a very specific subpopulation of black gay men. Those men not integrated as a function of social networks or residential housing/socialization patterns seldom become respondents in most AIDS studies. Men who do not identify themselves as gay or who wish not to be identified as having associations with the gay community often are not represented, in either the black or white community (Voeller, Chapter 19). What emerges is the importance of a working knowledge of various ethnic subgroups in order to design a study that reaches the greatest number of subpopulations.

Of particular concern are those studies that gather data using telephone methodology that does not take into account the characteristics of those individuals who are less likely to have a telephone or who share a phone, making it difficult to get accurate answers to sensitive questions about sexual behaviors. The greater tendency of blacks to be in institutional settings such as prison, the military, or board-and-care facilities, or to be homeless, influences the segment of the population reached by telephones. If the study were one of attitudes about AIDS, results are clearly confounded. On the other hand, attitudes of individuals in institutional settings may not be relevant because their behavior is frequently moderated by rules of the institution rather than their personal attitudes. For the prisoner who both desires to be sexually active and believes that condoms should be used, if prophylactics are unavailable his beliefs may have only limited translation into behavior (e.g., abstinence).

Measurement

READING ABILITY

Literacy levels are of particular concern with studies that employ paper-pencil measures (Mays & Jackson, in press). Many AIDS-related materials are beyond the average comprehension levels of individuals (Hochhauser, 1987). Equally important are those studies that seek to evaluate awareness, knowledge, or attitudes purely through exposure to written media. Communication researchers have devoted considerable resources to examination of how people use the media (Salomon & Cohen, 1978). One source of measurement error in some AIDS assessments is the measurement of factors dependent upon exposure to mainstream written sources (Allen, 1981; Allen & Bielby, 1979a, 1979b). In general, the media have been viewed as having played a tremendous role in making individuals aware of the AIDS epidemic without any consideration of differences in skills, access to, or abilities in relation to different types of media. Individuals with lower reading levels or less interest in mainstream white society are less likely to read mainstream daily newspapers or magazines such as *Time, Newsweek,* or the *Atlantic Monthly* (Mays, 1989).

Preliminary results of a study of young ethnic minority adults indicate that television is the source from which they get most of their information (Mays & Cochran, 1990b). However, special programs aside, most television coverage of AIDS involves short segments on the news or 30- to 60-second public service announcements. These sources can hardly impart the same amount of information as written material.

LANGUAGE DIFFERENCES

In measuring sexual behavior, it is important for researchers to be sensitive to the cultural nuances of black American life. These range from differences in expectations about sexuality to the particular language individuals use to refer to specific sexual behaviors (Mays & Jackson, in press). Clearly, to the extent that differences are understood, there will be better guidance in the choice of questions, the language used in assessing behavior, and strategies for subject recruitment.

Language differences are an important methodological concern. For example, among white gay men, oral-anal sex is referred to as rimming. Some black gay men have different terms (e.g., "tossing salad," "eating chocolate chip cookies"), but associate the term "rimming" with whites, even though they may also engage in exactly the same behavior. Therefore, if they are advised to avoid rimming, they may forego oral-anal sex with white men, but not black men. In many instances, body parts are referred to in terms associated with cultural folklore. "Boning," a term that has characteristically referred to a specific type of heterosexual vaginal intercourse, has become more popular in its use as a result of Spike Lee's recent movie "School Daze." Researchers attempting to assess the practice of vaginal intercourse may find they get a more accurate response from subjects using this term rather than more clinical descriptions (e.g., vaginal penetration with the penis).

RELEVANCE OF QUESTIONS

Questionnaires, surveys, and interviews are not without their intended outcomes. Participants in studies know this and will sometimes track the bias of researchers even while participating in the study (Mays & Jackson, in press). For gay men who are sensitive to possible homophobia in researchers, this can dramatically influence their willingness to respond accurately to questions. Similar issues exist for black Americans, who may distrust the motives of white researchers.

AVAILABILITY OF APPROPRIATE NORMS

Many standard measures have not been normed on black populations. Difficulties can arise when norms developed for white populations are applied inappropriately. For example, this controversy surrounds the use of intelligence and personality tests (Mackenzie, 1980, 1984). Researchers need to be sensitive to utilizing measures with appropriate norms when possible and, when this is not possible, to modifying their piloting procedures and data interpretations as needed.

Problems in Predicting the AIDS Epidemic Among Blacks

Efforts to predict both individuals' behaviors and the path of the HIV epidemic within the black community introduce additional issues related to cultural sensitivity.

Theoretical Models of Behavior

Many of the attitude-behavior models currently being used to predict AIDS-related behavior, such as the Health Belief Model (Becker & Maiman, 1975), include assumptions that are often rooted in Euro-American world views or social class values, which are inconsistent with the views and values of many of black Americans at risk. Most attitude-behavior models assume that people are *motivated* to pursue rational courses of action. They further assume that people have the *resources* necessary to proceed directly with these rational decisions. Barriers to a rational course of action are trivialized as "moderators" rather than viewed as the structure within which people may function. In addition, the influential effects of ubiquitous contradictory values experienced by individuals who are not from the dominant culture are ignored. For example, most individuals, no matter how poor, have access to television and thus are exposed to cultural values consistent with the majority culture. However, minority cultural values are also present, placing before the individual, perhaps, a more complex array of values from which to draw. The synergistic effects of this are unknown.

For inner-city poor black Americans confronting an environment in which much of their surrounding milieu is beyond their personal control, models of human behavior that emphasize individualistic, direct, and rational behavioral decisions overlook the fact that many blacks do not have personal control over traditional categories of resources—for example, money, education, and mobility. In this context, intention may not always lead to the desired behavior, as suggested by Ajzen and Fishbein's (1980) Theory of Reasoned Action. Instead, intention may lead one down an indirect path in which the behavioral outcome is jury-rigged from whatever resources are available.

For black Americans, social norms and extent of commitment to social responsibilities may be better predictors of future behavior than intentions lacking in resources for translation into effective action. Black Americans are less likely than whites to value the individualistic focus of white culture. It may be helpful to remember that Kwanzaa, an indigenous Afro-American holiday, celebrates the community's core values of unity, black community self-determination, collective work and responsibility, cooperative economics, purpose, creativity, and faith (Mays & Cochran, 1988b). For blacks, individualistically oriented behavior is frequently tempered by social responsibilities.

Statistical Models of Infection Rates

While much is known about the prevalence of HIV infection, projected trends of disease incidence are less reliable. Many of the surveys and studies used to estimate incidence are based on samples not representative of the general population (Centers for Disease Control, 1987). Currently the best incidence estimates of AIDS exist for white gay men. This subgroup has assisted in a relatively more accurate determination of the rate of seropositivity through their tireless participation in numerous studies of behavioral and biological status. Coupled with previously stockpiled frozen serum donated by gay men participating in earlier infectious disease studies, this has greatly advanced knowledge of HIV infection determinants in white gay men. Unfortunately data of this nature have not been available on black Americans.

The two highest risk categories for blacks are IV drug users and homosexual/bisexual men. Historically, neither of these groups have constituted the subject of focus to any great extent in biomedical research. Few research data exist from these groups for a variety of reasons, ranging from investigators' perceptions of them as a "hard-to-reach" population to fear of exposure of homosexual behavior to a lack of cultural relevance of the research conducted in the past.

Several techniques have been employed to estimate the prevalence and incidence of AIDS. Each has assumptions that may or may not be useful in the determination of AIDS in the U.S. black population.

ISSUES OF RELATIVE RISK

Relative risk refers to the comparison between two groups of disease incidence rates in the total at-risk population. Variations within each of the two populations separately as to level of risk for any individual are ignored. Since cultural factors mediate levels of risk, their inclusion in forecasting the infectious disease path may prove helpful, particularly in understanding what the future holds for black Americans. One important point to remember is that, unfortunately, traditional surveillance techniques report by groups and not by risk behavior. Thus, blacks are, in part, more at risk than other groups because of the higher incidence of IV drug use in the black community. Intravenous drug abuse and its related heterosexual spread contribute to a markedly higher risk rate for blacks and Hispanics. In fact, blacks account for 51% of heterosexual transmission cases associated with IV drug use, while Hispanics com-

prise another 27% (Centers for Disease Control, 1990). So it is not genetics, but patterns of behavior in environments in which HIV is likely
to be present that result in the higher than expected infection rates in
blacks and Hispanics.

In calculating the risk merely by ethnic group, it is important to determine whether this is truly a useful procedure for assessing risk. For
example, when the risk of AIDS for adult black Americans as a population is calculated, the relative risk as compared to whites is 3.1 (Curran
et al., 1988). Conversely, if one is interested in a specific group of blacks,
such as women IV drug users, the relative risk is 13.2 compared to whites
(Selik et al., 1988). Although the general population of heterosexuals
may be at low risk for AIDs, the odds of HIV infection by ethnic group
indicate a very higher probability for women in the black population.

Similar differences would be seen if we calculated the relative risk for
blacks when compared to whites by geographic region (Selik et al., 1988).
Black heterosexuals living in Ames, Iowa, are not at as high a risk for
contracting HIV as blacks living in New Jersey or Miami. These differences in the transmission pattern by ethnic group and geography call
into question the wisdom of modeling the disease for all cases rather
than by risk factor and ethnic group.

EXTRAPOLATION OF PREVALENCE FROM OBSERVED RATES

As indicated in the discussion of relative risk, in a general population
individuals vary in their level of risk for HIV infection. The bias in models
extrapolated from observed rates to predict the incidence of AIDS in the
black population is a function of the extent to which persons at high or
low risk were included or excluded, the geographic region from which
the sample was drawn, and the demographic composition of the sample
surveyed. Few of the past studies used to determine seropositivity or
seroconversion rates have had adequate numbers of black gay or bisexual men in their samples. The heterosexual seropositive samples have
relied heavily on the military screening program or adults attending
sexually transmitted disease clinics (Centers for Disease Control, 1987),
which, while providing a large pool of ethnic minority subjects, are
skewed samples of the black population.

Recent research suggests that seroconversion rates are slowing among
gay men as a result of changes in behavior (Centers for Disease Control,
1987), although there are indications that black gay men may not be
changing their behavior as rapidly as white gay men (Landrum, Beck-
Sague, & Kraus, 1988; Samuel & Winkelstein, 1987). Conversely, cases
due to IV drug use and heterosexual transmission, categories in which
blacks are dramatically overrepresented, are increasing (Centers for Disease Control, 1987). Thus, our estimates of the future of this epidemic
for black Americans may be least accurate.

Summary

We attempted here to raise some of the subtle and not so subtle issues that are involved in the study of AIDS risk–related sexual behavior in black Americans. There are several other issues equally important and compelling. For example, much of the public policy and funding allocation flows from our epidemiological data. Yet our lack of certainty about something as simple as the number of people in the true population of IV drug users hampers the accuracy of our predictions. Our ethnic category of "blacks" ignores the diversity of the black population at risk (Afro-Americans, black Cubans, black Caribbeans, and Africans residing in the United States), since each may be located sociobehaviorally within a different risk category (Mays & Jackson, in press). We use epidemiological models that assume exchangeability within and across ethnic groups. These models are flawed with confounders given that there are many indications that AIDS among black Americans is epidemiologically different.

In an epidemic, these issues can translate into the unnecessary loss of lives. It is essential that in our assessment and prediction of AIDS risk–related sexual behaviors among black Americans, we proceed with sensitivity to the obvious, as well as subtle, effects of ethnicity and culture. It is important that the attempt be made not to do this in a monolithic manner, ignoring the rich diversity of the black population. Methodological approaches must emerge out of the diversity of the population and not be determined by the structure imposed by funding. Assessment of AIDS and HIV infection must not flow from a position of cultural ignorance nor be designed in haste. To the extent that HIV infection is facilitated by our cultural discriminations, we as scientists cannot afford discriminatory approaches in looking for solutions to stopping the spread of HIV infection.

Acknowledgments

This chapter, which was originally prepared in December 1987 and revised in May 1988, was supported in part by awards from the National Institute of Mental Health (Contract No. 87MO198343, Grant No. 1 RO1 MH 42584-01) to both authors; a U.S. Public Health Service Biomedical Research Support Grant from the University of California at Los Angeles, a National Research Service Award (T32 HS 00007) from the National Center for Health Services Research and Health Care Technology Assessment to the first author; and a California State University/Northridge Foundation grant to the second author. Work on this chapter was completed while the first author was a National Center for Health Services Research Fellow at the Rand Corporation, Santa Monica, CA.

References

Ajzen, I., & Fishbein, M. (1980). *Understanding attitudes and predicting social behavior.* Englewood Cliffs, NJ: Prentice-Hall.

Allen R. L., & Bielby, W. T. (1979a). Blacks' attitudes and behaviors toward television. *Communication Research, 56,* 488–496.

Allen, R. L., & Bielby, W. T. (1979b). Blacks' relationship with the print media. *Journalism Quarterly, 6,* 437–462.

Allen, R. L. (1981). The reliability and stability of television exposure. *Communication Research, 8,* 233–256.

Barker, R. (1968). *Ecological psychology.* Stanford, CA: University Press.

Becker, M. H., & Maiman, L. A. (1975). Sociobehavioral determinants of compliance with health and medical care recommendations. *Medical Care, 13,* 10–24.

Beame, T. (1983). Racism from a Black perspective. In M. J. Smith (Ed.), *Black men/white men: A gay anthology.* San Francisco: Gay Sunshine Press.

Bell, A., & Weinberg, M. (1978). *Homosexualities: A study of diversity among men and women.* New York: Simon and Schuster.

Bell, A. P., Weinberg, M. S., & Hammersmith, S. K. (1981). *Sexual preference: Its development in men and women.* Bloomington: Indiana University Press.

Berk, M. C., Wilensky, G. R., & Cohen, S. B. (1984). Methodological issues in health surveys. *Evaluation Review, 8,* 307–326.

Blumstein, P., & Schwartz, P. (1983). *American couples.* New York: William Morrow and Company.

Brooks-Gunn, J., Boyer, C. B., & Hein, K. (1988). Preventing HIV infection and AIDS in children and adolescents: Behavioral research and intervention strategies. *American Psychologist, 43,* 958–964.

Byrne, D., & Fisher, W. A. (1986). In D. Byrne & K. Kelley (Eds.), *Alternative approaches to the study of sexual behaviors.* Hillsdale, NJ: Lawrence Erlbaum Associates.

Carrier, J. M. (1985). Mexican male bisexuality. In F. Klein & T. Wolf (Eds.), *Bisexualities: Theory and research.* New York: Haworth Press.

Centers for Disease Control. (1990). HIV/AIDS surveillance report. Atlanta, Georgia.

Centers for Disease Control. (1987). Human immunodeficiency virus infection in the United States: A review of current knowledge. *Morbidity and Mortality Weekly Report, 36* (suppl. no. S-6), 1–48.

Cochran, S. D. (1989). Women and HIV infection: Issues in prevention and behavior change. In V. M. Mays, G. W. Albee, & S. F. Schneider (Eds.), *Primary prevention of AIDS: Psychological approaches.* Beverly Hills, CA: Sage Publications.

Cochran, S. D., & Mays, V. M. (1986, August). *Sources of support in the Black lesbian community.* Paper presented at the meetings of the American Psychological Association, Washington, DC.

Cochran, S. D., & Mays, V. M. (1988a). Disclosure of sexual preference to physicians by Black lesbian and bisexual women. *Western Journal of Medicine, 149,* 616–619.

Cochran, S. D., & Mays, V. M. (1988b). Epidemiologic and sociocultural factors in the transmission of HIV infection in Black gay and bisexual men. In M. Shernoff & W. A. Scott (Eds.), *A sourcebook of gay/lesbian health care* (2nd ed.). Washington, DC: National Gay and Lesbian Health Foundation.

Curran, J. W., Jaffe, H. W., Hardy, A. M., Morgan, W. M., Selik, R. M., &

Dondero, T. J. (1988). Epidemiology of HIV infection and AIDS in the United States. *Science, 239,* 610–616.

Day, R. D., & Mackey, W. C. (1988). Children as resources: A cultural analysis. *Family Perspective, 20*(4), 251–264.

Flora, J. A., & Thorensen, C. E. (1988). Reducing the risk of AIDS in adolescents. *American Psychologist, 43,* 965–970.

Friedland, G. H., & Klein, R. S. (1987). Transmission of the human immuno-deficiency virus. *New England Journal of Medicine, 317,* 1125–1135.

Furstenberg, F. (1972). Attitudes toward abortion among young Blacks. *Studies in Family Planning, 3,* 66–69.

Gary, L. E., & Berry, G. L. (1985). Predicting attitudes toward substance use in a Black community: Implications for prevention. *Community Mental Health Journal, 21,* 112–118.

Gibbs, J. T. (1986). Psychosocial correlates of sexual attitudes and behaviors in urban early adolescent females: Implications for intervention. *Journal of Social Work and Human Sexuality, 5,* 81–97.

Ginzburg, H. M., MacDonald, M. G., & Glass, J. W. (1987). AIDS, HTLV-III diseases, minorities and intravenous drug abuse. *Advances in Alcohol and Substance Abuse, 6,* 7–21.

Gregersen, E. (1986). Human sexuality in cross-cultural perspective. In D. Byrne & K. Kelley (Eds.), *Alternative approaches to the study of sexual behavior.* Hillsdale, NJ: Lawrence Erlbaum Associates.

Guttentag, M., & Secord, P. F. (1983). *Too many women: The sex ratio question.* Beverly Hills, CA: Sage Publications.

Harding, T. (1987). AIDS in prison. *Lancet, ii,* 1260–1264.

Hochhauser, M. (1987, August). *Readability of AIDS educational materials.* Paper presented at the meetings of the American Psychological Association, New York.

Jackson, J. S., & Hatchett, S. J. (1985). Intergenerational research: Methodolog-ical considerations. In N. Datan, A. L. Green, & H. W. Reese (Eds.), *Intergenerational networks: Families in context.* Hillsdale, NJ: Lawrence Earl-baum Associates.

Kahn, E. (1987). Lesbians and AIDS: Everything you need to know about AIDS prevention. *On Our Backs,* 13–15.

Landrum, S., Beck-Sague, C., & Kraus, S. (1988). Racial trends in syphilis among men with same-sex partners in Atlanta, Georgia. *American Journal of Public Health, 78,* 66–67.

Mackenzie, B. (1980). Hypothesized genetic differences in IQ: A criticism of three proposed lines of evidence. *Behavior Genetics, 10,* 225–234.

Mackenzie, B. (1984). Explaining race differences in IQ: The logic, the meth-odology, and the evidence. *American Psychologist, 39,* 1214–1233.

Marmor, M. J., Weiss, L. R., Lyden, M., Weiss, S. H., Saxinger, W. C., Spira, T. J., & Feorino, P. M. (1986). Possible female to female transmission of human immunodeficiency virus. *Annals of Internal Medicine, 105,* 969.

Mason, J. O., Ogden, H. G., Berrett, D. A., & Martin, L. Y. (1986). Interpreting risks to the public. *American Journal of Preventive Medicine, 2,* 133–139.

Mays, V. M. (1989). AIDS prevention in Black populations: Methods of a safer kind. In V. M. Mays, G. W. Albee & S. F. Schneider (Eds.), *Primary prevention of AIDS: Psychological approaches.* Beverly Hills, CA: Sage Pub-lications.

Mays, V. M., & Cochran, S. D. (1986, August). *Relationship experiences and the perception of discrimination.* Paper presented at the meetings of the Amer-ican Psychological Association, Washington, DC.

Mays, V. M., & Cochran, S. D. (1987). Acquired immunodeficiency syndrome and Black Americans: Special psychosocial issues. *Public Health Reports, 102,* 224–231.

Mays, V. M., & Cochran, S. D. (1988a). The Black Women's Relationship Project: Surveying the lives of Black lesbians. In M. Shernoff & W. A. Scott (Eds.), *A sourcebook of gay/lesbian health care* (2nd ed.). Washington, DC: National Gay and Lesbian Health Foundation.

Mays, V. M., & Cochran, S. D. (1988b). Interpretation of AIDS risk and risk reduction activities by Black and Hispanic women. *American Psychologist, 43,* 949–957.

Mays, V. M., & Cochran, S. D. (1990a). *Acquired immunodeficiency syndrome and women.* Manuscript submitted for publication.

Mays, V. M., & Cochran, S. D. (1990b). *AIDS-related attitudes in a cohort of young, ethnic/minority heterosexuals.* Manuscript in preparation.

Mays, V. M., & Cochran, S. D. (1990c). *A survey of knowledge, attitudes and beliefs about AIDS by young black adults.* Manuscript in preparation.

Mays, V. M., & Jackson, J. S. (in press). AIDS survey methodology with Black Americans. *Social Science and Medicine.*

Peplau, L. A., Cochran, S. D., & Mays, V. M. (1986, August). *Satisfaction in the intimate relationships of Black lesbians.* Paper presented at the annual meeting of the American Psychological Association, Washington, DC.

Potterat, J. J., Phillips, L., & Muth, J. B. (1987). Lying to military physicians about risk factors for HIV infections. *Journal of the American Medical Association, 257,* 1727.

Salomon, G., & Cohen, A. A. (1978). On the meaning and validity of television viewing. *Human Communications Research, 4,* 265–270.

Samuel, M., & Winkelstein, W. (1987). Prevalence of human immunodeficiency virus in ethnic minority homosexual/bisexual men. *Journal of the American Medical Association, 257,* 1901–1902.

Scott, J. W. (1980). Black polygamous family formation: Case studies of legal wives and consensual "wives." *Alternative Lifestyles, 3,* 41–64.

Selik, R. M., Castro, K. G., & Pappaioanou, M. (1988). Racial/ethnic differences in the risk of AIDS in the United States. *American Journal of Public Health, 78,* 1539–1545.

Shilts, R. (1987). *And the band played on: Politics, people and the AIDS epidemic.* New York: St. Martin's Press.

Soares, J. V. (1979). Black and gay. In M. P. Levine (Ed.), *Gay men: The sociology of male homosexuality.* New York: Harper & Row Publishers.

Spanier, G. B., & Glick, P. C. (1980). Mate selection differentials between whites and blacks in the United States. *Social Forces, 58,* 707–725.

Spilka, B., Hood, R. W., & Gorusch, R. (1985). Religion and morality. In B. Spilka, R. Hood, & R. Gorusch, (Eds.), *The psychology of religion.* Englewood Cliffs, NJ: Prentice Hall.

Staples, R. (1973). *The Black woman in America.* Chicago: Nelson Hall.

Staples, R. (1981a). Race and marital status: An overview. In H. P. MacAdoo (Ed.), *Black families* (pp. 173–175). Beverly Hills, CA: Sage Publications.

Staples, R. (1981b). *The world of black singles.* Westport, CT: Greenwood Press.

Tucker, M. B., & Mitchell-Kernan, C. (1990). Sex ratio imbalance among Afro-Americans: Conceptual and methodological issues. In R. Jones (Ed.), *Black adult development and aging* (pp. 179–189). Berkeley, CA: Cobb & Henry.

U.S. Bureau of the Census. (1983). *General population characteristics, 1980.* Washington, DC: U.S. Government Printing Office.

U.S. Bureau of the Census. (1990). *United States population estimates, by age, sex, race, and Hispanic origin: 1980 to 1988.* Current Population Reports, Series P-25, No. 1045. Washington, DC: U.S. Government Printing Office.

U.S. Department of Health and Human Services. (1986). *Health status of the disadvantaged: Chartbook 1986.* DHHS Publ. No. (HRSA) HRS-P-DV86-2. Washington, DC: U.S. Government Printing Office.

Weinstein, N. (1987, October). *Perceptions of risk.* Paper presented at the Centers for Disease Control Conference on Behavioral Aspects of High Risk Sexual Behavior, Atlanta.

Wilson, P. (1986). Black culture and sexuality. *Journal of Social Work and Human Sexuality, 4,* 29–46.

Wyatt, G. E., Peters, S. D., & Guthrie, D. (1988). Kinsey revisited part II: Comparisons of the sexual socializations and sexual behavior of Black women over 33 years. *Archives of Sexual Behavior, 17,* 289–332.

6

Rates of Sexual Partner Change in Homosexual and Heterosexual Populations in the United Kingdom

Roy M. Anderson and Anne M. Johnson

The rate of spread of a sexually transmitted infection within a human community is determined by a variety of factors associated with the typical course of infection in an individual patient and the prevailing patterns of sexual behavior and activity within the population. A useful measure of the transmission potential of a sexually transmitted disease (STD) within a defined community is provided by a parameter called the *basic reproductive rate*. This epidemiological statistic, commonly denoted by the symbol R_0, measures the number of secondary cases of infection that, on average, are produced by one primary case in a susceptible population (Anderson, Medley, May, & Johnson, 1986; May & Anderson, 1987).

The magnitude of R_0 is determined by the average probability that an infectious person transmits the infection to a susceptible individual per partner contact, times the number of different sexual partners an individual typically experiences per unit of time, times the average duration of infectiousness of an infected individual (Anderson et al., 1986; Hethcote & Yorke, 1984; May & Anderson, 1987). For an infection to persist endemically in a given community the magnitude of R_0 must equal or exceed unity.

The definition of R_0 outlined above is based upon the assumption that individuals select sexual partners at random from the sexually active segment of a population. Published studies of sexual behavior, however, reveal marked heterogeneity in sexual activity within human communities, where most people have few partners (relative to the average) and a few have many partners (Anderson, 1988). In these circumstances the typical number of different sexual partners an individual experiences per unit of time should be interpreted as the *effective average rate* of sexual partner change. Effective average is defined as the mean rate of partner change (m) plus the variance-to-mean ratio (σ^2/m) (Anderson et al., 1986; May & Anderson, 1987). The definition is based upon the *assumption of proportionate mixing*: individuals in a given sexual activity group (defined by their rate of sexual partner change) choose partners from other activity groups in accordance with the proportional representations of the other groups in the population, weighted by the degrees of sexual activity in these groups.

This definition of the effective average rate of sexual partner change originates from the development of simple mathematical models of the transmission dynamics of sexually transmitted infections. It highlights the significance of quantitative measures of the statistical distribution of rates of partner change to an understanding of the transmission dynamics of the infection (the mean, m, and the variance, σ^2, are summary statistics of this frequency distribution). We need to know not just the average number of new sexual partners acquired per unit of time, but also the variation in this number. Those with high rates of acquiring new partners play a disproportionate role in the transmission of infection because they are both more likely to acquire infection and more likely to pass it on.

Rate of sexual partner change is but one facet of human sexual behavior of relevance to the transmission of human immunodeficiency virus (HIV), the etiological agent of AIDS. A large number of epidemiological studies of HIV transmission among male homosexuals, for example, suggest that certain types of sexual behavior, such as receptive anal intercourse, enhance the probability that a susceptible individual will acquire infection from his infectious partner (Detels, Chapter 1; Peterman & Curran, 1986; Stevens, Chapter 2; Voeller, Chapter 19; Winkelstein, Wiley & Nadian, 1986). Other studies suggest that a past history of other sexually transmitted infections, such as gonorrhea or genital ulcers (so-called cofactors), is an important determinant of the likelihood of acquiring infection in homosexual and heterosexual communities in developed and developing countries (Castro et al., 1988; Quinn, Mann, Curran, & Piot, 1986; Peterman & Curran, 1986; Piot & Carael, 1988; Voeller, Chapter 19; Winkelstein et al., 1986).

In all such studies of risk factors, however, the rate of partner change remains a highly significant variable. To assess the relevance of both the type and the frequency of sexual activity within a partnership plus the history of past STD infections to the risk of acquiring infection, it is necessary to carefully dissect out the relative contributions of these factors independent of the rate of partner change. In practice, this is often difficult since, for example, a history of past STD infection is invariably positively correlated with the rate of partner change. Similarly, the frequency of anal intercourse in male homosexual relationships is often related to partner change rates. Therefore, very careful statistical analyses are required to assess the relative risk of different types of sexual behavior. The exemplary studies of Moss et al. (1988) and Winkelstein et al. (1986) will illustrate the methods to be adopted and the problems arising in the interpretation of apparent associations between sexual behavior and the incidence of HIV infection or AIDS.

In this paper we focus on one attribute of sexual behavior of relevance to the study of HIV transmission: the rate of sexual partner change per unit of time. In particular, we focus on the statistical distribution of this attribute in homosexual and heterosexual communities in the United Kingdom as recorded in questionnaire and interview surveys carried out over the past 2 years. We recognize that these statistical distributions are very crude scores of the net risk of HIV transmission but argue that their measurement and analysis is an essential prerequisite to more detailed studies of the type and frequency of activity within individual partnerships. In broad terms, the distribution of partner change rates overlays finer distributional divisions based on the identification of the type, frequency, and duration of sexual activity within partnerships.

This chapter is organized as follows. The first section summarizes the major conclusions derived from simple mathematical models of HIV transmission in homosexual and heterosexual populations. Emphasis is placed on parameter definition and problems of interpretation arising from variability in factors such as rates of partner change, type and frequency of sexual activity within a partnership, and the infectiousness of infected persons over the long and variable incubation period of the disease AIDS. The second section reviews recent surveys in the United Kingdom of rates of sexual partner change, with emphasis on comparisons between homosexual and heterosexual populations, differences between age classes, and the impact of educational programs. The third section examines the risks of infection associated with different types and frequencies of sexual activity. The following section considers the question of cross-transmission between risk groups and "who mixes with whom" within a defined risk group. The paper ends with a discussion of future research needs and directions, in the context of furthering our quantitative understanding of the transmission of HIV via sexual activity.

Parameter Definition and Interpretation

Simple mathematical models of the transmission dynamics of HIV infection, based on the assumption of homogeneous mixing within sexually active male homosexual communities, define the basic reproductive rate of infection, R_0, as

$$R_0 = \beta m D \tag{1}$$

where β denotes the probability that a susceptible person acquires the infection from an infected partner (defined per partner duration), m denotes the mean number of different sexual partners (defined per unit of time), and D records the average duration of the period over which an infected person is infectious to others.

The equivalent definition for the transmission of HIV in a heterosexual community is

$$R_0 = \beta_1 \beta_2 m_1 m_2 D \tag{2}$$

where β_1 and β_2 are the transmission probabilities for contacts between infected females and susceptible males and between infected males and susceptible females, respectively. The parameters m_1 and m_2 are, respectively, the mean rates of partner change for females and males (May & Anderson, 1987).

These definitions arise from models with many simplifying assumptions. Most important, all three parameters, β, m, and D, are defined as average values for a given community and ignore the observed complications of variability in the infectiousness of infected persons (β) throughout the long and variable incubation period (D) and in rates of sexual partner change (m) whether in homosexual or heterosexual communities. Before examining these problems, however, it is informative to see how the definitions of R_0 are related to the doubling time of the epidemic in the early stages of the spread of infection. Simple analyses suggest that the relationship between the doubling time (the time interval required for a doubling in cases of AIDS or numbers seropositive), t_d, and R_0 is

$$R_0 = (D_1 n_2 / t_d) + 1 \tag{3}$$

With a doubling time in the early stages of the epidemic in the United States and the United Kingdom of roughly 9 months ($t_d = 0.75$ years) and an average infectious period of roughly 8 years (under the assumption that the average infectious period is approximately equal to current estimates of the average incubation period; see Medley, Anderson, Cox, & Billard [1987]), then equation (3) gives an R_0 estimate of roughly 8.4. In other words, these very simple calculations suggest that in the early stages of the epidemic each primary case of infection gave rise to eight secondary cases in male homosexual communities in developed countries. As the epidemic progresses the magnitude of the *effective* reproductive rate R will decline from its pristine potential (R_0), as a result of the decline in the number or density of susceptible men, to attain a state of stable endemic infection (in the absence of changes in sexual behavior) in which the effective reproductive rate is equal to unity.

Heterogeneity in the Rate of Sexual Partner Change

The simplest way in which to incorporate heterogeneity in sexual activity, as defined by rates of partner change, is via the assumption of proportionate mixing (see Anderson et al., 1986; Hethcote & Yorke, 1984). Suppose a population of N individuals consists of a series of subgroups of N_i individuals who have an average of i sexual partners per unit of time such that $N_i = Np(i)$ where $p(i)$ is the proportion of the population in the ith class. The proportionate mixing assumption assumes that individuals choose sexual partners in activity class i in proportion to their representation in the total population times their degree of sexual activity i. In a population where there are Y_i infected persons in activity class i, the rate of infection of a susceptible person is therefore the transmission probability per partner contact β times the sum of the contacts with infected individuals in all activity classes (the sum over i). More formally, the proportionate mixing assumption assumes that the per capita rate of infection, λ, is defined as

$$\lambda = \beta \Sigma_i i Y_i / \Sigma_i i N_i \tag{4}$$

The revised definition of the basic reproductive rate under this assumption is

$$R_0 = \beta(m + \sigma^2/m)D \tag{5}$$

where m is the mean rate of partner change and σ^2 is the variance of this rate. Similarly, for heterosexual transmission the parameters m_1 and m_2 in equation (2) must be replaced by $(m_1 + \sigma_1^2/m_1)$ and $(m_2 + \sigma_2^2/m_2)$, respectively. These revised definitions reveal much of importance to the influence of heterogeneity in sexual activity on the spread of HIV.

Consider the following simple numerical example. In a male homosexual community, suppose that 90% of individuals have 1 new sexual partner per year, while 10% have 20 per annum, such that the mean rate of partner change is 2.9 and the variance is 36.1. By reference to equation (5) it can be seen that the variance in sexual activity (σ^2) generated by a small proportion (10%) of highly active people makes a disproportionate contribution to the magnitude of the transmission potential of HIV (R_0) by comparison with the contribution of the mean activity level (e.g., $m = 2.9$, $\sigma^2/m = 12.4$). Roughly 80% of the transmission potential is generated by the variance induced by 10% of the population. This example highlights the importance of measuring variability in rates of sexual partner change.

Who Mixes with Whom

The assumption of proportionate mixing outlined above is made in the absence of detailed information on who mixes with whom. For example, do individuals in high activity classes predominantly choose sexual partners from their own partner change class, or do they mix more broadly with the other classes? To further refine models of transmission, information is urgently required in this area. The practical problems involved in the collection of such data, however, are considerable. It would be necessary not simply to ask how many partners an individual experienced over some defined interval, but to also ascertain who those partners were and to record their patterns of sexual partner change. Preliminary work of this nature in male homosexual communities has begun by constructing networks of sexual contacts (Pinching et al., 1986). An alternative approach is to interview the partners within a short or long-standing relationship to ascertain the degree of correlation between their partner change rates prior to entering the current relationship. Precise assessment of the influence of strong or weak correlation, within and between sexual activity classes, on the transmission dynamics of HIV must await the collection of more detailed information.

Probability of Transmission

The probability of transmission, β, reflects the average infectiousness of an infected individual throughout the average duration of an infected person's sexually active life span. It is often assumed that this period is approximately equal to the average incubation period of the disease. As knowledge accumulates of the typical course of infection in an infected person, the simple assumption that one can mirror the likelihood of transmission by some average value appears less realistic. Many factors influence the infectivity of an infected person. These include the type and intensity of sexual activity practiced with the susceptible partner, the genetic background of the person, the genetic strain or strains of the virus harbored by the infected person, the duration of time since the person first acquired the virus, and the person's past history of STD infection (Anderson, 1988; Hahn et al., 1986; May, Anderson & Johnson, 1988; Peterman, Stoneburner, Allen, Jaffe, & Curran, 1988; Srinivasan et al., 1987; Winkelstein et al., 1986).

Temporal variability in the infectiousness of an infected individual throughout the long and variable incubation period of AIDS is probably one of the most important determinants. A tentative hypothesis at present, based on observed fluctuations in HIV antigen concentrations in the serum of infected patients, is that there are two periods of peak infectivity. The first is shortly after infection and may last for a few months; the second occurs at the end of the incubation period as AIDS-related complex (ARC) develops. This latter period may last for a year or more. The interval between these two phases of infectivity appears to vary greatly as a result of high variation in the duration of the incubation period (Goudsmit & Paul, 1988; May et al., 1988; Pederson et al., 1987). In some infected patients the incubation period may be as short as 18 months, whereas others show no signs of disease after 8 to 9 years of infection. Current evidence, based on the study of transfusion-associated cases, suggests an average incubation period of 8 to 9 years (Medley et al., 1987). However, this estimate is likely to increase as the duration of study of infected persons lengthens. If there are two distinct phases of infectivity (to match peaks in antigen abundance in blood sera and body excretions plus secretions), separated by a period of low to negligible infectiousness, the definition of the basic reproductive rate must be modified to take account of this temporal variability. Very simple multistage models suggest the following definition (see Blythe & Anderson, 1988; May et al., 1988):

$$R_0 = (\beta_1 D_1 + \beta_2 D_2)(m + \sigma^2/m) \tag{6}$$

where β_1 and β_2 denote the transmission probabilities for the first and second phases of infectivity, respectively, and D_1 and D_2 denote their average durations. The parameters m and σ^2 are as defined in equation (5).

Much of the recent epidemiological research on transmission probability has focused on the relevance of the type and intensity of sexual activity. This work is often based on the study of seroconversion in partnerships in which one of the partners is seropositive and the other seronegative for HIV antibodies (Peterman et al., 1988). The results from such studies have, in general, been inconclusive. In some instances partners have seroconverted after a single or small number of sexual contacts, whereas in other cases repeated sexual contact over long intervals of time has not resulted in transmission (see "Types of Sexual Activity," p. 142). One possible explanation for such variability is that the stage of infection at which sexual contact occurs (i.e., at the beginning or the end of the incubation period) is of greater significance than the type, intensity, and duration of sexual contact. In future studies of transmission probability it is essential to relate seroconversion with both the time since the infected partner acquired infection and the time scale on which he or she will develop the disease AIDS. Ideally, longitudinal data are required on viral concentrations in secretions, such as semen and vaginal fluids, linked with information on sexual activity within the partnerships under study (Goudsmit & Paul, 1988).

The relative significance of other sources of variability in infectiousness, such as the risk group to which the patient belongs (e.g., intravenous drug user, male homosexual, heterosexual, hemophiliac), host genetic background, and the genetic strain of the virus, is very poorly understood at present.

Rates of Sexual Partner Change

A variety of studies of homosexual men in developed countries (Winkelstein et al., 1986) and heterosexuals in developing and developed countries (Evans, McCormack, Bond, MacRae, & Thorp, 1988; France et al., 1988; Quinn et al., 1986) reveal that rates of sexual partner change are positively correlated with the likelihood that a person is either infected with HIV or shows symptoms of AIDS. Over the past few years a number of studies have been completed of variability in partner change rates within defined populations. These include studies of male homosexuals, intravenous (IV) drug users, female prostitutes, and heterosexuals (Anderson, 1988; May and Anderson, 1987; McManus & McEvoy, 1985; Piot, Plummer, Rey, & Ngugi, 1987; Winkelstein et al., 1986, 1987a and b). In this section we focus on recent work in the United Kingdom but we refer to published studies in the United States for comparison.

The methods employed in these surveys vary greatly and include quota sampling in the general population, random location quota sampling, and blanket sampling of cooperative groups of people such as students and attenders of health clinics. The surveys are invariably based on an interview plus the completion of a confidential questionnaire.

Homosexual Men

Over the past 5 years much attention has been focused on rates of sexual partner change in male homosexual communities (see Anderson, 1988; Carne et al., 1986; May & Anderson, 1987; McKusick, Harstman, & Coates, 1985; McManus and McEvoy, 1986; Winklestein et al., 1986). The majority of studies have revealed high mean rates of partner change per unit of time and great variability between individuals. For example, a study of McManus and McEvoy (1986) of male homosexual behavior in the United Kingdom in 1984, based on the completion of a postal questionnaire circulated by gay magazines and in gay clubs and public houses, revealed that approximately 5% of the respondents ($N = 1,292$) reported more than 100 different sexual parterns in the previous year (Fig. 6-1). Variability between individuals, however, is high, as indicated by a survey by Carne et al. (1986) carried out on an STD clinic population in London during 1986. This study reported a mean rate of partner change of 4.7 per month (m) and a variance (σ^2) of 56.7. Similar trends have been demonstrated in the United States, as shown in the study by Winkelstein et al. (1986) of male homosexuals in the city San Francisco in 1984, where 26.8% of the sample ($N = 641$) reported 10 or more partners in the previous year (Fig. 6-2).

The most detailed recent study in the United Kingdom is that carried out by the British Market Research Bureau (BMRB) on behalf of the

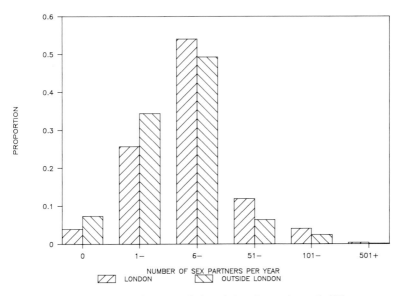

Figure 6-1. Frequency distribution of the claimed number of different sexual partners per year in a sample of male homosexuals from London (n = 523) and outside London (n = 744) in 1984. (Data from McManus & McEvoy, 1986.)

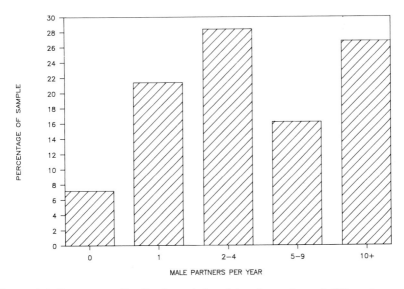

Figure 6-2. Frequency distribution of the claimed number of different sexual partners per year in a sample of male homosexuals from San Francisco in 1984. (Data from Winkelstein et al., 1986.)

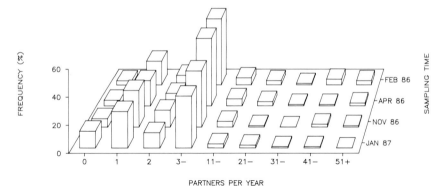

Figure 6-3. Frequency distributions of the claimed number of different sexual partners per year of male homosexuals in England in four sampling periods in 1986–1987. (Data from BMRB 1987.)

government in 1986 and 1987 to evaluate the effectiveness of an educational campaign (Anderson, 1988; BMRB, 1987). The sampling program was aimed at gay clubs and pubs listed in the publication *Gay Times* and was done over four periods: February 28–March 8, 1986; April 14–26, 1986; November 16–21, 1986; and January 26–February 7, 1987 (N = 156, 298, 284, and 251, respectively, for the four time periods). The recorded frequency distributions of stated number of different sexual partners at each sampling time period over the previous year are presented in Figure 6-3. The distributions reveal high means and high variability (long "tails" to the distributions). Of major interest, however, is the trend for the mean partner change rate to decline over the year in which sampling took place. The mean declined from 10.5 partners in February 1986 to 4.8 in January 1987. Similar trends have been recorded for male homosexuals attending STD clinics in London (Carne et al., 1986) and in surveys in the United States (McKusick et al., 1985). Changes have not been restricted to rates of sexual partner change but have also included higher percentages of men adopting safer sex practices, such as the use of condoms during anal intercourse.

Bisexual Men

In the light of the high levels of seropositivity to HIV antigens in male homosexual communities in Europe and the United States, many epidemiologists have argued that a major route of transmission of HIV from male homosexual communities to heterosexual populations is via bisexual men (Table 6-1). Studies of rates of sexual partner change strat-

Table 6-1
HIV Seroprevalence in Samples of Homosexual
Men in Different Regions of the United States
(CDC, 1987b)

Area	Seroprevalence (%)
Seattle–Tacoma	37
Denver	40
Madison–Milwaukee	28
Chicago	41
Boston	49
New York City	36
Philadelphia–Pittsburgh	50
Baltimore	35
Atlanta	44
Lexington–Louisville	23
Arkansas	12
Kansas City–St Louis	35
Albuquerque	12
Long Beach	28
Los Angeles	48
San Francisco	52

ified by the sex of the partner are therefore of some importance in the study of the spread of AIDS.

The proportion of homosexual men who claim to have regular sex with female as well as male partners varies widely in published surveys. In the United Kingdom survey by McManus and McEvoy (1986), 16% of respondents claimed that their sexual orientation was 4 or less on the Kinsey Scale. In contrast, the BMRB (1987) survey in England in 1986–1987 revealed that approximately 30% of the gay men sampled also claimed to have sex with women. Of these 30%, roughly one-third reported sex with two or more female partners over the previous 12 months (Fig. 6-4A). These proportions remained approximately constant over the four periods of sampling during the year of study.

A more detailed study by Winkelstein et al. (1986) reported claimed numbers of female and male sexual partners in a sample of 173 bisexual men in San Francisco in 1984. Interestingly, this study also reported that approximately 30% of male homosexuals claimed to be bisexual (Fig. 6-4B). The majority of those interviewed who claimed to be bisexual reported greater numbers of male than female sexual partners. However, both studies, suggest that bisexual men tend, on average, to have quite high numbers of female sexual partners, although the variance in partner change rates also appears to be large.

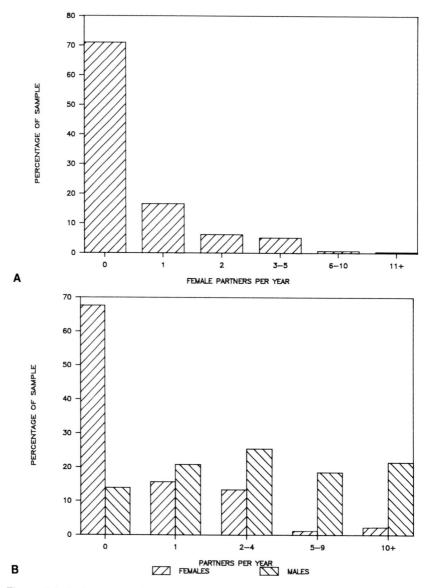

Figure 6-4. A: frequency distribution of the claimed number of different female sexual partners per year in a sample of bisexual men in England in March-April 1986. (Data from BMRB [1987].) B: frequency distributions of the claimed number of female and male sexual partners per year in a sample of bisexual men in San Francisco in 1984. (Data from Winkelstein et al., 1986.)

Heterosexuals

Relatively few surveys of heterosexual activity in defined populations have as yet been completed. Over the past few years, however, a small number of studies in Europe and North America have begun to focus attention on rates of sexual partners change. In the United Kingdom three surveys have recently been completed, although each study has employed different sampling methods. The first of these was conducted by the Harris Research Organization in England in November 1986 and employed a quota sampling method ($N = 823$) stratified by age and sex, based on interview and the completion of a confidential questionnaire. The recorded frequency distributions of claimed sexual partners per 1-month, 2-month, 1-year and 5-year periods is displayed (Figure 6-5). Over these four time periods the mean rates of partner change were 1.0, 1.1, 1.5, and 2.3 respectively. Variability was high, however, and the respective variances (males and females combined) for these means were 0.6, 1.09, 4.99, and 13.2. Note that the variance rises rapidly as the time period over which the number of sexual partners is recorded increases. This study also revealed marked changes in the number of different sexual partners with age (Fig. 6-6). Over the previous year the mean number of different partners was 1.94, 1.51, and 1.15, respectively, in the 18- to 24-year, 25- to 34-year, and 35- to 44-year age groups. The variance also changed with age, being 9.29, 5.37, and 1.22, respectively, in the three groups. A more recent study by the BMRB over the period April 1986 to January 1987 employed a random location quota sampling method (see Anderson, 1988; BMRB, 1987). Following the initial sam-

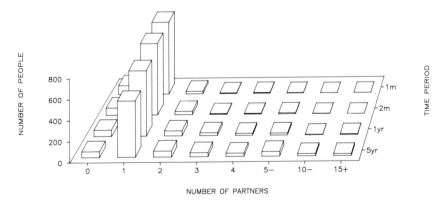

Figure 6-5. Frequency distributions of the claimed number of sexual partners of the opposite sex in heterosexual men and women samples in November 1986 in England. Numbers of different partners are recorded for time intervals of 1 month, 2 months, 1 year, and 5 years. (Data from Anderson, 1988.)

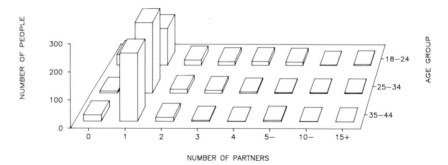

Figure 6-6. Frequency distributions of the claimed number of sexual partners of the opposite sex in a sample of heterosexual men and women sampled in November 1986 in England. Numbers per year are stratified by age group (males and females combined). (Data from Anderson, 1988.)

pling period, three additional periods of sampling were employed, in April 1986, November 1986, and January 1987, to assess the impact of educational programs about AIDS. The recorded frequency distributions of numbers of different sexual partners are displayed in Figure 6-7. The means of these distributions were 0.92, 0.91, and 0.94, respectively, in sampling periods 1 (April 1986), 2 (November 1986), and 3 (January 1987) of the study. Note that these figures reveal no change in the pattern of

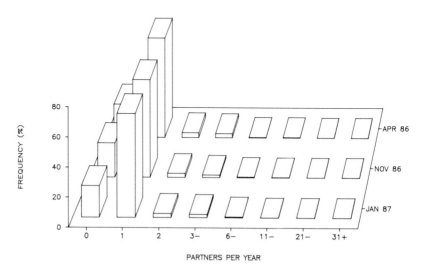

Figure 6-7. Frequency distributions of the claimed number of different sexual partners of the opposite sex in samples of heterosexual men and women drawn in three sampling periods in 1986–1987. (Data from BMRB, 1987.)

partner change rate over a time period in which an intensive educational campaign was launched ($N = 702$, 713 and 708, respectively, in periods 1 to 3). This trend is in marked contrast to the patterns recorded in samples of male homosexuals that were discussed in the previous section (see Fig. 6-3).

These two surveys show encouraging agreement in the claimed mean number of partners over a 1-year period (1.5 and 0.9), but the quota sampling methods employed pose many problems in data interpretation and analysis. In an attempt to address some of the problems inherent in quota sampling, a pilot study of sexual life-style is in progress in the United Kingdom based on a random sample of the population.

The majority of studies of sexual life-style have been based on clinic, volunteer, or quota samples. None of these studies can therefore be regarded as necessarily representative of the general population. While quota sample surveys have the advantage that they can be executed more rapidly and cheaply than random sample surveys, they have a number of inherent methodological disadvantages. In quota samples, interviewees are instructed to interview a sample of persons, specifying the numbers to be of a certain age, sex, and social class or educational level, which can lead to substantial bias. The four main problems associated with quota sampling as compared with probability sampling are:

1. Because there is no random selection it is not possible to attach estimates of standard errors to the sample results. There is no measurement of response rate in a quota sample. In a random sample both sampling error and sampling bias are measurable.
2. Within quota groups, interviewers may not obtain a representative sample of respondents.
3. The social class control, which is widely used in British surveys, is based on a hazardous statistical foundation and may rely to some extent on the interviewer's judgment.
4. It is difficult to strictly control field work (largely as a result of interviewer bias).

In the current U.K. survey, a three-stage probability sampling design was employed to obtain a random sample of adults in England, Scotland, and Wales:

1. Electoral enumeration districts were randomly selected, with probability proportional to metropolitan, urban, and rural. One hundred ten enumeration districts were sampled, with probability proportional to population size in each stratum.
2. From the electoral register of each enumeration district, an equal

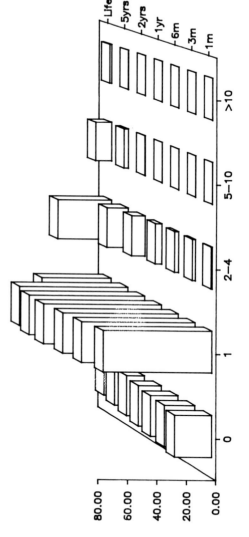

136

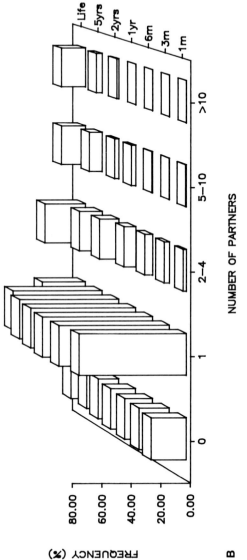

TIME PERIOD

NUMBER OF PARTNERS

FREQUENCY (%)

B

Figure 6-8. Frequency distributions of the claimed number of female (A) and male (B) sexual partners in samples of male and female heterosexuals drawn from the general population in the United Kingdom in 1987. Numbers of partners are recorded for time intervals of 1 month, 3 months, 6 months, 1 year, 2 years, 5 years, and lifetime. (Data from Johnson et al., 1988.)

number of names was randomly selected, and these were con-
verted to an address.

3. Interviewers were instructed to visit each address, enumerate
    the residents at that address between the ages of 16 and 64, and
    then select one person for interview by a random numbers pro-
    cedure.

The interview comprised two separate components. First, interviewees were interviewed using a structured face-to-face interview concerning their demographic characteristics and attitudes to AIDS. They were
then asked to complete a self-completion questionnaire about their sexual life-style. This was completed within the presence but not within
visibility of the interviewer.

The essential components of the self-completion interview were:

1. Numbers and sex of sexual partners, over a series of different
    time periods, in order to obtain rates of partner change.
2. Duration and approximate dates of sexual partnerships.
3. Use of condoms (for women, current contraception use).
4. Use of prostitution.
5. Experience of injection drug use and needle sharing.

Questionnaires were sealed in a plain envelope with no identifying information that could be traced back to the individual interviewee. Of
the 2,074 addresses sampled in the pilot study, the response rate after
exclusion of ineligible addresses (defined as the percentage of interviews
completed) was approximately 50% (Johnson et al., 1988). This figure
is disappointingly low and hence great caution must be exercised in the
interpretation of the results of the pilot study (a further pilot study is
planned in an attempt to raise the compliance rate). Of particular concern is the representative nature of the sample with respect to the known
demographic, economic, and social divisions of the population of Great
Britain (Johnson et al., 1988).

With these caveats in mind, Figure 6-8, records the claimed numbers
of different sexual partners for males and females for seven different
time intervals (lifetime, 5 years, 2 years, 1 year, 6 months, 3 months,
and 1 month). Over all age groups the mean number of different sexual
partners (of the opposite sex) in the previous year was 1.11 for males
and 0.89 for females. Interestingly, these figures are in good agreement
with those obtained in the Harris and BMRB surveys, which employed
quota sampling methods (see Figs. 6-5 and 6-7). A more detailed description of the results of the random sampling pilot survey are contained in Johnson et al. (1988).

Aside from attempts to obtain a picture of heterosexual habits in the
general population, a variety of studies have focused on particular

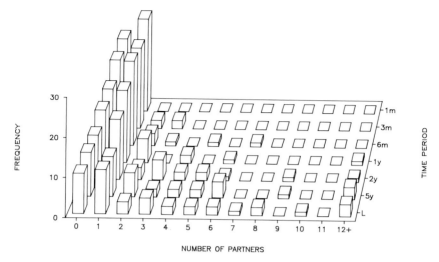

Figure 6-9. Frequency distributions of the claimed number of different sexual partners of the opposite sex in a sample of heterosexual students interviewed in October 1987 in London. Numbers of different partners are recorded for time intervals of 1 month, 3 months, 6 months, 1 year, 2 years, 5 years, and lifetime. (Data from Anderson, 1988.)

groups of individuals, such as those volunteering for study or those attending STD clinics. The results of one such study, of a student population attending a university in England in October 1987 (age range 20 to 22 years) are presented in Figure 6-9 (Anderson, 1988). The mean numbers of partners (males and females combined) over seven time intervals (lifetime, 5 years, 2 years, 1 year, 6 months, 3 months and 1 month) are 3.85, 3.45, 2.27, 1.51, 1.02, .073 and 0.56, respectively. Variability, however, was high and the respective variances for the same time intervals were 42.13, 38.91, 15.90, 8.51, 1.82, 0.60, and 0.25. Interestingly, the mean for the past 12-month period (1.5) was in good agreement with the equivalent figures for the same age group in the Harris, BMRB, and pilot random sampling surveys.

A rather different pattern emerges from studies of heterosexual patients attending STD clinics in London. These individuals probably lie in the tails of the distributions recorded in surveys of the general population. The results from one such survey carried out in March 1987 (Johnson and Sonnex unpublished) are recorded in Figure 6-10 (N = 114 females and 108 males). The mean number of different partners over the preceding 12 months are roughly three times greater than the equivalent figures from samples drawn from the general population (means: 3.36 for males and 3.4 for females). Similar patterns have been recorded

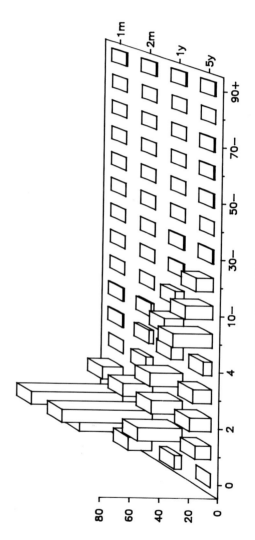

TIME PERIOD

NUMBER OF FEMALE SEX PARTNERS

FREQUENCY

A

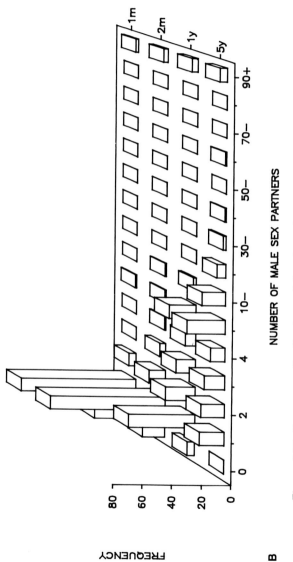

FREQUENCY

NUMBER OF MALE SEX PARTNERS

B

Figure 6-10. Frequency distributions of the claimed number of male (A) and female (B) sexual partners in samples of female and male heterosexuals attending an STD clinic in London in 1987. Numbers of partners are recorded for time intervals of 1 month, 2 months, 1 year, and 5 years. (A. M. Johnson & M. Sonnex, unpublished data.)

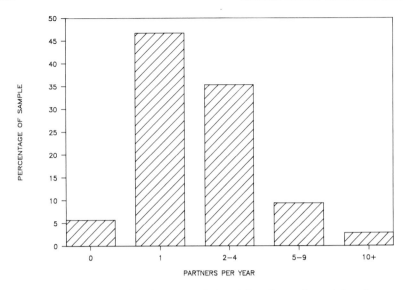

Figure 6-11. Frequency distributions of the claimed number of female sexual partners per year in a sample of male heterosexuals from San Francisco in 1984. (Data from Winkelstein et al., 1986.)

in the United States by Winkelstein et al. (1986) in a sample of heterosexual men ($N = 212$) attending a clinic in San Francisco in 1984 (Fig. 6-11).

These studies highlight the importance of obtaining random samples if accurate conclusions are to be drawn concerning sexual activity in the general population. Of equal importance, however, is the comparison between male homosexuals and heterosexuals attending STD clinics. Even though the rates of partner change of the latter group are three to four times greater than average levels in the general heterosexual population, they are still less than those recorded for homosexual males attending clinics or volunteering for participation in sexual life-style surveys.

Types of Sexual Activity

While rates of partner change have an important influence on the risk of HIV infection, the type of sexual activity occurring between partners also influences the probability of virus transmission. Personal perceptions of what constitutes a sexual partner may vary considerably. It is therefore essential to collect data on types of sexual activity and, for the purposes of behavioral research, to define what sexual activities constitute a "partnership."

Homosexual Practices

Studies among homosexual men have shown that receptive anal inter-course is the sexual act associated with the greatest risk of infection (Detels, Chapter 1; Darrow et al., 1987; Kingsley et al., 1987; Moss et al., 1988; Stevens, Chapter 2; Voeller, Chapter 19; Winkelstein et al., 1987a). The multicenter AIDS cohort study analyzed behavioral risk fac-tors in persons seroconverting to HIV positivity in the course of the study. Data indicate that the anal receptive partner has a much higher risk of infection than the insertive partner. Ninety-one of 1,817 (5%) men who had practiced receptive anal intercourse in the year of the study seroconverted, while only 3 of 334 (0.9%) who reported only in-sertive anal intercourse seroconverted. Risk of seroconversion increased with the number of partners with whom receptive, but not insertive, intercourse was practiced. Of 220 men who abstained from anal inter-course, but reported other sexual practices (such as passive oral inter-course), none seroconverted (Kingsley et al., 1987).

Heterosexual Practices

The relative risks for infection through different types of heterosexual activity have been much less extensively studied. Studies of the het-erosexual partners of HIV-positive individuals indicate that HIV can be transmitted from men to women and women to men through vaginal intercourse without obvious evidence of traumatic or other sources of breaks in the epithelium of the genital tract (Cabane, Thiberge, Godeau, Moreau, & Wattiaux, 1984; Calabrese & Gopalakrishna, 1986; Harris et al., 1983; L'age-Stehr et al., 1985; Padian, Marquis, et al., 1987; Padian, Wiley, & Winkelstein, 1987; Redfield et al., 1985; Staszewski, Tauris & Black, 1987). Case reports provide evidence of both male-to-female and female-to-male transmission of HIV through only one or two sexual exposures to an infected individual (Cabane et al., 1987; Staszewski et al., 1987). Case cluster reports indicate that some individuals remain infectious to successive heterosexual partners over several years (Clu-meck et al., 1987).

While case reports describe possibilities for transmission, they can tell us little about the probability that an infected individual will transmit HIV to his or her uninfected partner given certain constraints of type and duration of contact.

MALE-TO-FEMALE TRANSMISSION

A number of studies have investigated female partners of male AIDS patients and seropositives in an attempt to measure transmission risks (Tables 6-2 and 6-3). These studies give widely differing results. Among the female partners of hemophiliacs the highest proportion of infected

Table 6-2
Studies of Male-to-Female Sexual Transmission of HIV

Source	Clinical Status of Male Index Case	Risk Group of Index Case	n Seropositive Female Partners/ n Studied	Percentage Seropositive Partners
Peterman et al. (1988)	Not given	Transfusion associated	5/23	22
Padian et al. (1987a, 1987b)	AIDS/ARC asymptomatic	Mixed	22/97	23
Redfield et al. (1985, 1986)	Not given	Army personnel	6/18	33
Steigbigel et al. (1987)	AIDS/ARC	IV drug users	41/88	48
Fischl et al. (1987)[a]	AIDS	Mixed	14/28	50
Taelman et al. (1987)	Not given	Patients from Africa	28/38	73

[a] Includes 10 seroconversions observed over median follow-up of 24 months.

partners is 17%, but in other risk groups anywhere from 18 to 73% of heterosexual partners have been found to be seropositive. The varying results from these studies may in part be due to methodological differences between them. The studies vary in the risk group and disease status of the index case (several examined only partners of AIDS cases), in the method of case ascertainment, and in the definition of the contact case (some examined only "regular" partners). Most studies have examined partners at only one point in time, and few follow-up studies linked to details of sexual behavior have been published.

There is some evidence that risk of transmission increases with the

Table 6-3
Studies of Sexual Transmission of HIV to Female Partners of Seropositive Male Hemophiliacs

Source	n Seropositive Female Partners/ n Tested	Percentage Seropositive Partners
Miller et al. (1987)	1/30[a]	3
Jones et al. (1985)	3/77	4
Allain (1986)	10/148	7
Kreiss et al. (1985)	2/21	10
Goedert et al. (1987)	4/21	17

[a] Intravenous drug users.

duration of the sexual relationship and frequency of sexual exposure. Fischl et al. (1987) observed 10 seroconversions among 24 previously seronegative female partners of AIDS patients followed over a median of 24 months, even though these women had had a sexual relationship for several years with the index case without apparent transmission, prior to the diagnosis of AIDS. Padian and colleagues (Padian, Marquis, et al., 1987; Padian, Wiley, & Winkelstein, 1987; Padian, 1990) found an increased risk of infection in female partners with more than 100 sexual exposures to the index case, but in univariate analysis found no relationship between stage of disease in the index case and probability of transmission. From their data, they estimate a risk of infection of 1:1,000 episodes of sexual contact. Such summary statistics should perhaps be viewed with caution, because of the current uncertainty of the relationship between degree of exposure and risk of transmission. Transmission risk may be a function of as-yet unmeasurable biological variables, such as levels of viral excretion in the index case, rather than behavioral differences between couples. It has been suggested that infectivity may increase as infected individuals progress towards AIDS. Goedert et al. (1987) studied the female partners of 24 hemophiliacs and documented three seroconversions in the sexual partners more than 4 years after the date of seroconversion in the male partner. Transmission was associated with a low T4 lymphocyte count in the index cases.

Few studies have documented the risk of different sexual practices. In the study by Fischl et al. (1987), seroconversion was associated with lack of use of barrier contraception. Padian, Marquis, Francis, Anderson, & Rutherford (1987) reported a relative risk of 2.3 for those ever engaging in anal intercourse, but this was not necessary for transmission of HIV.

In Africa, there is evidence of an increased risk of seropositivity with general ulceration, and this could be a cofactor in transmission. However, it may equally be a marker of high numbers of sexual partners.

FEMALE-TO-MALE TRANSMISSION

In Western nations studies of female-to-male transmission (Padian, 1990) to date have been based on small samples, so it is difficult to assess the relative risks of female-to-male and male-to-female transmission. Studies of female-to-male transmission give a range of values similar to those for male-to-female transmission (Table 6-4).

Some of these studies are unpublished and details of sexual behavior are not available. In the study by Fischl et al. (1987), 9 of the 17 male partners of female AIDS patients were seropositive at entry to the study and 3 subsequently seroconverted. In many couples both partners had no risk factors for infection other than being Haitian immigrants. It is not, therefore, always clear whether infection occurred from male to female or from female to male.

Table 6-4
Studies of Female-to Male Sexual Transmission

Source	Risk Group of Index Cases	n Seropositive Male Partners/ n Studied	Percentage Seropositive Partners
Peterman et al. (1988)	Transfusion associated	1/13	8
Redfield et al. (1985)	Army personnel	2/6	33
Taelman et al. (1987)	Patients from Africa	4/10	40
Steigbigel et al. (1987)	IV drug users	7/12	58
Fischl et al. (1987)[a]	Mixed	14/17	71

[a] Three seroconversions observed over median follow-up of 24 months.

Summary

While our understanding of heterosexual transmission risks is incomplete, types of behavior and use of barrier contraception may be important variables in determining transmission probabilities. Studies of heterosexual behavior must include information not only about numbers of partners but also about types of sexual behavior within those partnerships.

Sexual Networks

While rates of partner change and transmission probabilities will be important determinants of the AIDS epidemic, the patterns of mixing between higher and lower risk groups may also substantially alter transmission dynamics. This requires that studies of sexual life-styles answer the question "Who mixes with whom?"

In a homosexual community there can be direct sexual contact between any member of that community. Early studies of sexual networks among persons with AIDS demonstrated the importance of sexual interaction within small groups for viral transmission (Pinching et al., 1986). Similar case clusters have been described among heterosexuals (Clumeck et al., 1987; Franzen, Jertborn, & Biberfield, 1986), although the dynamics in this situation are different. For heterosexuals there are two distributions of rates of partner change, one for men and one for women. While studies of rates of sexual partner change show marked variability between individuals, we know very little about the extent to which men and women mix with partners with similar sexual life-styles. If mixing occurs within relatively discrete risk groups, viral transmission might be largely confined to those groups with the highest rates of partner change. However, if there is direct or indirect mixing between groups with high and low rates of partner change, there may be potential for greater viral spread.

Consider a society characterized by relative female chastity and extensive male use of prostitution. The mean number of opposite sex partners for men and women in a given time period must be equal but the frequency distribution would be quite different. For women, a high proportion of the variance would be accounted for by a small proportion of prostitute women with very large numbers of partners who mix with a population of men with a less skewed distribution of numbers of sexual partners. The men in turn mix with their regular female partners, who have low rates of partner change. Such a situation is described by a study in Nairobi, Kenya. Piot et al. (1987) have documented the rapid spread of HIV into the prostitute population (Table 6-5). In 1981, 3% of male STD clinic attenders and 4% of prostitutes sampled were infected. By 1985, the figures had risen to 15% and 61%, respectively. In the seropositive prostitute population the average number of partners was estimated at 123 per month, whereas among seropositive male STD clinic attenders the average number of partners was only 1.5 per month, although 44% admitted to contact with prostitutes. In turn, while none of 111 women in antenatal clinics were seropositive in 1981, 2% of 735 were infected by 1985. Presumably infection occurred in part as a result of men taking the infection home to their wives.

The situation in the West appears to be somewhat different. Studies of prostitutes in Europe and the United States have found infection among prostitutes to be closely associated with a history of intravenous drug use (Table 6-6). Thus, for example, in the Centers for Disease Control multicenter study (1987a), of 568 prostitutes 50% were intravenous drug users, 62 (11%) were HIV seropositive, and three quarters of these

Table 6-5
Selected Characteristics of Prostitutes Seropositive and Seronegative for HIV in Nairobi, Kenya, 1981[a]

	Seropositive (n = 5)	Seronegative (n = 11)
Mean (years)	21.5	24.9[b]
Number of sex partners per month	180[c]	54
Mean duration of prostitution (years)	0.7	2.1
Gonorrhea (%)	80[d]	14
Genital ulcer (%)	60	28

[a] Reprinted with permission from Piot et al. (1987).
[b] $p < .05$.
[c] $p < .01$.
[d] $o = .03$.

Table 6-6
Studies of HIV Antibody Prevalence in Female Prostitutes
in Europe and the United States

	Europe	Number Tested	Seropositive N (%)	% Drug Use
Barton et al. (1985)	London	50	0 (0)	N.A.
Brenky-Faudeux & Fribourg-Blanc (1985)	Paris	56	0 (0)	N.A.
Smith & Smith (1986)	Nuremberg	99	0 (0)	N.A.
Krogsgaard et al. (1986)	Copenhagen	01	0 (0)	N.A.
Schultz et al. (1986)	West Germany	≈2000	20 (1)	50
Papaevangelou et al. (1985)	Athens	200	12 (6)	0
Tirelli et al. (1986)	Pordenone, Italy	14	10 (71)	100
	United States			
Centers for Disease Control (1987a, 1987b)	Multicenter	568	62 (11)	76

N.A., not applicable.

gave a history of drug use. In this situation, it is necessary to understand the connections between drug use and prostitution. In contrast with studies in Africa (Mann et al., 1986), prostitutes in the West report high levels of condom use with clients, which may have had a role in preventing sexual infection. However, relative protection could equally be due to a lower prevalence of prostitution as well as lower rates of partner change among prostitutes. Unpublished data from a study of 52 prostitutes in London indicate a wide variation in numbers of sexual partners per week, with a median of 3 to 6 partners, dependent on the place of work, and a range of 0 to 40 (H. Ward, personal communication). However, approximately one quarter of these clients may be regular clients, thus reducing the overall rate of partner change.

To assess the potential role of male use of female prostitutes in viral spread in different societies, information is required on the prevalence of prostitution, the sexual life-style of prostitutes, and the prevalence and frequency of use of prostitutes by the male population.

In the studies carried out in the 1930s and 1940s by Kinsey and his colleagues, 29.6% of white college males reported ever having contact with prostitutes, the vast majority reporting premarital contact only (Gebhard & Johnson, 1979). However only 3.9% reported extensive contact of 20 partners or more. More recent data on this subject are urgently required if we are to compare the role of prostitution in the spread of HIV between countries.

One area of concern in the development of a heterosexual epidemic in the West has been potential spread from bisexual men and heterosexual drug users (May's & Cochran, Chapter 5; Reinsch, et al., Chapter 10; Voeller, Chapter 19). An analysis of heterosexual contact cases of AIDS in the United States (excluding the non-U.S. born) indicates that more than two thirds of the cases are contacts of intravenous drug users and only 14% are contacts of bisexual men (Chamberland, White, Nelson, & Dondero, 1987). Part of the explanation appears to be that many bisexual men have relatively few heterosexual contacts and certainly fewer than do their homosexual contacts (Winkelstein, Samuel, Padian, & Wiley, 1987b) (see Fig. 6-4). Winkelstein et al. (1987b) reported that while 21% of bisexual men in San Francisco reported more than 10 *male* partners in 6 months in 1984 but only 2% reported more than 10 *female* partners, 83% reported 0 or 1 female partners over the same time period. However, without adequate assessment of the size of the bisexual and drug-using population and their sexual behavior, it becomes almost impossible to determine the potential for spread by these two routes. One further question is the extent to which drug users choose other drug users as their sexual partners.

One useful research method in the study of sexual networks might be to study current sexual partners from a representative population sample and compare their sexual life-styles in terms of rates of partner change.

Discussion

The studies reviewed in this chapter highlight the need for more quantitative research on the relationship between sexual behavior and the risk of transmission of HIV from infected to susceptible partner. Many practical problems surround the collection of accurate information in this field, such as the choice of a representative sample from a population, the individual's interpretation of what constitutes a sexual relationship or a sexual preference, and the reliability of information on rates of sexual partner change and type of sexual activity.

Information on rates of sexual partner change in both homosexual and heterosexual populations is beginning to accumulate. The available data reveal patterns of high variability within samples, where the variance of the number of different sexual partners per unit of time invariably exceeds the mean. Such heterogeneity is an important factor in the spread of HIV, since the small fraction of highly sexually active individuals are both more likely to acquire infection and more likely to transmit infection to their susceptible partners. The quantification of variability in defined populations is a clear priority in future research.

A further feature of the data is the trend for male homosexuals to have much higher rates of sexual partner change than heterosexual men or women. In the BMRB study in England, data from the March–April

1986 sampling period revealed that the mean rate of claimed sexual partner change for male homosexuals was eight to nine times greater than the equivalent mean in the sample of heterosexuals. This difference could, in part, help to explain why HIV has spread more rapidly in homosexual than heterosexual populations in developed countries, irrespective of other differences in sexual behavior. However, it is clear that marked changes in behavior have occurred in many male homosexual communities over the past few years. Mean rates of partner change have declined (see Fig. 6-3) and an increasing proportion of individuals claim to be adopting safer sex practices. These changes have resulted in rates of HIV seroconversion falling to low levels at present in male homosexual communities in cities such as London, New York, and San Francisco. Similar changes in sexual activity do not appear to have occurred in heterosexual populations in developed countries.

Many other factors aside from rates of sexual partner change influence the spread of HIV. However, our quantitative knowledge of the importance of, for example, the duration and type of sexual activity to transmission is extremely limited at present. The major problem in the interpretation of the available data is that the infectiousness of an infected person appears to vary greatly between individuals and over the long and variable incubation period of AIDS. In future studies of transmission between infected and susceptible sexual partners it is of great importance to attempt to quantify levels of viral abundance in secretions and excretions and to link such measures with the likelihood of transmission per sexual act.

Finally, our focus on rates of sexual partner change ignores much detail concerning who mixes with whom in sexually active populations. Networks of sexual interrelationships are clearly of great importance to any assessment of the degree to which HIV will spread and persist in homosexual and heterosexual communities. Very detailed surveys will be required in future research to assess the degree to which individuals in different sexual activity classes (defined by their rates of sexual partner change) choose partners from their own or other activity groups.

Acknowledgment

We gratefully acknowledge financial support from the Medical Research Council, United Kingdom.

References

Allain, J. P. (1986). Prevalence of HLTV/LAV antibodies in patients with haemophilia and other sexual partners in France. *New England Journal of Medicine, 315,* 517.

Anderson, R. M. (1988). The epidemiology of HIV infection: Variable incubation plus infectious periods and heterogeneity in sexual activity. *Journal of the Royal Statistical Society, Series B, 151,* 66–93.

Anderson, R. M., Medley, G. F., May, R. M., & Johnson, A. M. (1986). A preliminary study of the transmission dynamics of the human immunodeficiency virus (HIV), the causative agent of AIDS. *IMA Journal of Mathematics in Medicine and Biology, 3,* 229–263.

Barton, S. E., Underhill, G. S., Gilchrist, C., Jeffries, D. J., & Harris, J. R. W. (1985). HTLV III antibody in prostitutes. *Lancet, ii,* 1424.

Blythe, S. P., & Anderson, R. M. (1988). Distributed incubation and infectious periods in models of the transmission dynamics of the human immunodeficiency virus (HIV). *IMA Journal of Mathematics in Medicine and Biology, 5,* 1–19.

BMRB. (1987). *AIDS Advertising Campaign, report on four surveys during the first year of advertising, 1986–87.* London: British Market Research Bureau Limited.

Brenky-Faudeux, D., & Fribourg-Blanc, A. (1985). HTLV III antibody in prostitutes. *Lancet, ii,* 1424.

Cabane, J., Thibierge, E., Godeau, P., Moreau, A., & Wattiaux, M. J. (1984). AIDS in an apparently risk-free woman. *Lancet, ii,* 105.

Calabrese, L. H., & Gopalakrishna, K. V. (1986). Transmission of HTLV III infection from man to woman and woman to man. *New England Journal of Medicine, 324,* 987.

Carne, C. A., Weller, I. V. P., Johnson, A. M., Loveday, C., Pearce, F., Hawkins, A., Smith, A., Williams, P., Tedder, R. S., & Adler, M. W. (1987). Prevalence of antibodies to human immunodeficiency virus (HIV), gonorrhoea rates and altered sexual behaviour in homosexual men in London. *Lancet,* I, 656–658.

Castro, K. G., Lieb, S. I., Jaffe, H. W., Narkunas, J. P., Calisher, C. H., Bush, T. J., Witte, J. J., & the Belle Glade Field Study Group. (1988). Transmission of HIV in Belle Glade, Florida: Lessons for other communities in the United States. *Science, 239,* 193–197.

Centers for Disease Control. (1987a). Antibody to human immunodeficiency virus in female prostitutes. *Morbidity and Mortality Weekly Report, 36,* 157–161.

Centers for Disease Control. (1987b). *A review of current knowledge and plans for expansion of HIV surveillance activities.* Atlanta, GA: Centers for Disease Control.

Chamberland, M., White, C., Nelson, A., & Dondero, T. J. (1987, June). *AIDS in heterosexual contacts: A small but increasing group of cases.* Paper presented at the Third International Conference on AIDS, Washington, DC.

Clumeck, N., Hermans, P., Taelman, H., Roth, O., Zissis, G., & De Wit, S. (1987, June). *Cluster of heterosexual transmission of HIV in Brussels.* Paper presented at Third International Conference on AIDS, Washington, DC.

Darrow, W. W., Echenberg, D. F., Jaffe, H. W., O'Malley, P. M., Byers, R. H., Getchell, J. P., & Curran, J. W. (1987). Risk factors for human immunodeficiency virus (HIV) infections in homosexual men. *American Journal of Public Health, 77,* 479–483.

Evans, B. A., McCormack, S. M., Bond, R. A., MacRae, K. D., & Thorp, R. W. (1988). Human immunodeficiency virus infection, hepatitis B virus infection, and sexual behaviour of women attending a genitourinary medicine clinic. *British Medical Journal, 296,* 473–475.

Fischl, M. A., Dickinson, G. M., Scott, G. B., Klimas, N., Fletcher, M., & Parks, W. (1987). Evaluation of heterosexual partners, children and household contacts of adults with AIDS. *Journal of the American Medical Association*, 257, 640–644.

France, A. J., Skidmore, C. A., Robertson, J. R., Brettle, R. P., Roberts, J. J. K., Burns, S. M., Foster, C. A., Inglis, J. M., Galloway, W. B. F., and Davidson, S. J. (1988). Heterosexual spread of human immunodeficiency virus in Edinburgh. *British Medical Journal*, 296, 526–529.

Franzen, C., Jertborn, M., and Biberfield, G. (1986, June). *Four generations of heterosexual transmission of LAV/HTLV III in a small Swedish town*. Paper presented at the Second International Conference on AIDS, Paris.

Gebhard, P. H., & Johnson, A. B. (1979). *The Kinsey data: Marginal tabulations of the 1938–1963 interviews conducted by the Institute for Sex Research*. Philadelphia: W. B. Saunders Company.

Goedert, J. J., Eyster, M. E., & Biggar, R. J. (1987, June). *Heterosexual transmission of human immunodeficiency virus (HIV). Association with severe T4 cell depletion in male haemophiliacs*. Paper presented at Third International Conference on AIDS, Washington, DC.

Goudsmit, J., & Paul, D. A. (1988). Circulation of HIV antigen in blood according to stage of infection, risk group, age and geographic origin. *Epidemiology and Infection*, 99, 701–710.

Hahn, B. H., Shaw, G. M., Taylor, M. E., Redfield, R. R., Markham, P. D., Salhuddin, S. Z., Wong-Staal, F., Gallo, R. C., Parks, E. S., & Parks, W. P. (1986). Genetic variation in HTLV-III/LAV over time in patients with AIDS or at risk for AIDS. *Science*, 232, 1548–1553.

Harris, C., Butkus Small, C. B., Klein, R. S., Friedland, G. H., Moll, B., Emeson, E. E., Spigland, I., & Steigbigel, N. H. (1983). Immunodeficiency in female sexual partners of men with the acquired immunodeficiency syndrome. *New England Journal of Medicine*, 308, 1181–1184.

Hethcote, H. W., & Yorke, J. A. (1984). Gonorrhoea; transmission dynamics and control. *Lecture Notes in Biomathematics*, 56, 1–105.

Johnson, A. M., Wadsworth, J., Elliott, P., Blower, S., Wallace, P., Prior, L., Miller, D., Adler, M. W., & Anderson, R. M. (1988). A pilot study of sexual lifestyles in a random sample of the population of Great Britain. *AIDS*, 3, 135–142.

Jones, P., Fearns, M., McBride, L., & Hamilton, P. (1987, June). *Continuing surveillance of HIV associated morbidity and mortality in a well-defined population*. Paper presented at Third International Conference on AIDS, Washington, DC.

Jones, P., Hamilton, P. J., Bird, G., Fearns, M., Oxley, A., Tedder, R., Cheingsong-Popov, R., & Codd, A. (1985). AIDS and haemophilia: Morbidity and mortality in a well-defined population. *British Medical Journal*, 291, 695–699.

Kingsley, L. A., Detels, R., Kaslow, R., Polk, B. F., Rinaldo, Jr., C. R., Chmiel, J., Detre, K., Kelsey, S. F., Odaka, N., Ostrow, D., VanRaden, M., & Visscher, B. (1987). Risk factors for seroconversion to human immunodeficiency virus among male homosexuals. *Lancet*, I, 345–348

Kreiss, J. K., Kitchen, L. W., Prince, H. E., Kasper, C. K., & Essex, M. (1985). Antibody to human T-lymphotrophic virus type III in wives of haemophiliacs. *Annals of Internal Medicine*, 102, 623–626.

Krogsgaard, K., Gluud, C., Pedersen, C., Neilsen, J. O., Juhl, E., Gerstort, J., & Neilsen, C. M. (1986). Widespread use of condoms and low prevalence of sexually transmitted diseases in Danish non-drug addict prostitutes. *British Medical Journal, 293*, 1473–1474.

L'Age-Stehr, J., Schwarz, A., Offermann, G., Langmaak, H., Bennhold, I., Niedrig, M., & Koch, M. A. (1985). HTLV III infection in kidney transplant recipients. *Lancet, ii*, 1361–1362.

Mann, J. M., Quinn, T., Francis, H., Miatudila, M., Piot, P., & Curran, J. (1986, June). *Sexual practices associated with LAV/HTLV III seropositivity among female prostitutes in Kinshasa, Zaire.* Paper presented at Second International Conference on AIDS, Paris.

May, R. M., & Anderson, R. M. (1987). Transmission dynamics of HIV infection. *Nature, 326*, 137–142.

May, R. M., Anderson, R. M., & Johnson, A. M. (1989). The influence of temporal variation in the infectiousness of infected individuals on the transmission dynamics of HIV. In R. Kulstad (Ed.), *AIDS 1988* AAAS Science Symposium Papers, 1989, pp. 75–83.

McKusick, L., Harstman, W., & Coates, T. J. (1985). AIDS and sexual behavior reported by gay men in San Francisco. *American Journal of Public Health, 75*, 493–496.

McManus, T. J., & McEvoy, M. B. (1986). A preliminary study of some aspects of male homosexual behaviour in the United Kingdom. *British Journal Sex Medicine, 14*, 110–120.

Medley, G. F., Anderson, R. M., Cox, D. R., & Billard, L. (1987). Incubation period of AIDS in patients infected via blood transfusions. *Nature, 328*, 719–721.

Miller, E. J., Miller, R. R., Goldman, E., Griffiths, P. D., & Kernoff, P. B. A. (1987, June). *Low risk of anti-HIV seroconversion in female sexual partners of haemophiliacs and their children.* Paper presented at Third International Conference on AIDS, Washington, DC.

Moser, C., & Kalton, G. 1971. *Survey methods in social investigation.* London: William Heinemann.

Moss, A. R., Osmond, D., Bachetti, P., Chermann, J-C., Barre-Sinoussi, F., & Carlson, J. (1988). Risk factors of AIDS and LAV/HTLV-III seropositivity in homosexual men. *American Journal of Epidemiology, 125*, 1035–1047.

Padian, N. (1990). Heterosexual transmission: infectivity and risks. In N. J. Alexander, H. L. Gabelnick, J. Spierler (Eds.), *The heterosexual transmission of AIDS.* New York: Alan R. Liss.

Padian, N., Marquis, L., Francis, D. P., Anderson, R. E., Rutherford, G. Q., O'Malley, P. M., & Winkelstein, W. (1987). Male to female transmission of human immunodeficiency virus. *Journal of the American Medical Association, 258*, 788–791.

Padian, N., Wiley, J., & Winkelstein, W. (1987, June). *Male to female transmission of HIV current results, infectivity rates, and San Francisco seroprevalence estimates.* Paper presented at Third International Conference on AIDS, Washington, DC.

Papaevangelou, G., Roumeliotou-Karayannia, A., Kallinkos, G., & Paoutsakis, G. (1985). LV/HTLV III infection in female prostitutes. *Lancet, ii*, 1018.

Pedersen, C., Nielsen, C. M., Vestergaard, B. F., Gerstoft, J., Krogsgaard, K., & Nielsen, J. O. (1987). Temporal relation of antigenaemia and loss of antibodies to core antigens to development of clinical disease in HIV infection. *British Medical Journal, 295*, 567–569.

Peterman, T. A., & Curran, J. W. (1986). Sexual transmission of human immunodeficiency virus. *Journal of the American Medical Association, 256*, 2222–2226.

Peterman, T. A., Stoneburner, r. L., Allen, J. R., Jaffe, H. W., & Curran, J. W. (1988). Risk of human immunodeficiency virus transmission from heterosexual adults with transfusion—associated infections. *Journal of the American Medical Association, 259,* 55–57.

Pinching, A. J., et al. (1986). Networks of AIDS patients. Personal communication.

Piot, P., & Carael, M. (1988). Epidemiological and sociological aspects of HIV-infection in developing countries. *British Medical Bulletin, 44,* 66–88.

Piot, P., Plummer, F. A., Rey, M. A., & Ngugi, E. N. (1987). Retrospective seroepidemiology of AIDS virus infection in Nairobi prostitutes. *Journal of Infectious Diseases, 155,* 1108–1112.

Quinn, T. C., Mann, J. M., Curran, J. W., & Piot, P. (1986). AIDS in Africa: An epidemiologic paradigm. *Science, 234,* 955–963.

Redfield, R. R., Markham, P. D. Salahuddin, S. Z., Wright, D. C., Sarngadharan, M. G., Gallo, R. C. (1985). Heterosexually acquired HTLV III/LAV disease (AIDS-related complex and AIDS). Epidemiologic evidence for female-to-male transmission. *Journal of the American Medical Association, 254,* 2094–2096.

Redfield, R. R., Wright, D. C., Markham, P. D., Gallo, R. C., Salahuddin, S. Z., & Burke, D. S. (1986, June). *Frequent bidirectional heterosexual transmission of HTLV III/LAV between spouses.* Paper presented at Second International Conference on AIDS, Paris.

Schultz, S., Milberg, J. A., Kristal, A. R., & Stoneburner, R. L. (1986). Female to male transmission of HTLV III. *Journal of the American Medical Association, 255,* 1703–1704.

Smith, G. L., & Smith, L. K. F. (1986). Lack of HIV infection and condom use in licensed prostitutes. *Lancet, ii,* 1392.

Srnivasan, A., York, D., Rangathan, P., Ferguson, R., Butter, D., Fearino, P., Kalyanaraman, V., Jaffe, H., Curran, J., & Anand, R. (1987). Transfusion-associated AIDS: Donor-recipient human immunodeficiency virus exhibits genetic heterogeneity. *Blood, 69,* 1766–1770.

Staszewski, S., Schieck, E., Rehmet, S., Helm, E. B., & Stille, W. (1987). HIV transmission from male after only two sexual contacts. *Lancet, ii,* 628.

Steigbigel, N. H., Maude, D. W., Feiner, C. J., Harris, C. A., Saltzman, B. R., Klein, R. S., et al. (1987, June). *Heterosexual transmission of infection and disease by the human immunodeficiency virus (HIV).* Paper presented at Third International Conference on AIDS, Washington, DC.

Taelman, H., Bonneux, L., Cornet, P., et al. (1987, June). *Transmission of HIV to partners of seropositive heterosexuals from Africa.* Paper presented at Third International Conference on AIDS, Washington, DC.

Tirelli, U., Vaccher, E., Sorio, R., Carbone, A., & Monfardini, S. (1986). HTLV III antibodies in drug-addicted prostitutes used by US soldiers in Italy. *Journal of the American Medical Association, 256,* 711–712.

Winkelstein, W., Lyman, D. M., Padian, N., Grant, R., Samuel, M., Wiley, J. A., Anderson, R. E., Lang, W., Riggs, J., & Levy, J. A. (1987a). Sexual practices and risk of infection by the human immunodeficiency virus: The San Francisco Men's Health Study. *Journal of the American Medical Association, 257,* 321–325.

Winkelstein, W., Samuel, M., Padian, N. S., & Wiley, J. A. (1987b). Selected sexual practices of San Francisco heterosexual men and risk of infection by the human immunodeficiency virus. *Journal of the American Medical Association, 257,* 1470–1471.

Winkelstein, W., Wiley, J. A., & Nadian, N. (1986). Potential for transmission of AIDS-associated retrovirus from bisexual men in San Francisco to their female sexual contacts. *Journal of the American Medical Association, 255,* 901–902.

7

Changing Sexual Behavior

Leon McKusick

Imagine someone emerging from a singles' bar, slightly inebriated and slightly in love with a person whom the individual has just met. As the two exit, they are met by the intense stares of two young research assistants, each eager to offer the couple a questionnaire. They are saying something about AIDS, saying "Would you please fill this out, drop it in the mail to us?" saying something about "the University of California." Then off the couple strolls down the street, curiously opening the questionnaire, only to leaf through pages of inquiry about how much they normally drink, the details of their sexual practices, whether or not they use condoms, whether or not they are married, their attitudes toward AIDS, and finally, whether or not they care about infecting each other with human immunodeficiency virus (HIV). That's a good conversation starter, as well as quite a way to begin a relationship—a picture of coping with life in the nineties.

As we look at how expediently people have responded to AIDS, we must first look at AIDS in the context of other health threats. Bear in mind that the behavioral response to AIDS has been quicker than *any* response to *any* other sexually transmitted disease. Behavior change in the AIDS epidemic has been quicker than *any* behavior change in response to *any* other *illness*—quicker than that to the illnesses caused by smoking or alcoholism, quicker than that to the threat of lethal heart disease (Coates, Stall, & Hoff, 1987). AIDS has probably had the strongest singular impact upon sexual behavior and the psychology of sex since Kinsey began measuring sexual behavior (Kinsey, Pomeroy, & Martin, 1948). Many middle-aged gay men in San Francisco can be overheard bemoaning the absence of the "good old days." This is something that middle-aged men may do in all communities. However, in the San Francisco gay community, these men do not grieve for what has not been lost.

AIDS has caused profound changes. According to a report prepared for the Office of Technology Assessment (Coates, Stall, & Hoff, 1987), by 1987 risky sexual behavior had changed more in San Francisco (Communication Technologies, 1987; Doll et al., 1987; McKusick, Horstman, & Coates, 1985; McKusick, Wiley, et al., 1985; Winkelstein et al., 1987) than anywhere else it has been measured and reported in the United States (Communication Technologies, 1986; Fox et al., 1987; Jones et al., 1987; Joseph et al., 1987; Juran, 1987; Kelly et al., 1987; Klein et al., 1987; Martin, 1987; Siegel et al., 1987). We had the opportunity to watch this mobilization since 1981, to observe patients and respondents go through the stages from denial to acceptance of disease threat; from fear of HIV being transmitted to them to fear that they may have transmitted HIV to others; from complete disease ignorance to a debate-level expertise fueled by concern for oneself and for others.

Why was this response so rapid in San Francisco? Three possible explanations emerge from observation. First, because the disease was still new and not well understood, the alarm had been sounded. The government had been mobilized, money was generated, and agencies had sprung up to deal with all aspects of the epidemic, particularly transmission and public health concerns. Condom use is either debated or satirized daily. This promotes awareness of behavioral health guidelines. The undeniable presence of AIDS—in a doctor's office waiting room, on the street, at a dinner party, in the supermarket, in the mall—visualizes, humanizes, actualizes the problem, defining the alacrity of behavioral response. In research, we measured this awareness of disease presence by asking our respondents the number of their friends or acquaintances diagnosed, and whether they have visual memory of what it looks like to have a severe form of AIDS. Indeed, in early analyses this awareness was shown to correlate with lowered risk (McKusick, Horstman, & Carfagni, 1983).

Second, the disease was perceived as terminal—the threat of a potential death sentence was a strong deterrent. In the following years, as the disease became better known, as some have survived, as treatment has become available that prolongs life, as people are getting *used* to this disease, as they have gotten *used* to other behaviorally induced diseases, perceived threat has become a less powerful motivator for change. Indeed, findings from a study of smoking show that threat stimulates short-term change, but further reinforcement is needed to sustain change (Surgeon General of the United States, 1986).

Third, on the whole, the San Francisco gay community is cohesively organized and very intercommunicative, much more so than other risk groups. Do not underestimate the power of 20 gay men on the phone in the afternoon; 4,000 friends will have the word by evening. On the whole, the gay community is emotionally and psychologically supportive, as well. When people feel supported, they have increased personal efficacy in threatening situations.

The three possible explanations for the speed of response to risky sexual behavior have been codified into variables that correlate with presence or absence of sexual behavioral risk reduction in three research projects conducted by the University of California San Francisco Center for AIDS Prevention Studies Psychosocial Research Component (Thomas Coates, Ph.D., principal investigator, Stephen Morin, Ph.D., Ron Stall, Ph.D., James Wiley, Ph.D., Joseph Catania, Ph.D., and Susan Kegles, Ph.D.). It is likely that the same or similar response factors apply to the gay community as a whole. They may be less relevant to other AIDS risk groups.

The three studies mentioned involved a sample of 450 gay men in San Francisco surveyed once a year from November 1983 to 1986; a stress reduction intervention group conducted in 1987 for 60 seropositive men; and a sample of 2,774 California high school students surveyed in mid-1987 as part of an AIDS education evaluation (Table 7-1). Based on preliminary results from these data, a hypothetical model emerges of individuals who may continue to be at risk for transmitting HIV. Knowing who is at risk can help us to design effective and compassionate interventions that may contain the spread of HIV.

Table 7-1
Description of Three Samples

Group	N =	Behavior Change	Time Frame	Data Collection
Gay men in San Francisco	454	Dramatic reduction in number of partners and risk acts; increased condom use	1983–86	Yearly mail survey of cohort
Stress group for seropositive gay men	60	Experimentals reduced number of sexual partners more than controls	1987 (3 months)	Pre- and post-test experimental
California high schoolers	2,774	No effect	1987 (1 month)	Pre- and post-test program evaluation

Variables Hypothesized to Influence Sexual Risk Activity
(Table 7-2)

Demographic Variables

Race appears to be associated with use of condoms during intercourse
in California high school students. Hispanic students were least likely
to use condoms during intercourse, then blacks, whites, and finally
Asians. High school girls were less likely to have and use condoms than
high school boys.

Younger gay men (under 30) appear to be behaviorally more flexible
than older gay men (over 40), and are therefore more easily able to
reduce sexual risk activity (McKusick, Wiley, et al., 1985). In the same
sample, although the percentage of men in a primary gay relationship
has not changed in each period of measurement during the time of the
AIDS epidemic, the number of monogamous gay men in relationships
has increased, as has the number of abstinent single men (McKusick &
Coates, 1987).

Table 7-2
Correlates of Risk Activity

Demographic variables
 Race[a]
 Age[b]
 Relationship status and monogamy[b]

Health Awareness Variables
 Knowledge of HIV antibody status[b]

Health Belief Variables
 Personal efficacy[a,b]
 Agreement with behavioral health guidelines (response efficacy)[a,b]
 Perceived threat[a,b]

Behavioral Variables
 Use of drugs or alcohol[b]
 Use of drugs or alcohol during sexual activity[b]
 Stress reduction activity[c]

Psychological Variables
 Depression[b]
 Stress[b]
 Denial[b]
 Self-esteem[b]

Social Variables
 Perception of peer norms[a]

[a] Significant interaction with behavioral risk levels in California high school students.
[b] Significant interactions with behavioral risk levels in gay men in San Francisco.
[c] Significant interaction with behavioral risk levels in seropositive gay men in stress reduction seminar.

Health Awareness Variables

An individual's knowledge of being antibody positive for HIV has been associated with lowered sexual risk activity in studies of gay men (Coates, Morin, & McKusick, 1987). Strength of agreement with behavioral health guidelines also has been correlated with degree of sexual risk reduction (McKusick, Wiley, et al., 1985).

Health Belief Variables

Aspects of the health belief model have been adapted to the AIDS epidemic (McKusick, Conant, & Coates, 1985). In the study of gay men, the strongest correlations were found between personal efficacy, a measure of the respondent's belief in his capability to reduce risk, and level of risk activity. Personal efficacy also appears to be crucial in the maintenance of lowered risk over time. A measure of perceived threat, based on an individual's agreement that AIDS is an immediate and feared threat in the environment, also correlated with level of sexual risk activity in cross-sectional analyses of the San Francisco gay men's sample (Morin & Charles, 1985).

Behavioral Variables

Stall, McKusick, Wiley, Coates, and Ostrow (1986) found that use of drugs or alcohol overall, and specifically use of drugs or alcohol during sexual activity, led to more HIV-transmitting sexual activity in gay men. HIV-positive gay men who participated in a 10-week stress reduction seminar were more likely to reduce numbers of sexual partners during a 3-month period than controls who did not participate in the group (Coates, McKusick, Kuno, & Stites, 1990). Reduction of stress may have been a factor in this reduction of risk activity, although Dr. Donald Francis has pointed out that perhaps these men were too busy going to a group to spend time looking for sexual partners.

Psychological Variables

In early analyses of gay men in San Francisco, anonymous sex was reported as a means of reducing stress (McKusick et al., 1983). Further investigation has shown that stress is correlated with high-risk sex, and stress reduction with reduction of numbers of sexual partners (Coates, McKusick, Kuno, & Stites, 1988).

Psychological depression, as measured by the Brief Symptom Inventory (BSI), was correlated with both lowered risk and sustained risk reduction over time. AIDS is depressing and, in this instance, depression may be an adaptive reaction to the presence of AIDS in the environment of the individual, or may simply signal the absence of motivation to have sex of any kind in response to the situation. Weber, Coates, and

McKusick (1988), using the denial subscale of the Minnesota Multiphasic Personality Inventory (MMPI) and Weber's own AIDS denial scale, found strong positive correlations between amount of psychological denial and level of risky sexual activity. Morin and Charles (1985) found self-esteem to be inversely predictive of level of risk activity.

Social Variables

High school students who reported that they perceived their peers to be engaged in condom use also were likely to engage in condom use, as well as in other forms of sexual risk reduction activity.

Hypothetical Profiles of Persons at Risk of HIV Transmission

Target Profile 1: A Gay Man (Table 7-3)

Homosexual activity continues to be the primary way HIV is being transmitted, even though massive behavior change has already occurred among gay men. In absolute numbers, we can assume that more gay men seroconvert to HIV in a given week than members of any other risk group. Consequently, behavioral health education is still most crucial and most needed in this population, to promote further change and to sustain what changes have been made.

The gay man at risk of HIV transmission is likely to be single and over 40. He may use drugs or alcohol during sex. He may experience low self-esteem and stress; he may be swimming in denial. (As a physician once remarked in reference to certain members of the gay community, "You know, denial is not just a river in Egypt.") He is not likely to agree that AIDS is an immediate and personal threat, nor is he likely to have a high level of personal efficacy.

Table 7-3
High-Risk Profile 1: Gay Male

20–30 years old
In a shorter term primary relationship
Does not know his antibody status
Less likely to agree with behavioral health guidelines
Lower personal efficacy
Lower perceived threat
Less depressed
Drinks or does drugs during sex
Swimming in denial
Stressed, not reducing stress

Table 7-4
High-Risk Profile 2: Heterosexual High Schooler

Hispanic or Black
Female
Less likely to use condoms than males
Less likely to agree with behavioral health guidelines
Lower personal efficacy
Greater perceived threat
Lower perceived peer norms

Target Profile 2: A High School Student (Table 7-4)

Since high school students are still behaviorally flexible as a result of
their youth, are experimenting with sex, and are accessible to education,
we must teach them how to avoid HIV transmission. Based on research
findings, the high school student at risk is more likely to be Hispanic
or black than white or Asian, more likely to be a girl than to be a boy,
less in agreement with behavioral health guidelines, less likely to perceive the threat of AIDS in her or his environment or to experience
personal efficacy, and less likely to consider that her or his peers are
engaged in safer sex practices.

Intervention Guidelines

Based on our predictive data and hypothetical profiles, some tenets can
be offered of an effective behavioral intervention strategy for those involved in counseling and/or education of persons in high risk groups.

**1. As much as possible, understand the psychology motivating your
clients' behavior.** In order to do so, you may need to pay particular
attention to the literature on gay psychology, on racial and ethnic minority issues, and, given the link with substance and alcohol use, on
behavioral addiction.

**2. Be explicit, in the language of your client, when describing and
asking about sex behavior.** This is necessary for two reasons: first, so
that you know that the two of you are talking about the same thing.
Regarding oral-anal contact, for example, one must be aware of collo-
quialisms such as "rimming" in order to obtain accurate reportage. Second, to establish trust and derive honest responses, the therapist or
educator must speak in the syntax of sex. Specific syntax varies from
culture to culture, and is usually highly valued by the members of each
culture. Knowledge, or lack of knowledge, of such syntax on the part
of the interviewer is seen by the respondent as grounds for interpersonal
trust or distrust. Also, highly clinical terms, which communicate the

specifics of sexual activity clearly and discreetly in the halls of science, may be a real turnoff in the field.

3. Intervene in a nonjudgmental fashion. An instrument has not been designed yet to measure the lightning speed with which a person will reject a recommendation out of hand if he or she perceives that he or she is being judged, particularly if the topic is sex.

4. Be precisely informed and *authoritative* about behavioral health guidelines. In epidemiological discussions, equivocation often occurs about the relative risk of transmission of HIV via certain sexual behaviors. Although one must be aware of the unanswered questions about what exactly constitutes transmitting sexual activity, in order to intervene one must also speak with authority. This is particularly true when the therapist or educator is confronting a client's resistance to changing a known high-risk practice, such as anal intercourse without a condom. The client may defend his behavior by saying, "No one agrees on exactly how HIV is transmitted." The authoritative educator would respond, "Nonetheless, your risk of infection will remain extremely high if you continue having anal intercourse without a condom." If the client retorts, "If that's the case, I'm probably already infected, so what's the difference," the educator would reinforce the point by saying, "If you were already infected, you would then need to protect your immune system from any *additional* viruses or bacteria, which means the same: you would need to wear a condom during anal intercourse."

5. Call upon the social responsibility of your client to protect others. Some people can be motivated to help others more easily than to protect themselves. Our studies have shown a high level of concern for others, particularly in those gay men who know their HIV antibody status is positive. The STOP AIDS Project, a very effective risk reduction campaign executed in San Francisco and instituted elsewhere, is partly founded on this idea, drawing people into participation in a fight against HIV transmission for the protection of the community.

6. Support and enhance psychological predictors of risk reduction, particularly self-esteem and personal efficacy, then support psychological factors that maintain lowered risk. Awareness of recommendations, with the backing of efficacy, constitute the "one-two punch" against AIDS transmission. The educator or therapist can enhance the client's belief in his or her ability to reduce behavioral risk by framing behavior change in a positive context (e.g., that it is a good thing to master); by simplifying the issues (e.g., that, once mastered, it is a relatively easy thing to do); and by supporting the whole personality in change (e.g., reinforcing that whatever the client does to reduce risk at whatever pace is a demonstration of self-worth and efficacy).

7. Be aware of social support networks (particularly the role and influence of the primary partner). This is where subculture sensitivity comes in handy. Many at-risk individuals are in primary relationships that are not institutionally sanctioned, but that are extremely supportive and viable sources of support and reinforcement. Perception of peer norms has also been shown to be an important factor. If the Joneses are using condoms, and keeping it up, it may be time, once again, to keep up with the Joneses.

8. **Create a risk hierarchy and encourage your client to move at least toward "lowered risk" if not to "no risk."** In many instances an individual cannot immediately leap from risky sexual activity into abstinence from such activity. This can be a source of pure frustration to the exuberant educator or therapist who desires to stop HIV transmission immediately, and can stimulate negative attitudes toward the client. One solution is to move the client in obtainable stepped goals to lesser risk activity. This process is more likely to promote efficacy, and consequently to result in the desired change. Two overall goals could be: first, to stop activity that transmits HIV to others, invoking social responsibility; and second, to stop activity that leaves one vulnerable to infection, while reinforcing self-esteem.

Two Clinical Examples

To bring the reader a step closer to a psychology of risk reduction, and the human beings involved, two psychotherapy cases are presented. The first case, although not a complete success, describes stress and self-esteem and their impact on risk activity, and encourages understanding of gay psychology and homosexual behavior. The second case highlights the idea of efficacy and the use of the HIV antibody- test to support behavior change.

Stress, Self-Esteem, and Gay Psychology

A 34-year-old gay hemophiliac, diagnosed with AIDS-related complex, has been engaging in anonymous sex. Regarding HIV transmission, therapy with this man has been a partial success: He moved from behavior in his anonymous contacts that might infect others to behavior that only hurts him. He was very reluctant to stop his favorite behavior: oral-anal activity with strangers, more coloquially known as "rimming" by white gay men. He is currently on azidodeoxythimidine (AZT). His repeated rimming has caused reinfection with amebic parasites, which is, his physician lectures, "very dangerous" for him given his immune status. The behavior persists in spite of medical recommendations and his awareness of risk.

He and I agreed on the following goal hierarchy: it is better for him to perform fellatio and to swallow semen than to rim; it would be even better to perform fellatio without swallowing semen; and, ultimately, it would be good for him to get away from anonymous contact altogether, where old patterns of behavior tend to be reinforced. Although the hierarchy involves practices that are equally risky or of unknown risk, he believes that he can follow it, and its overall direction is toward less risk. Moreover, he reports that he would be *extremely* embarrassed and ashamed to think that word might get out to his friends that he engages in unsafe sex, a good example of the influence of peer norms.

Looking at the situation, here are the man's liabilities. He has low self-esteem, his anger and stress often result in self-destructive sexual behavior, and he is reluctant to change his more habitual sexual patterns. Distrustful, it took him several sessions to talk to me at all about his sexual risk activity, since he anticipated my negative judgment.

He also has several assets. Although an adult child of an alcoholic, he does not drink. He has sought and is accessible to psychotherapy, which is promoting his self-esteem. He has elected to go on AZT, which, from a psychological standpoint, further enhances his self-esteem and hope for the future. Finally, turning a liability into an asset, he is compulsive. With the right encouragement, he can be just as compulsively interested in risk reduction as he is about unsafe sex. He and I continue together in laborious and lascivious discussions of the frank details of his sexual feelings and fantasies and what to do about them, a kind of talk that he relishes.

Efficacy and Test Results

Two gay men were in a relationship of 1 year within which they had strived to have safer sex. They came for couples' counseling together to discuss nesting stage (McWhirter & Mattison, 1984) issues (buying a house and moving together, merging assets, establishing agreements regarding commitment and fidelity) and, the proverbial psychotherapeutic bottom line, to discuss their ambivalences about getting closer. One of their shared concerns was each others' HIV antibody status.

For this example, I'll call them Peter and Norman. Peter knew he was seropositive for HIV but had no ARC symptoms. Norman did not know his antibody status when they appeared for counseling. The process of Norman's testing became a principal focus of relationship commitment. In the 2 weeks between the drawing of his blood and the receipt of the results, Norman expressed a great deal of fear and anxiety. Characteristically loyal and devoted, Norman had provided the glue that had held these two together. However, during this waiting period he expressed his ambivalence toward the relationship in the form of a cold lack of empathy for Peter.

Peter, a man who, by character, is subject to withdrawal under stress, needed help to stay engaged. This 2-week period was a test of his ability to "be there" for Norman and wish him the best, in spite of his own anxiety, or even his possible envy, about the results. If Norman's test came back positive, Peter might then feel guilty for possibly having infected him. If Norman were negative, Peter, then the only seropositive in the relationship, might feel alienated, alone, and in need of reassurance about feared rejection or desertion.

Norman's test was seronegative. He immediately felt celebratory, and then, as immediately, he experienced a kind of "survivor's guilt," knowing he had escaped a painful situation that Peter had not. Soon thereafter came a wellspring of empathy and expression of love for Peter. Peter immediately reciprocated with his own affirmations of Norman, a successful resolution of a very unstable time for the two of them. After a year in a blooming relationship that was considerably sexual (including anal intercourse, both ways, with condoms), Norman's seronegative result was a simple reinforcement of their safer sex couple efficacy. It was then an easy step for the three of us to generalize this efficacy to the quality and viability of their relationship on the whole. In this sense, the antibody test was a challenge, and ultimately an aid, to them, as the relationship passed into the next stage of couples' development.

Conclusion

The heart of truth about AIDS is being told by whole people, who are individually and uniquely coping with disease and sex in the larger context of their lives and life meanings. As we move to reduce high-risk sexual behavior, we will be continually frustrated and crippled in our efforts unless we respectfully observe, as Kinsey did, this delightful variety.

Acknowledgment

This chapter was supported in part by two grants: National Institute of Mental Health (NIMH) grant #MH39553, and NIMH/National Institute on Drug Abuse AIDS Center grant #42459.

References

Coates, T. J., McKusick, L., Kuno, R., & Stites, D. (1990). Stress reduction training reduced number of sexual partners but did not affect immune function in HIV positive gay men. *American Journal of Public Health, 79,* 885–887.

Coates, T.J., Morin, S. F., Lo, B., Stall, R. D., & McKusick, L. (1987, August). *AIDS antibody testing: Will it stop the AIDS epidemic? Will it help people infected with HIV?* Paper presented at the meetings of the American Psychological Association, New York.

Coates, T. J., Morin, S. F., & McKusick, L. (1987). Behavioral consequences of AIDS antibody testing among gay men. *JAMA, 258,* 1889.

Coates, T. J., Stall, R. D., & Hoff, C. C. (1988, June). *Changes in sexual behavior among gay and bisexual men since the beginning of the AIDS epidemic.* Report prepared for the Office of Technology Assessment. Washington, DC.

Communication Technologies. (1986). *Designing an effective AIDS prevention campaign strategy for Los Angeles County.* San Francisco: Communication Technologies, Inc.

Communication Technologies. (1987). *A report on designing an effective AIDS prevention campaign strategy for San Francisco: Results from the fourth probability sample of an urban gay male community.* San Francisco: Communication Technologies.

Coll, L. S., Darrow, W. W., O'Malley, P., Bodecker, T., & Jaffe, H. (1987, June). *Self-reported changes in sexual behaviors in gay and bisexual men from the San Francisco City Clinic cohort.* [Abstract] The Third International Conference on AIDS, Washington, DC.

Fox, R., Ostrow, D., Valdiserri, R., Van Raden, M., Visscher, B. & Polk, B. F. (1987, June). *Changes in sexual activities among participants in the Multicenter AIDS Cohort Study.* [Abstract] The Third International Conference on AIDS, Washington, DC.

Jones, C. C., Waskin, H., Gerety, B., Skipper, B. J., Hull, H. F., & Mertz, G. J. (1987). Persistence of high risk sexual activity among homosexual men in an area of low incidence of AIDS. *Sexually Transmitted Diseases, 14,* 79–82.

Joseph, J., Montgomery, C., Kessler, R. C., Ostrow, D. G., Emmons, C. A., & Phair, J. P. (1987, June). *Behavioral risk reduction in a cohort of homosexual men: Two year follow-up.* [Abstract] The Third International Conference on AIDS, Washington, DC.

Juran, S. (1987, April). *Sexual concern and behavioral change as a result of fear of AIDS.* Paper presented at the Society for the Scientific Study of Sex, Eastern Region Conference, Philadelphia, PA.

Kelly, J. A., St. Lawrence, J. S., Hood, H. V., et al. (1989). Behavioral interventions to reduce AIDS risk activities. *Journal of Consulting and Clinical Psychology, 57,* 60–67.

Kinsey, A. C., Pomeroy, W. B., & Martin, C. R. (1948). *Sexual behavior in the human male.* Philadelphia: W.B. Saunders Company.

Klein, D. E., Wolcott, D. L., Landsverk, J., Namir, S., & Fawzy, F. I. (1987). Changes in AIDS risk behaviors among homosexual male physicians and university students. *American Journal of Psychiatry, 144,* 742–747.

Martin, J. L. (1987). The Impact of AIDS on gay male sexual patterns in New York City. *American Journal of Public Health, 77,* 578–581.

McKusick, L., & Coates, T. J. (1987). HTLV-III transmitting behavior in San Francisco gay men. In D. Ostrow (Ed.), *Biobehavioral control of AIDS.* New York: Irvington Press.

McKusick, L., Conant, M. A., & Coates, T. J. (1985). The AIDS epidemic: A model for developing intervention strategies for reducing high risk behavior in gay men. *Sexually Transmitted Diseases, 12,* 229–234.

McKusick, L., Horstman, W. R., & Carfagni, A. (1983, August). *Report on AIDS and the sexual behavior of San Francisco gay men.* Paper presented at the Annual Convention of the American Psychological Association, Anaheim, CA.

McKusick, L., Horstman, W., & Coates, T. (1985). AIDS and the sexual behavior reported by gay men in San Francisco. *American Journal of Public Health, 75(5),* 493–496.

McKusick, L., Wiley, J. A., Coates, T. J., Stall, R., Saika, G., Morin, S., Charles, K., Horstman, W. R., & Conant, M. A. (1985). Reported changes in the sexual behavior of men at risk for AIDS: San Francisco, 1982–1984: The AIDS Behavioral Research Project. *Public Health Reports, 100,* 622–629.

McWhirter, D. P., & Mattison, A. M. (1984). *The male couple.* Englewood Cliffs, NJ: Prentice-Hall.

Morin, S. F., & Charles, K. (1985, August). *Health belief and sexual behavior change.* Paper presented at the annual conference of the American Psychological Association, Toronto.

Siegel, K., Mesagno, F., Chen, J. Y., & Christ, G. (1987, June). *Factors distinguishing homosexual males practicing safe and risky sex.* [Abstract] The Third International Conference of AIDS, Washington, DC.

Stall, R., McKusick, L., Wiley, J., Coates, T. J., & Ostrow, D. G. (1986). Alcohol and drug use during sexual activity and compliance with safe sex guidelines for AIDS. *Health Education Quarterly, 13(4).*

Surgeon General of the United States. (1986). *The health consequences of involuntary smoking.* Washington, DC: U.S. Department of Health and Human Services.

Winkelstein, W., Samuel, M., Padian, N., Wiley, J. A., Lang, W., Anderson, R. E., & Levy, J. A. (1987). The San Francisco Men's Health Study: Reduction in human immunodeficiency virus transmission among homosexual/bisexual men: 1982–1986. *American Journal of Public Health, 76,* 685–689.

III
HISTORICAL
PERSPECTIVES

8

Social History: Disease and Homosexuality

John E. Boswell

Moral problems relating to AIDS are among the most difficult and complex that American society has faced in the 20th century. Civil libertarians and lawyers have barely begun to untangle the knotted cords of individual and public rights; the medical community has been forced to confront its own fears, inadequacies, and ethical shallowness in unprecedented ways; and the public at large is divided and torn by conflicts between compassion for the suffering and fear of contagion.

Disease has held a mirror to the cultures and people of the world for as long as literature has recorded the fears and follies of humans. Writings from Exodus to Camus' *La Peste* have chronicled or pictured the reactions of individuals and nations to biological catastrophe as a means of exploring human souls and behavior. This is partly because the terrifying effects of plague—sickness and death—are themselves staples of both history and literature, being the chief plot devices nature uses in setting the length and shape of human lives. They are at once natural and unnatural, universal and singular, absolutely predictable and totally unnerving: ideal for both literature and history, which must decipher the constants of the human heart by exploring its reactions to particular and extraordinary events.

Plague adds another dimension to the revealing power of sickness and death by transforming an essentially individual experience into a communal one. Long ago, the deceased were "laid to rest" on family land and thought of as the members of the family or community who had gone before; in the modern world the dead are all gathered together away from the areas used by the living and hidden behind cemetary

171

walls. The sick—although we are all sometimes sick—are kept away from the healthy, often in special buildings for this purpose. These distinctions and their social embodiments are created by profound anxieties about sickness and dying, which in the 20th century have taken on the air of the bizarre and macabre rather than normal events in the lives of humans. Medicine and its amelioration of the frailty of the body have contributed to this, but it is also in large measure the result of loss of faith in personal or social values beyond good health and a pleasant life.

Plague demolishes these barriers and looses on the living and healthy the anxieties supposed to be contained behind hospital and graveyard walls. Disease and mortality become communal experiences and worries. Healthy people must think about disease, must experience sickness other than their own, must face and cope with death. An unusual increase in the number of sick and dying, especially if related to a communicable illness, can so terrify the living and well that they cease to believe in the distinctions that formerly protected them. Whole societies can begin to think of themselves as diseased and dying, even when the majority of them are not, because the horror of epidemic breaks through the cultural barriers that had protected the ordinary person from seeing the inevitable, or so disorders the natural plot line for individuals that notions of "lifetime" collapse.

The most haunting images of plague in the Western tradition are those of the Black Death of 1348–1350, which killed about one third of the entire population of Europe in 2 years.[1] Its effects were so devastating that the political and religious structures of Europe were profoundly transformed by it.[2] Astonishingly, much literature composed at the time takes little notice; some seems utterly oblivious.[3] Huge numbers of people apparently went on with "business as usual" or pretended to do so even when it was actually impossible. The cynical and ribald tales told in the *Decameron* by aristocrats who fled the plague and sealed themselves off from a dying world are the best known literary legacy of this. The nursery rhyme "Ring around the rosey" (a description of one of the symptoms) is an incongruously light-hearted reminder of a later

---

[1] For a broad overview of the role of epidemic disease in world history, see McNeill (1976). On the Black Death in particular, the standard studies are, for Europe, Ziegler (1969) and, for the Middle East, Dols (1977). Dols (1974) also published a very enlightening comparison of Eastern and Western responses to the plague. Of more recent literature one might note Gottfried (1978, 1983).

[2] On this see the classic study by Huizinga (1924), and more recently (*inter alia*) Mollat and Wolff (1972), Leff (1967), and Kieckhefer (1984).

[3] But see Barasch (1976). By contrast, leprosy, although actually a very small medical problem, had a profound impact on the medieval imagination (see, e.g., Brody, 1974). Delumeau (1978) treated fear in general as a cultural factor.

occurrence of the same plague, and one of the few cultural muniments to record the magnitude of its horror: "all fall down."

But the Black Death was by no means the most disruptive plague on record. Most of the native population of South America died within 50 years of the arrival of Europeans from the plague of smallpox the Spanish brought with them. (These facts, widely known and treated in most histories of the conquest of South America, are briefly covered in McNeill, 1976, pp. 176–207.) Because the conquistadores were accustomed (and more immune) to smallpox themselves, because they neither understood nor cared about the Indians dying by the millions, and—most poignantly, perhaps—because the culture that might have recorded the revelations of this plague perished along with its creators, nothing survives to share the secrets of the human heart that might have been learned from what was perhaps the most devastating single pestilence in history. (The fate of the North American Indians was similar, but the destruction of native culture took somewhat longer in North America, and the diversity of the invading cultures both ameliorated and aggravated the cultural conflict.)

All plagues have a shattering impact on the community they affect, but if the community is a disenfranchised one the dominant elements of society may well ignore the situation, and decline to share the experience or the wisdom wrung from it. A few references in *The Color Purple* are probably the first literary expression of the horror of sickle cell anemia to reach the American reading public, although the disease has terrorized the black community for centuries.

Like other plagues, the AIDS epidemic is both universal and particular. In the communities most affected by it so far, one sees the classic responses to plague. Some find that the walls that protected them from death and sickness have collapsed, and they regard themselves as members of the community of the dying even if they are apparently healthy. Others—like Boccaccio's storytellers—shore up the old walls, build new ones, or escape to fantasy.

To some extent the reaction of the public could be compared to that of the white community to sickle cell anemia (medical interest but little empathy), or to that of the conquistadores (indifference or inattention to the suffering, some concern about demographic consequences, in this case whether it will become more prevalent in the general population). The latter comparison is poignant since it is now believed that the virus that causes AIDS spread to the gay population of the New World from the heterosexual population of Africa.

However, in several ways the AIDS epidemic and public reaction to it are unparalleled. The walls around disease and death have been greatly buttressed in the 20th century not only by preoccupation in industrial societies with the health of the body but also with the concept of being "normal"—a concept that has replaced "moral" or "good" in politically based, religiously plural societies. In moralistic cultures everyone is conscious of being a sinner, a fact that mitigates the degree of alienation visited upon or felt by those who transgress the rules. "Normality" is a much more alienating concept, since it frequently applies to someone's being—something he or she is unlikely to be able to change—rather than to behavior that could be altered. Serious illness is abnormal in health-obsessed America, and few things are as alienating as being seriously unhealthy. Cancer, and the shame its victims often feel or are made to feel, is the clearest example of this, but any physical abnormality causes embarrassment and discomfort to most Americans and exposes its possessor to ostracism and isolation from a public trying to maintain the walls against disease and abnormality.

This blaming and punishing of victims for not being "normal" is particularly devastating in the cases of the largest groups of people with AIDS in America, gay men and drug addicts. Even without fear of contagion, having a fatal illness would isolate them from American society. When communicability and social stigma are added, the rejection, callousness, and indifference become overwhelming. It is not too difficult to see why, in a society deeply troubled by drug abuse (particularly cocaine) on the part of political, artistic, and even sports elites, the public might choose to view heroin addiction as a reprehensible moral choice rather than a pitiful form of suffering. The circumstances of the poorest and most alienated Americans are quite different from those of the rich and powerful, and one might more charitably regard their addiction as a desperately wrong choice of remedies than as a malicious crime, but it is still understandable that people entertain ambivalence about problems of drug abuse. It is less easy to understand why the pathetic, excruciating consequences this often entails are then conflated with these individuals' bad choices, inducing at best utter indifference to their fates and at worst active revulsion and hostility.

The plight of middle-class persons afflicted with lung cancer after years of smoking, with heart disease after years of being overweight or failing to exercise, with injuries sustained in an automobile accident while intoxicated, with financial ruin after fiscal recklessness—these prompt compassion in most humans, and soften judgments about erstwhile failings. Even if we see that in some way their misery is related to their previous actions, we are reminded that misfortune is as common to all of us as are unwise or selfish choices. We know that we would expect our families, friends, neighbors, doctors, and civic officials to care for us in need, even if we had been less than perfectly circumspect in our lives, and we are usually willing to extend a hand to any fellow citizen without inquiring into the extent to which he or she may be responsible for some of the difficulties.

For some reason the unwise choices of poor blacks and Hispanics in America's urban centers seem to place them out of reach of compassion or empathy from most Americans, including, tragically, the government. It is as bad for gay men, which is more surprising in some ways, because they are not usually separated from the indifferent majority by emotional, social, or physical distance. They come from ordinary families of all social classes. The wall dividing them from the majority is not demarcated by a ghetto or a color or economic standing.

The notion that Western society has been characterized by generally increasing progress and tolerance on social issues such as the position of gay people is a misprision. Although many of the people reading this may live in social enclaves where there is considerable tolerance, probably at no time and place in Western history have attitudes toward gay people been more hostile than in the United States and Europe throughout most of the 20th century. This is often as true among liberal, well-educated people as among the population at large, and it is related to the historical vagaries and inconsistencies of the transition from moral values to the pseudoscientific ones noted above.

In the ancient world "norms" for human beings were largely social and behavioral: there was public agreement and expectation about how to be a good citizen, a good parent, a good child, a good friend, and so on, based on codes of conduct and behavior that anyone could fulfill. There was little or no consciousness of gay people as a distinct category of human being; they could and did fulfill these duties as well as anyone else (including being a good parent: marriage and parenthood were not thought to be restrictively coterminous with love). Men known to have erotic interest in other men occupied high positions, were extremely influential, and often were much admired. Although there was some awareness of sexual preference, it was unrelated to important social "norms" and not a matter of any more significance than one's choice of food or housing would be today.[4]

Christianity introduced a different set of norms, worse for gay people, but only marginally worse, contrary to popular opinion (for Christian attitudes see Boswell, 1980; Brundage, 1987; and Noonan, 1965). From

---

[4] On homosexuality in the ancient world see bibliography in Boswell (1980) for references up to 1980. It omitted Veyne (1978) and Sullivan (1979). Works by Buffière, Bremmer, and Kempter all appeared in 1980; publications appeared in 1981 by Veyne and Barrett, the latter of which includes a thorough review of the literature on homosexuality in Homer. Subsequent materials (to 1982) are cited in Boswell (1982-83). Foucault (1984) offered a superficial but challenging overview of Greek and Roman sexual constructs in his *Histoire de la Sexualité,* especially volumes 2 and 3. None of these works takes into account the chapter on Roman homosexuality in Boswell (1980); for criticism of it see MacMullen (1982), and for general agreement Lilja (1982). Scroggs (1983), although addressing religious issues, provided a useful overview of sexual practices in the Mediterranean during the first centuries of the Christian era.

about the fourth century of the Christian era to the Renaissance the predominant public norm was holiness: a complex concept derived from Christian scriptures and teaching, social taboos and decorum, and personal sentiment. This norm could be applied in two ways to gay people. They could be viewed as "separate but equal" (i.e., bound by the same rules of holiness except for the variable of gender). A Christian ceremony of union for same-gender couples performed widely in the early Middle Ages; a genre of debates in high medieval literature about the relative merits of homosexual versus heterosexual love (in which the gay side wins two out of three); complaints in 12th- and 13th-century literature that gay clerics enjoy special advantages—all are traces of the "separate but equal" approach. (I am preparing a study of the ceremony of union for publication; for further information on the other topics see Boswell, 1980.)

A second, better-known strand of thought opposed homosexual behavior categorically. It held that to meet the standard of "holiness" a sexual act must not only occur within a marriage, but also be procreative. Since homosexual acts, even within the context of a permanent union, are not procreative, they could not be holy. This narrower view was mostly limited to ascetic strains in the early church, but gradually gained ground in Europe from the 12th to the 14th century, at a time when many other minorities (e.g., Jews) were also incurring greater social stigma and ostracism, and it eventually swept the other, more tolerant view before it. By the end of the Middle Ages homosexuality was considered a serious sin everywhere in Europe, and there were penalties for it in most civil law codes (Boswell, 1980; Brundage, 1987; Noonan, 1965).

This put gay people in an inferior category, but it is crucial to note that it was a category everyone else also occupied sometimes: what was wrong with gay people was that they were sinners, but it was "normal" to be a sinner. Every human being on earth since the Fall (except Jesus and the Virgin Mary) had been or would be a sinner. Even the people who promulgated this view of sexuality admitted that most conjugal acts performed by most couples did not meet its standards; it was, therefore, not so disturbing that gay people failed them as well (see, e.g., Augustine).

It was the modern world that created the barriers now isolating gay people so effectively and making them pariahs in many Western cultures. Beginning in the 18th century, having for the most part lost interest and faith in the transcendental values underlying the idea of "holiness," European society increasingly replaced that idea with the concept of "normality." As medicine has advanced and the residents of prosperous industrialized cultures have become more and more focused on their bodies and health and less and less interested in nonmaterial values, a paramount arena for assessing the "normal" in Western thought has come to be "health"—physical and psychological.

The unholiness of homosexuality has been transformed into "unhealthiness."[5] Although one might suppose that being "unhealthy" would provoke sympathy rather than hostility, and that gay people would therefore be better off violating the standard of "normality/ health" than that of holiness, this is not the case. In the Middle Ages most people would have acknowledged that they stood, at least occasionally, on the outside of the norm of holiness, and were not therefore categorically different from any other sinner. However, most Americans do not think of themselves as standing outside "normality." They are not conscious of having anything in common with "abnormal" or diseased persons, either physically or mentally. Such people belong in a category utterly distinct, one the average person not only disapproves of but fears. Much obsession with health is in fact focused on protection from contamination or things that cause disease or unhealth (exemplified, e.g., in the concern at many levels in this country that homosexuality will "spread" if not actively restrained).

It was, by a bitter twist of fortune, not many years after the American Psychiatric Association had removed homosexuality from its diagnostic list of illnesses, signaling the widespread conviction on the part of the scientific community that homosexuality was part of the "normal" range of human sexuality, that AIDS became prominent in the gay community, reestablishing in the imaginations of many people the link between homosexuality and disease.

A diagnosis of AIDS is in fact often the first public indication of someone's homosexuality, and may expose him to forms of ostracism he had managed to avoid during much of his life, since gay people, unlike most minorities, can often pass undetected in society. Bruce Niles in "The Normal Heart," Larry Kramer's play about AIDS, evinces this double terror: the prospect of losing his job (along with medical benefits and other essentials for survival in America) is just as terrifying to him as the possibility of contracting AIDS.

Yet another element of alienation results from the fact that large numbers of gay men cannot turn to their families for support. Most victims of plague or any serious disease rely on their families if the public rejects them, and most minorities can find strength and comfort in the bosom of the family—at least some of whom usually experience the same pain—if society oppresses or demeans them. Gay men do not come from gay families, and have often not disclosed their homosexuality to their

---

[5] On the "medicalization" of homosexuality, see the seminal article by Chauncey (1982-83) and the literature cited there. This theme has since become standard in discussions of 20th-century sexuality.

relatives out of fear of hurting them or being rejected. They are alone with their suffering in a way matched by few victims of any such disaster in human history.

The utter exclusion of gay people from the map of human types results in their being completely disengaged from normal standards of morality in the eyes of the public. American boys are led to believe, explicitly or by implication, that there is nothing worse than being a "queer." An enormous amount of social banter and ritual among preadolescent and adolescent boys centers on identifying—and avoiding being identified as—"queer," on the basis of mannerisms, social interaction, clothing, and other superficial criteria, as if being "queer" were a more serious and important character failing than dishonesty, cruelty, hypocrisy, or cowardice. Indeed, American society subtly conveys to all its members that homosexuality is worse than any other heinous act, including murder, because these at least can be named, while until relatively recently homosexuality could not even be discussed in most public settings. That one could casually mention murder even now in a polite conversation in many circumstances where intimate same-sex physical acts could not be openly discussed without causing acute anger or discomfort implies to almost all members of the society, directly or indirectly, that the latter are not only worse, but much worse, than the former. It requires exceptional perspicacity and insight to resist this implicit value scale, to discern that it is not the official moral teaching of any Western ethical system, and not even necessarily what American society actually wishes to convey, but an accidental by-product of the interaction of residual and dimly understood moral teachings, pseudomedical values, and taboos about sexuality and speech. Few come to this realization.

What is not spoken can be much more powerful than what is. A mother who reports that she punished her son for playing with his "thing," or a man who suggests to his girlfriend that they "do it," run no risk of being misunderstood. Although "thing" could refer to hundreds of body parts or objects in a child's life, the refusal to mention precisely what it is conveys with absolute clarity exactly what is not being mentioned; and although "do it" could refer to any conceivable physical or mental operation, its import is unmistakable: because the activity is not being named, we know exactly what it is. Likewise, the silence about homosexuality in American culture is deafening: no one is misled about what is not being mentioned, or why it is not. Everyone knows what acts are alluded to in insults, gestures, and jokes, but never named or discussed openly; what "morals charges" are not published in the paper; who the people who wear green on Thursday are. And everyone concludes from the impossibility not only of discussing them in a nice way but even of derogating them in candid and open discourse that this sin is grave indeed: worse than all the mentionable sins and failings. It is, literally, unspeakable, and therefore off the map of normal, understandable human foibles and failings.

The relegation of this vast area of human sexuality to an unmappable moral wasteland leaves both the heterosexual majority and the gay minority to wander aimlessly without signposts, markers, or boundaries when confronted with moral dilemmas focused on sexuality. Other participants in the AIDS and Sex conference have cited instances of blood donors with AIDS who freely admitted giving contaminated blood, but denied to their deaths that they were gay, even though the evidence was irrefutable. What moral system would suggest that the giving and getting of erotic pleasure with one's own gender is a graver failing than willfully causing the agonized deaths of other human beings? None would officially, but the unspoken popular morality of America does so in clear and unmistakable terms, allowing the latter to be discussed in graphic detail anywhere in the culture but forbidding mention of the former in most public settings except in very oblique and censored terms. Large numbers of drug users and homosexual men will risk death for the pleasure of certain substances or acts, but find the stigma attached to homosexuality so terrifying that the former will insist that drug abuse, not eroticism, is the source of their infection, and the latter, having risked death with apparent equanimity for the sake of sex, will often lie about their sexual orientation to avert the blame attached to it.

Moral confusion obscures the view of the heterosexual majority on the subject even more thoroughly. Of many examples one might note, perhaps the most chilling is the way in which the promiscuity of gay men, probably a factor in the spread of the disease, has been sensationally exploited in the media, debated in the medical community, and fulminated against in Congress. Many writers have suggested, overtly or by implication, that gay men have not only brought the catastrophe on themselves through their promiscuous sexual behavior, but also endangered the (implicitly more monogamous and restrained) heterosexual majority. None of the shocked fulminators ask themselves a question about the majority's role in the lives of gay people clearly exposed by this very tragedy: How can society blame gay men for promiscuity when, for nearly a millennium, it has systematically denied social, legal, and religious acceptance to gay couples? Promiscuous encounters can be hidden from view; permanent relationships cannot. No one questions where an unmarried man goes at night, but every level of American society questions the position of an unrelated person of the same gender as a lifetime partner. With every passing year it becomes more difficult for most gay men to explain to family, friends, coworkers, bosses, land-

lords, and others who the "friend" in the house is.[6] A culture that oppresses, penalizes, or stigmatizes all forms of homosexuality can hardly
expect gay men to form visible and permanent unions. The occasional
and casual aspect of their relationships was often their only safeguard
against severe social sanctions. Blaming AIDS victims for their life-style
is a classic case of blaming the victim, not unlike the conquistadores'
conclusion that the Indians were an inferior people because they died
in such numbers under European dominance.

Such moral myopia is not only inimical to the well-being of individual
humans, heterosexual and homosexual, but to society as a whole. American society conditions itself to react with horror and disgust to many
ordinary aspects of human sexuality, or refuses to recognize them at all,
driving them underground or rendering them conceptually invisible.
And it does so with a pervasive covertness, offering no articulation that
could be challenged. As long as it remains so difficult even to question,
much less to change, popular aversions and antipathies in these crucial
areas, all efforts to cope with the problems of human existence that
depend directly or indirectly on sexuality—from birth to sickness to
death—will be hindered, distorted, and baffled.

Most human disasters elicit surprises, good and bad, from the char-
acters of those drawn into them. The courage and optimism of one young
Dutch girl can be as lasting and influential an image as the cruelty of the
Nazis. While many institutions of American life, most notably the federal
government, have evinced a spectacular callousness and indifference to
AIDS and its toll of human suffering, other communities not traditionally
known for social responsibility or humane concern have assumed the
burdens of caring for the sick and raising money and consciousness to
prevent, combat, and treat the disease. The arts community, the gay
community, and black and Hispanic groups have demonstrated unprecedented organization, effort, wisdom, and strength in dealing with
their own and others' tragedies and losses, and reflections of these
changes in American society may persist as long as those of neighbors
burning down homes or hounding children out of school.

Confronted with a man blind from birth, Jesus was asked by judgmental bystanders whose sin had caused the malady, the blind man's
own or that of his parents. He replied that it was neither: that the man
was blind not as a punishment but for a purpose, that the works of God
might be manifest in him. Then Jesus restored his sight (John 9:1–4).
From a scientific point of view there is no ulterior motive to the action

---

[6] It has been less difficult for women, because of the patronizing assumption that unmarried women had no choice in the matter, and may have decided to live with another
woman *faute de mieux*. This may be one (but only one) of the reasons there has been less
promiscuity among lesbians.

of viruses. They do not afflict people to show forth any glory; it is simply a biochemical process at work. And there is, as yet, no miracle cure. But this does not mean that a society's reaction to the infection may not reveal to future observers much, both good and bad, about that society's character and moral values. Although many of the lessons about human behavior taught by any epidemic could have been learned from earlier plagues or writings about them, most were not. Meanness and cruelty, compassion and heroism seem to be reinvented in the face of each new disaster. It remains to be seen whether American society will learn more than earlier societies did, whether it can pass to posterity the wisdom it acquires, and how it will acquit itself morally during this time of trial.

References

Augustine. (1955). *De bono conjugali* [On the good of marriage]. p. 13. New York: Catholic University of America Press.

Barasch, M. (1976). *Gestures of despair in Medieval and early Renaissance art.* New York: New York Univesity Press.

Barrett, D. S. (1981). The friendship of Achilles and Patroclus. *Classical Bulletin, 57,* 87–93.

Bremmer, J. (1980). An enigmatic Indo-European rite: Paederasty. *Arethusa, 13(2),* 279–298.

Brody, S. (1974). *The disease of the soul: Leprosy in Medieval Literature.* Ithaca, NY: Cornell University Press.

Brundage, J. (1987). *Law, sex, and Christian society in Medieval Europe.* Chicago: University of Chicago Press.

Boswell, J. (1980). *Christianity, social tolerance and homosexuality: Gay people in western Europe from the beginning of the Christian era to the fourteenth century.* Chicago: University of Chicago Press.

Boswell, J. (1982-83, Fall-Winter). Revolutions, universals and sexual categories. In "Homosexuality: Sacrilege, vision, politics." *Salmagundi, 58–59,* 89–113.

Buffière, F. (1980). *Èros adolescent: La pédérastic dans la Grèce antique.* Paris: Belles Lettres.

Chauncey, G. (1982-83, Fall-Winter). From sexual inversion to homosexuality: Medicine and the changing conceptualization of female deviance. In "Homosexuality: Sacrilege, vision, politics." *Salmagundi, 58-59,* 114–146.

Delumeau, J. (1978). *La peur en Occident.* Paris: Fayard.

Dolls, M. (1974). Comparative communal responses to the Black Death in Muslim and Christian societies. *Viator, 5,* 269–288.

Dolls, M. (1977). *The Black Death in the Middle East.* Princeton, NJ: Princeton University Press.

Foucault, M. (1984). *Histoire de la sexualité. Vol. 2: L'usage des plaisirs; Vol. 3: Le souci de soi.* Paris: Gallimard.

Gottfried, R. (1978). *Epidemic disease in fifteenth-century England: The medical response and demographic consequences.* New Brunswick, NJ: Rutgers University Press.

Gottfried, R. (1983). *The Black Death: Natural and human disaster in Medieval Europe.* New York: Free Press.

Huizinga, J. K. (1924). *The waning of the Middle Ages*. London: Edward Arnold and Company.

Kempter, G. (1980). *Ganymed: Studien zur typologie, ikonographic und ikonologie*. Cologne: In Kommission bei Bohlau.

Kieckhefer, R. (1984). *Unquiet souls: Fourteenth-century saints and their religious milieu*. Chicago: University of Chicago Press.

Leff, G. (1967). *Heresy in the later Middle Ages*. Manchester, England: Manchester University Press.

Lilja, S. (1982). *Homosexuality in republican and Augustan Rome*. Helsinki: Societas Scientarium Fennica.

MacMullen, R. (1982). Roman attitudes to Greek Love. *Historia, 31(4)*, 484–502.

McNeill, W. (1976). *Plagues and peoples*. Garden City, NJ: Doubleday.

Mollatt, M., & Wolff, P. (1972). *The popular revolutions of the late Middle Ages*. London: Allen & Unwin.

Noonan, J. (1985). *Contraception: A history of its treatment by the Catholic theologians and canonists*. Cambridge, MA: Harvard University Press.

Scroggs, R. (1983). *The New Testament and homosexuality*. Philadelphia: Augsburg Press.

Sullivan, J. P. (1979). Marital sexual attitudes. *Philologus: Zeitschrift für klassiche Philologie, 123*, 288–302.

Veyne, P. (1978). La famille et l'amour sous le Haut-Empire romain. *Annales E.S.C., 33*, 3–23.

Veyne, P. (1981). L'homosexualité à Rome. *L'Histoire, 30*, 76–78.

Ziegler, P. (1969). *The Black Death*. London: Harper & Row.

IV
CROSS-CULTURAL
PERSPECTIVES

9

Acquired Immune Deficiency Syndrome in Africa

J. O. Ndinya-Achola, F. A. Plummer, P. Piot, and A. R. Ronald

Since its recognition in 1981, AIDS has now become a global problem and is likely to affect all parts of the world. In most African countries where AIDS has been diagnosed, the syndrome was first recognized between 1982 and 1984 (Buchanan, Downing, & Tedder, 1986; Obel et al., 1984), but even before then individuals originating from Africa had diagnoses made in some European countries (Clumeck, Sonnet, Taelman, et al., 1984). Thus, although the official recognition of the occurrence of the syndrome in Africa came later, the disease may have been spreading quietly in Africa for some time. It is not easy to determine the size of infected population worldwide, but going by information obtained from limited serology surveys, it is estimated that the number of individuals infected in Africa could be several millions, and the infection rate is going up. The causative agent(s), human immunodeficiency virus (HIV) Type 1 or 2, both occur in Africa, although their distribution seems to have some geographical limits, HIV-1 being more common in East and Central Africa while HIV-2 is more common in West Africa (Denis, Barin, Gershy-Damet, et al., 1987; Gurtler, Zonlek, Frosner, & Deinhardt, 1987).

Determinants of HIV Occurrence in Africa

The most important modes of transmission of HIV in Africa are sexual transmission and vertical transmission. Transmission through transfusion of blood and blood products may be important, but there are not many studies conducted to determine its magnitude. In Kenya, HIV seroconversion was associated with transfusion in 30% of patients with hemophilia A (Kitonyi, Bowry, & Kasili, 1987). Screening of transfusion blood against HIV in African countries is a recent activity. Many African countries only started such screening late in 1986 or 1987, a time when certain African populations were showing seropositivity rates of between 1% in blood donors and 88% in some high-risk groups (Kreiss, Koech, Plummer, et al., 1986; Mann, Francis, Quinn, et al., 1986; Van de Perre, Clumeck, Carael, et al., 1985). The role of unsterile needles is less well investigated, but it was shown in Zaire that seropositivity rates were greatly influenced by increased number of injections received within the past year (Piot, Quinn, Taelman, et al., 1984).

As discussed by Dr. Malcolm Potts in this volume (Chapter 12), numerous sexual partners remains the single most important risk factor for HIV transmission in Africa, and heterosexual transmission is the most important mode of HIV transmission. Table 9-1 shows the prevalence of HIV seropositivity among prostitutes in Africa (Denis et al., 1987; Durand, Garigue, & Booulomie, 1986; Gurtler et al., 1987; Mann, Quinn, Piot, et al., 1984; Mhalu, Mbena, Bredberg-Reden, et al., 1987; Plummer, Sinonsen, Ngugi, Cameron, & Ndinya-Achola, in press; Van de Perre et al., 1985). Between 1984 and 1985, the average annual incidence of AIDS in Kinshasha was estimated at 550 to 1,000 cases per million population with a current male-to-female ratio of 1.1:1 (Mann et al., 1986). In Nairobi between 1981 and 1986, a marked rise in seropositivity was noted among a group of prostitutes. A similar rise, although not so dramatic, was demonstrated in men with genital ulcer disease attending the Special Treatment Clinic in Nairobi (Table 9-2) (Piot, Plummer, Ray, et al., 1987). These rates of seroconversion suggest a link with promiscuity, sexually transmitted diseases, and HIV seroconversion. Further evidence is available from a study conducted in

Table 9-1
Prevalence of HIV Antibody in Prostitutes in Africa

City/Area	Year	Number Studied	Percent Positive
Meiganga, Cameroon	1985	221	8
Abidjan, Côte d'Ivore	1986	101	20
Nairobi, Kenya	1985	286	61
Blantyre, Malawi	1986	265	56
Butare, Rwanda	1984	33	88
Arusha, Tanzania	1986	42	0
Dar-es-Salaam, Tanzania	1986	225	29
Kinshasha, Zaire	1985	377	27
Equateur, Zaire	1986	283	11

Table 9-2
Prevalence of LAV/HTLV-III Antibodies in Selected Populations in Nairobi
Between 1980 and 1985[a]

Population Studied	Year of Study (Number of Positives/Number Tested)					
	1980	1981	1982	1983	1984	1985
Prostitutes		5/116 (4%)		32/39 (82%)	45/76 (59%)	126/215 (59%)
Women with STD[b]		2/49 (4%)	6/74 (8%)			
Men with STD	0/118	2/70 (3%)	4/68 (8%)	13/93 (14%)		19/107 (18%)
Pregnant women		–0/111				17/735 (2%)

[a] Adapted from Piot et al. (1987).
[b] STD, sexually transmitted disease.

Nairobi between March and December 1986, involving 340 men. HIV seroconversion was associated with frequent contacts with prostitutes, past history of genital ulcer disease, and lack of circumcision (Tables 9-3 and 9-4) (Simonsen, Cameron, Gakinya, et al., 1988). In that study, lack of association between injections and transfusion was noted (Table 9-5). This should not give an impression that transfusion and use of contaminated needles is not an important route of transmission.

The role that genital ulcers, particularly syphilis and herpes simplex, play in transmission of HIV has been demonstrated elsewhere (Handsfield, Ashley, & Rompalo, 1987). As to whether certain etiological agents of genital ulcers are more efficient in facilitating this transmission is not known, but if this was the case it would partially explain the difference

Table 9-3
Sexually Transmitted Diseases and HIV Infection[a]

	HIV Positive (N = 38)	HIV Negative (N = 302)
Past history of urethritis	22	140 (pNS)
Past history of genital ulcers	24	58 ($p = 10.6$; OR = 10.2, CI95% = 4.7–22.1[b])
Current diagnosis Urethritis	17	163 (pNS)
Present diagnosis Genital ulcers	24	138 ($p = .028$)

[a] Adapted from Simonsen et al. (1988).
[b] p = probability; OR = odds ratio; CI = confidence interval.

Table 9-4
Interaction of Circumcision and Past Genital Ulcer Disease[a]

	HIV Positive (N = 37)[b]	HIV Negative (N = 301)[b]
Circumcised		
Past genital ulcers	15	35
No past genital ulcers	5	196 (OR = 16.8; CI95% = 5.3–57.0; p = 10.6)
Uncircumcised		
Past genital ulcers[c]	8	23 (OR = 1.8)
No past genital ulcers	9	47 (pNS)

[a] Adapted from Simonsen et al. (1988).
[b] Circumcision status not recorded in two cases.
[c] Uncircumcised men with no past genital ulcers were also at increased risk of HIV infection compared to circumcised men with no genital ulcer disease (OR = 7.51; CI95% = 2.17–27.22; p = .00029).

in HIV seroprevalence between West Africa and East and Central Africa. During the 1987 meeting of the African Union Against Venereal Diseases and Treponematoses (AUVDT), it was noted that chancroid was more common in East, Central, and Southern Africa, but less common in West Africa. Instead, the more common genital ulcer diseases in West Africa are herpes and syphilis (AUVDT, 1988). Whereas in Europe and North America HIV infection in women is still primarily a problem in high-risk groups (prostitutes, intravenous drug users, and women living in areas of high heterosexual transmission) (Peckham, Senturia, & Ades, 1987), in Africa, where heterosexual transmission is common, the ratio of infected women of child-bearing age is relatively high. In Nairobi, the seropositivity rates among antenatal clinic attenders was shown to be about 2.7% (Braddick et al., 1987). This implies a perinatal trans-

Table 9-5
Lack of Association between Injections and Transfusions
and HIV Infection

	HIV Positive (N = 38)	HIV Negative (N = 340)
Infections during the past year		
0	22	163 (pNS)
1	4	54
2–3	7	73
4	5	12
Blood transfusions in the past 5 years	0	7 (pNS)

mission rate of about the same magnitude, and the figures are likely to increase. Studies are in progress in Nairobi, Kamplal, and Lusaka to determine the net vertical transmission rates and the effect of HIV infection in both the mother and the neonate.

The incidence of AIDS has increased dramatically over the last 2 years. Based on figures reported to the World Health Organization as of May 30, 1987, 34 countries in Africa reported a total of 4,343 out of a total 49,132 cases reported from all over the world (*Weekly Epidemiological Record*, 1987). Thus, African cases represented about 8.8% of all AIDS cases. In September of the same year, the number of cases reported from Africa had increased to 5,823 out of a total 60,653 cases representing about 9.6% of total cases (*Weekly Epidemiological Record*, 1987). This represents an increase of 1,480 cases, or a contribution of 12.8% of all new cases (Fig. 9-1). During the same period, U.S. cases contributed to 71.7% in July and 69% in September. Overall, the percentage contributed from Africa is increasing, although during the same period the number of reporting African countries also increased, from 34 to 41. These figures do not reflect the true incidence; for example, figures for Uganda and Tanzania in September were not revised, and it is also unlikely that the Central African Republic has 254 cases, while its neighbor Chad has only one reported case.

Many countries are underreporting. This underreporting may be due to inadequate facilities for case reporting or poorly coordinated report-

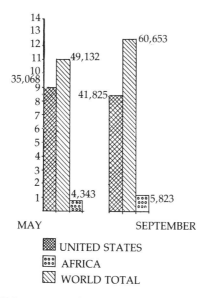

Figure 9-1. AIDS cases in Africa and the rest of the world (1987).

ing, or may be deliberate,due to political reasons. It may also be true that case definition as originally suggested by the U.S. Centers for Disease Control (CDC) and adopted with modifications may have been difficult to apply in Africa. Thus, cases of "slim" disease in Uganda may not initially have been reported as AIDS. With the revised CDC case definition (*Morbidity and Mortality Weekly Reports,* 1987), provision has been made for the possibility of local names being used for AIDS; thus "slim" disease is accounted for.

Clinical Presentation of AIDS in Africa

The clinical picture of AIDS in Africa is very similar to that of AIDS elsewhere in the world. Minor differences may occur with regard to opportunistic infections in view of the prevalence of communicable disease or infectious diseases in Africa.

Weight loss, diarrhea, and lymph node enlargement are the commonest presenting symptoms. The dramatic weight loss has resulted in the nicknaming of AIDS as a "weight-losing agent" or "slim disease" in some parts of Africa. In a series of clinically and laboratory diagnosed AIDS cases, diarrhea was the presenting symptom in over 75% of cases (D. M. Owili, personal communication). The microbiological agents of diarrhea have not been well defined, but it is suggested that *Salmonella* (other than *S. typhi* and *S. paratyphi*), shigella, and *Cryptosporidium* may be the most important.

Other opportunistic pathogens include mycobacteria. *Mycobacterium tuberculosis* was demonstrated in two cases out of 17 HIV-positive patients investigated for chest infections in Zimbabwe. Other mycobacteria may be equally important. *Cryptococcus neoformans* infections have been increasing since the advent of AIDS. In Kenyatta National Hospital, Nairobi, cryptococcal meningitis was seen at the rate of 1 to 2 cases per year, but during 1986 two cases were recorded, both associated with AIDS, and in 1987 up to October two cases have again been recorded associated with AIDS. In all, since 1984 when the first AIDS case was reported in Kenya, there have been seven cases of AIDS-associated recorded cryptococcal meningitis and all have ended fatally. The possible association of other commonly occurring infectious diseases, such as malaria and other protozoal infections, with AIDS is the subject of much investigation.

Sociological Aspects of AIDS

AIDS will no doubt greatly influence social aspects of Africa's life-styles. Such aspects as property inheritance rules, some of which were closely associated with inheritance of the remaining spouse, are likely to be most severely affected. It is unlikely that a man, knowing that the former husband of a widow died of AIDS, will want to marry her. Polygamy will probably be equally affected. Other traditional practices that will undoubtedly lead to special problems with AIDS in Africa are: the practice of a father-in-law having sex with the bride, sharing of a wife by brothers, sex practiced early after childbirth, and the belief that sexually transmitted diseases can be cured by intercourse with a virgin. On the other hand, there are practices that may help minimize the spread of AIDS. It has been noted that where bride price is high, there is less extramarital sexual contact. Other African ethnic groups have some taboos that prohibit early sex. All these diverse traditions have their roles to play, which are not necessarily beneficial. In the presence of AIDS, it is necessary to review traditional practices and to promote the good ones and discard outmoded or bad practices. In urban areas where traditional life-styles may have disappeared, it will be necessary to restore moral codes using the influence of religion and family ties.

Have the Traditional Healers Any Role To Play?

The traditional doctors have not been left out in the story of AIDS. While the majority of traditional doctors interviewed in Kenya accept that their methods have very little influence on AIDS, a few claim that some of their herbal medicines prolong the lives of AIDS victims. One traditional healer has claimed that the extract of a common East African tree locally known as Mwarobaini (*Azadiratcha indica*), in combination with another herbal extract, is effective in reducing symptoms related to AIDS. None of these claims have been scientifically tested or proved. However, a local East African herb, *Ketodrostris foetidissima*, has been used for a long time to treat or modify the course of measles in children. The extract of this herb was shown in vitro to have inhibitory effects on measles virus as well as certain bacteria, including *Staphylococcus aureus* and *Salmonella typhimurium* (Odero, 1985). Indeed, the number of compounds so far tested for their effect on HIV and related viruses is very large, but some of these medicinal values of certain African plants should be tested.

Whenever people are ill and suffering from an incurable disease, they become desperate and will seek help from any source. AIDS has made its sufferers feel very desperate. At the moment, when chemotherapy against AIDS is not entirely satisfactory, what modern medicine offers may not differ significantly from what traditional medicine offers. Thus, African populations are in a position that is highly prone to abuse and exploitation. Traditional healers have been known to claim that they have a cure for such diverse diseases as cancer, diabetes, chronic alcoholism, tuberculosis, and psychiatric disorders. When such claims are made, the people who fall victims to them end up paying large sums of money for treatment that does not work. This will no doubt happen or is already happening in AIDS. It is a duty of the governments and other relevant bodies to counsel and educate traditional healers so that they refrain from exploiting AIDS sufferers by offering ineffective remedies.

The Problem of Prostitution and AIDS

The word "prostitute"has been used in this text before, but its definition is not universally accepted. The sixth edition of the *Concise Oxford Dictionary* defines the word prostitute as "a woman who offers her body to promiscuous sexual intercourse especially for payment or as religious rite; man who undertakes homosexual actions for payment." In both cases, the man or woman is offering sex to somebody, who is apparently left out of the definition. It is suggested that the clients of "prostitutes" should also be included in this definition.

In almost all African countries prostitution is illegal, yet despite its illegal status, prostitution is a widespread practice. It is not easy to identify prostitutes. In many instances, the person practicing prostitution is employed in a fairly decent job, but they practice prostitution to earn extra money because salaries are low. Married men who have left their wives in the rural areas but live and work in urban areas have resorted to casual sex with prostitutes. Women have been known to have occasional or alternative husbands. Despite such situations, the only group of individuals recognized as prostitutes are the ones with no other label to hide behind. Usually they are unmarried and unemployed women. In this context, prostitutes then stand out as undesirable groups, and once designated undesirable they conduct their business in a manner that conceals what they are. Therefore, prostitution is a widespread and diffuse practice in Africa. The people practicing it are difficult to characterize and their only common denominator is increased promiscuity. They have been identified as reservoirs of sexually transmitted diseases in Africa, and serological surveys have indicated that they have higher rates of HIV infection than ordinary populations.

What steps can be taken to minimize prostitution in Africa? This is indeed a very difficult question. If prostitution is legalized, there will be a serious uproar from religious organizations, and governments are likely to resist that suggestion because it will "lower moral standards" of the population. At this time of the AIDS pandemic, what is needed is to "raise" rather than "lower" moral standards. Should laws be enacted to give severe punishment to prostitutes? Again, very large numbers of people are involved and the lawmakers themselves are not necessarily exempt.

At the moment, the bigger problem of prostitution is restricted to the urban centers. It has even been suggested in Uganda that AIDS was originally an urban disease that was affecting mostly the prostitutes and their clients, but that is was transmitted to rural areas by people who frequently visited urban centers. It was suggested at one time that African urban prostitutes should be repatriated and rehabilitated in rural areas. Here again, we are faced with a nearly impossible task. Already, prostitutes have very high HIV seropositivity rates and there is no guarantee that they will completely abandon prostitution, so this suggestion may only transfer more infection from urban to rural areas.

In prostitution, Africa is faced with a big problem with little or no ready solution. The best that can be done is to identify those groups of individuals who are considered as prostitutes. Health education, with strong emphasis on the use of protective devices (e.g., condoms and spermicides) during sex, should be targeted to them. Condoms should be made generally available with proper instruction on their use and disposal.

Control Strategies

Using a hypothetical country with a population of 100 million people, and a moderate estimate of 1% of the population being infected but asymptomatic, the total number of individuals infected is 1 million. Given such a large infectious pool and the lack of effective chemotherapy or vaccine, the most logical options for control are measures that minimize or stop transmission.

Transmission through transfusion of blood and blood products is the easiest to influence through strict screening of all blood and its products against HIV. Although an expensive undertaking initially, for a disease like AIDS the expense is justified.

Transmission through contaminated needles and other equipment is the next type that can be influenced, but it is best discussed together with sexual transmission.

Sexual transmission is the most urgent and most worrying issue in Africa. Sexual contact with multiple partners is a widespread phenomenon in many African countries. Where there is no strict religious control in sexual behavior, most communities regard sex outside marriage as socially acceptable. In a study conducted in Nairobi, most African males had their first sexual contact between 14 and 16 years of age, and by age 30 they had had up to 15 or more partners (Simonsen et al., 1988). Therefore, any efforts directed at influencing sexual behavior must be initiated very early in life, preferably from 10 years onward.

The best approach would be intense sex education starting at primary school level and continued at all ages up to age 65. The nature of the problems of sexually transmitted disease, with strong emphasis on AIDS, its causative agent, the natural history of illness, and the nonavailability of drugs or vaccine must be stressed. Those aspects of modified or altered sexual behavior must also be clearly spelled out, as must the use of protective methods (e.g., the condom and spermicides). It has been shown among a Nairobi group of prostitutes that where AIDS is of concern the prostitutes insist on the use of condoms (Ngugi, Plummer, Bosire, & Ndinya-Achola, 1987). If health education programs are to succeed, there must be synchronized activity. It is no use giving intensive health education in one country when the adjoining country is doing nothing.

It was shown that the occurrence of certain sexually transmitted diseases was associated with higher tendency to seroconvert to HIV positivity. In particular, genital ulcer disease (notably chancroid) and genital chlamydia facilitated female-to-male transmission of HIV (Simonsen et al., 1988). It is postulated that either the presence of a sexually transmitted disease agent leads to genital ulceration–facilitated transmission by attracting lymphocytes locally at the genital lesions, or the process of ulceration itself creates a portal of entry for HIV. Whatever the mechanism, measures directed at eradicating or minimizing genital ulcer diseases such as chancroid, syphilis, lymphogranuloma venereum, and gonorrhea should be studied in countries with high prevalence of these diseases.

Effective control of the spread of AIDS can only be achieved when the threat of the AIDS epidemic is acknowledged by governments and medical authorities worldwide. In their paper, Seale and Medvedev (1987) concluded that "The initial reactions to AIDS of public health authorities, clinicians, and medical scientists all over the world is to deny that there are any serious problems, and when they are so obvious that they can no longer be denied, to blame groups of people who are perceived to be undesirable, such as drug addicts, homosexuals, foreigners, prostitutes, the promiscuous and the Pentagon." This statement is unfortunately true, and in the African context is probably the most important factor that has facilitated the rapid spread of AIDS in Central and Eastern Africa. Whereas early recognition of the magnitude of the AIDS problem, coupled with honest case recording and reporting, may initially harm a country in terms of adverse publicity, in the long run it will save the country from a devastating AIDS epidemic.

References

Acquired immunodeficiency syndrome global data: AIDS cases reported to World Health Organization as of 1987 (1987). *Weekly Epidemiological Record*, 40, 137, 301.

Braddick, M., Kreiss, J. K., Quinn, T., Ndinya-Achola, J. O., Veracauteren, G., Plummer, F. A., et al. (1987, June). *Congenital transmission of HIV in Nairobi, Kenya.* [Abstract] The Third International Conference on AIDS, Washington, DC.

Brunell, P., Daum, R., Cooper, L., Oleske, J., Luban, N., et al. (1987). Classification system for human immunodeficiency virus (HIV) infection in children under 13 years of age. *Morbidity and Mortality Weekly Report, 36*, 15.

Buchanan, D. J., Downing, R. G., & Tedder, R. S. (1986). HTLV III antibody positivity in Zambian Copper belt. *Lancet, 1*, 155.

Clumeck, N., Sonnet, J., Taelman, H., Mascart-Lemone, F., De Bruyere, M., Vandeperre, P., Dasnoy, J., Marcellis, L., Lamy, M., Jonas, C., Eyckmans, L., Noel, H., Vanhaeverbeek, M., & Butzler, J. P. (1984). Acquired immunodeficiency syndrome in African patients. *New England Journal of Medicine, 310*, 492–497.

Denis, F., Barin, F., Gershy-Damet G., Rey, J. L., Lhuillier, M., Mounier, M., Leonard, G., Sangare, A., Goudeau, A., M'Boup, S., Essex, M., & Kanki, P. (1987). Prevalence of human T-lymphotrophic retroviruses type III (HIV) and type IV in Ivory Coast. *Lancet, 1*, 408–411.

Durand, J. P., Garigue, F. P., & Booulomie, J. (1986, June). *AIDS in Cameroon.* [Abstract] The Third International Conference on AIDS, Paris.

Gurtler, L. G., Zonlek, G., Frosner, G., & Deinhardt, F. (1987, June). *Prevalence of HIV (1) and HIV (2) in selected Malawian populations.* [Abstract] The Third International Conference on AIDS, Washington, DC.

Handsfield, H. H., Ashley, R. L., & Rompalo, A. M. (1987, June). *Association of anogenital diseases with human immunodeficiency virus infection in homosexuals.* [Abstract] The Third International Conference on AIDS, Washington, DC.

Kitonyi, G. W., Bowry, T., & Kasili, E. G. (1987, June). *AIDS studies in Kenyan hemophiliacs.* [Abstract] The Third International Conference on AIDS, Washington, DC.

Kreiss, J. K., Koech, D., Plummer, F. A., Holmes, K.K., Lightfoote, M., Piot, P., Ronald, A. R., Ndinya-Achola, J. O., D'Costa, L. J., Roberts, P., Ngugi, E. N., & Quinn, T. C. (1986). AIDS virus infection in Nairobi prostitutes: Spread of the epidemic to East Africa. *New England Journal of Medicine, 314*, 414–418.

Mann, J. M., Francis, H., Quinn, T. C., Kapita, B., Asila, P. K., Bosenge, N., Nzilambi N., Jansegers, L., Piot, P., Ruti, K., & Curran, J. W. (1986). HIV seroprevalence among hospital workers in Kinshasha: Lack of association with occupational exposure. *Journal of the American Medical Association, 256*, 3099–3102.

Mann, J. M., Francis, H., Quinn, T., Pangu, K. A., Ngaly, B., Nzila, B., Nzila, N., Kapita, B., Muyembe, T., Kalisa, R. S., Piot, P., McCormic, J., & Curran, J. W. (1986). Surveillance for AIDS in a Central African City, Kinshasha, Zaire. *Journal of American Medical Association, 255*, 3255–3259.

Mann, J.M., Quinn, T., Piot, P., Ngaly, B., Nzila, N., et al. (1987). *HIV seroprevalence among prostitutes in Kinshasha, Zaire: Sexual practices associated with seropositivity.* Manuscript submitted for publication.

Mhalu, F., Mbena, E., Bredberg-Reden, U., Kiango, J., Nyamuryekunge, G., Biberfeld, G., et al. (1987, June). *Prevalence of HIV in healthy subjects and groups of patients in some parts of Tanzania.* [Abstract] The Third International Conference on AIDS, Washington, DC.

Ngugi, E. N., Plummer, F. A., Bosire, M., & Ndinya-Achola, J. O. (1987, June). *Effect of an AIDS educational program on increasing condom use in a cohort of Nairobi prostitutes.* [Abstract] The Third International Conference on AIDS, Washington, DC.

Obel, A. O. K., Sharif, S.K., McLigeyo, S.O., Gitonga, E., Shah, M. V., & Gitau,

W. (1984). Acquired immunodeficiency syndrome in an African. *East African Medical Journal, 61*, 724–726.

Odero, B. (1985). *Medicinal aspects of Ketodrastris foetidissima*. Unpublished Masters thesis, University of Nairobi, Nairobi, Kenya.

Peckham, C. S., Senturia, Y. D., & Ades, A. E. (1987). Obstetric and perinatal consequences of human immunodeficiency vurus (HIV) infection. *British Journal of Obstetrics and Gynaecology, 94*, 403–407.

Piot, F., Plummer, F. A., Ray, M. A., Ngugi, E.N., Ndinya-Achola, J. O., Veracauteren, G., D'Costa, L. J., Laga, M., Nsanze, H., Fransen, L., Haase, D., van der Groen, G., Brunham, R. C., Ronald, A. R., Brun-Vezinet, F. (1987). Retrospective seroepidemiology of AIDS in Nairobi populations. *Journal of Infectious Diseases, 155*, 1108–1112.

Piot, P., Quinn, T. C., Taelman, H., Feinsod, F. M., Kapita, B., Wobin, O., Mbendi, N., Mazebo, P., Ndangi, K., Stevens, W., Kalambayi, K., Mitchell, S., Bridts, C., & McCormick, J. B. (1984). Acquired immunodeficiency syndrome in a heterosexual population in Zaire. *Lancet, 2*, 65–69.

Plummer, F. A., Simonsen, J. N., Ngugi, E. N., Cameron, D. W., & Ndinya-Achola, J. O. (In press). Incidence of human immunodeficiency virus (HIV) infection and related disease in a cohort of Nairobi prostitutes.

Seale, J. R., & Medvedev, Z. A. (1987). Origin and transmission of AIDS. Muthuse hypodermics and the threat to the Soviet Union (discussion paper). *Journal of the Royal Society of Medicine, 80*, 301–304.

Simonsen, J., Cameron, W., Gakinya, M., Ndinya-Achola, J. O., D'Costa, L. J., Karasira, P., Cheang, M., Ronald, A. R., Piot, P., & Plummer, F. A. (1988). Human immunodeficiency virus infection in men with sexually transmitted diseases. *New England Journal of Medicine, 319*, 274–278.

Van de Perre, P., Clumeck, N., Carael, M., Nzabihimana, E., Robert-Guroff, M., De Mol, P., Freyens, P., Butzler, J. P., Gallo, R. C., & Kanyamupira, J. B. (1985). Female prostitutes: A risk group for infection with human T-cell lymphotropic virus type III. *Lancet, 2*, 524–597.

10

The AIDS Epidemic in Brazil

Maria Eugenia Fernandes

The health situation in Brazil is a reflection of the poor educational, economic, and social conditions in general. About 70% of the population suffers from different degrees of malnutrition. In order to better evaluate the quality of life in an underdeveloped country, we have to consider three variables:

Child Mortality, Maternal Mortality, and Life Expectancy

In Brazil, 67 out of 100 children die during the first 5 years of life. In the northeast, the poorest Brazilian region, the infant mortality rate reaches 300:1,000. Meanwhile, in the First World, infant mortality rate is 10 to 15 deaths per 1,000 children born alive. In my country, the primary cause of death before 5 years of age is the combination of malnutrition and infection. In São Paulo State, the most industrialized area in the country, 50% of elementary-grade children attending public schools have some degree of malnutrition.

Most developed countries consider vaccine-preventable infectious diseases an almost-solved problem. In 1986, no cases of diphtheria were reported in the United States to the Centers for Disease Control (CDC) in Atlanta, whereas 1,800 cases were reported to the Brazilian Ministry of Health. Four thousand cases of whooping cough occurred in the United States in 1986, as compared to 24,000 cases in Brazil. In the same year, there were 400 polio cases in Brazil and 10 in the United States. Measles is still a great problem: the Pan American Health Organization and the São Paulo surveillance team estimated that in 1983 113,000 cases of measles occurred in São Paulo State, with 1,700 deaths. We still have to consider that there is an underreporting of notifiable diseases in the country. The actual number of cases is about threefold the reported numbers. Medical doctors are predicting an increase in pediatric tuberculosis in the near future because mothers are refusing bacille Calmette-Guérin vaccination; after being passed through a flame, the same needle is used for all children. Mothers fear their child's contamination by human immunodeficiency virus (HIV).

High maternal mortality is directly related to ignorance, lack of resources, and poverty. In Brazil, maternal mortality is about 20 times higher than that in developed countries and far higher than that verified in some countries where the per capita income is lower than that of Brazil. One out of 16 women dies between 15 and 40 years of age, when they are most needed and productive. The World Health Organization estimates that there are 3 to 5 million abortions in Brazil every year, which are responsible for 400,000 deaths. Abortions are widely known as one of the most common causes of sterility and mortality in the country. Comparatively, in the United States, less than 1 woman per 100,000 dies as a consequence of abortion; in Cuba, 1:100,000; in Denmark, 0:100,000; and in Brazil, 10 to 20 per 100,000. Because abortion is an illegal practice, all abortion clinics are clandestine, and most have very poor sanitary conditions—no trained personnel; no adequate cleaning, disinfection, or sterilization procedures. In these clinics, HIV could eventually be transmitted from patient to patient.

Brazil is well provided with all kinds of diseases—the infectious diseases typical of the Third World and cardiovascular illnesses, neoplasias, trauma, and hospital infections, which also afflict industrialized nations. In terms of life expectancy, since the early 1980s, Brazilians can hope to live around 63 years, versus Canadians (77 years) and Cubans (73 years on the average).

Emergence of AIDS

Amid this panorama of sickness AIDS emerged. The first cases were reported in São Paulo and Rio de Janeiro in 1982. The upper-class Brazilian homosexual men that often traveled to New York and San Francisco to enjoy themselves and were less swayed by the prejudice of society were those who were infected with HIV and brought it back with them to Brazil, where the virus spread. As a much praised vacation resort among American homosexuals, especially Carnival time, when permissiveness and sexual promiscuity are tolerated, Brazil soon rose to second in the world in absolute numbers of AIDS cases according to the latest data from the World Health Organization. To date, at the end of 1987, we have 2,237 cases reported to the AIDS National Program in

Brazil. Most cases reported in Brazil are concentrated in São Paulo State, with 1,185 cases; in São Paulo city alone, there are 902 cases reported to the State Health Department.

The notification of patients has run into technical, social, and ethical problems. At present, we estimate that we probably have 3,000 cases of AIDS, 9,000 to 15,000 cases of AIDS-associated conditions, and 150,000 to 300,000 of symptomless carriers. During the next 5 years, we will have at least 75,000 cases, which means approximately 15,000 cases/year until 1993.

Most AIDS cases reported are from urban areas like Rio de Janeiro and São Paulo. Patients from all over the country go to these two cities seeking better care and treatment. What makes the hospitalization problem even worse now is that we have only 164 hospital beds in São Paulo State, while 350 beds are needed. The National AIDS Program estimates that through December 1987 it will be necessary to have a total of 735 hospital beds in the country. A patient at risk now has to wait at least 1 month to have an appointment with the Public Health Service.

Epidemiology

HIV is spreading fast in Brazil. Of the reported cases, 77.4% were transmitted through sexual relationships, 48.5% through homosexual contact, 23.7% through bisexual contact, and 5.4% through heterosexual contact. Since 20 to 30% of AIDS cases occurred among bisexual men, the data point out the risk of a progressive dissemination of HIV in the heterosexual population and an increasing number of pediatric cases.

Transmission of HIV through blood transfusions and blood products, including intravenous drug use, is responsible for 14.9% of the cases, of which 5.8% are through the use of intravenous drugs, 5.0% through blood and blood product transfusions, and 4.1% in hemophiliacs. It is important to emphasize that only one third of the blood collected in Brazil is tested for HIV, according to the National Division of Health and Blood Products.

In the industrialized countries, blood donors have been tested since 1985, whereas it was only in April of 1987 that the Brazilian Ministry of Health and Ministry of Social Welfare promulgated guidelines requiring that all blood donors be tested for HIV. In an important survey undertaken on 22,200 samples of blood from donors living in São Paulo city, collected between June 1985 and November 1986, the prevalence of antibodies for HIV was 0.18%. In contrast, in 51,000 blood donors from Atlanta, GA, the seropositivity was 0.21%. In 51 samples from patients with hemophilia A collected in Rio de Janeiro from 1983 to 1984, 98% presented positive results. In São Paulo, the seropositivity of 56 patients with hemophilia A was 73%. Several studies undertaken in the United

States with patients with hemophilia A indicated that the seropositivity for HIV was as high as 80% in 1984.

A research study of the prevalence of HIV antibodies was undertaken in four Brazilian tribes. Two tribes were from the Xickrin group and two tribes from the Paracanã group. The total Indian population from those four tribes was 673. Among these Indians, 307 (45%) were tested for HIV antibodies with the use of ELISA and Western blot techniques. Of these 155 men and 152 women, all were seronegative. There is no available information about antibody prevalence in intravenous drug users in our country.

As in Africa, we have a lot of unnecessary blood transfusions, and Brazilian patients often demand an injection instead of pills. The manufacturing rate of disposable needles and syringes is not enough for adequate procedures, and unfortunately they can only be found in large cities.

Another point is that AIDS patients frequently look for alternative treatments such as acupuncture, during which most needles are not properly sterilized. There were two cases reported to the State Health Department in São Paulo, pointing out acupuncture as the only risk factor. One of the cases was a child whose Japanese parents were not infected, and the epidemiological survey did not identify any other risk factor.

Pediatric cases contributed to 2.7% of the total number of cases.

Effects of AIDS on the Brazilian Health Care System

The identification of new cases every day in Brazilian cities and the fast growth of epidemics show a totally unprepared health system. AIDS has dramatically shown us the reality of our hospitals with their

Lack of physical space for hospitalization

Lack of trained personnel in quantity and quality in the care of AIDS patients

Lack of medicines such as antibiotics, antiparasitic drugs, pentamidine, azidodeoxythimidine, and so on

Lack of basic resources such as gloves, gowns, and germicides to make proper procedures possible

Lack of adequate and updated laboratories for diagnosis of opportunistic infections and for HIV antibody tests

Lack of equipment for such procedures as bronchoscopy, endoscopy, and tomography, which are so necessary for diagnosis of opportunistic infection

All this, added to the refusal of surgeons to operate on these patients as well as of pathologists to perform the much-needed necropsies for fear of contamination, hamper the collection of accurate data for establishing a reliable picture of the AIDS epidemic in Brazil.

In order to illustrate these difficulties, we can analyze research undertaken with 100 patients during the period from 1985 to 1986, by the AIDS National Reference Center for hospitalization and treatment at Emilio Ribas Hospital in São Paulo. The average stay of these patients was 60 days. In the clinical history, the main signs and symptoms were: fever (in 68 cases), loss of weight (67 cases), diarrhea (50 cases), dyspnea (37 cases), candidiasis (64 cases), lymphadenopathy (43 cases), and neurologic signs (35 cases). Tuberculosis was the most important opportunistic infection, occurring in 43% of the cases. This can be explained by the high prevalence of tuberculosis in Brazil.

The clinical and radiological diagnosis of *Pneumocystis carinii* was made in 32% of suspected cases. This low rate, compared to that in the United States and Europe, is due to the lack of pulmonary biopsies. *Cryptococcus neoformans* was found in 7% of our cases, cerebral toxoplasmosis in 15% of the cases, *Mycobacterium avium* in 1%, *Histoplasma capsulatum* in 2%, herpes zoster in 5%, and herpes simplex in 8 cases. Kaposi's sarcoma was present in 22% of the cases. In this sample of 100 patients, 50% showed diarrhea, although no *Cryptosporidum* species was found in anyone because of laboratory difficulties.

During the period between 1983 and 1987, the average rate of nosocomial infection in AIDS patients was 50%, with most cases due to gramnegative bacteria. This high rate is mainly due to immunosuppression, high permanence, and catheterization.

Can Brazil support the onus of this epidemic? In terms of costs, we have to consider two basic points: first, the unmeasurable cost of pain, suffering, anguish, and the shame of being sick with AIDS. Second, the measurable costs, such as hospitalization and treatment, must be considered. In 1987, the Ministry of Health budget was set at US$ 16 million for medical assistance and treatment and US$ 1 million for educational purposes.

In Brazil the AIDS patient is only hospitalized when he or she cannot be treated at home. In our country, there is an increasing number of AIDS patients with social problems who cannot afford treatment outside of hospitals. Thus the majority of Brazilian AIDS patients are treated in public hospitals, since health insurance does not cover costs of hospitalization due to infectious diseases, and only the very rich patients can afford treatment in a private hospital. An AIDS patient usually has an average of two hospital stays, each of them about 20 days, costing a total of US$ 17,000.

In addition, Brazil suffers from a public health system that is underpaid. In the richest Brazilian state, a physician who works full time earns US$ 6,000 per year in the Public Health System. A nurse working during the same period earns US$ 2,400 per year, and a nurse's aide earns US$ 600 per year (as of October 1986). Considering the poor conditions of patient care for AIDS patients, it becomes extremely difficult for health care providers to work with this epidemic. Yet the impact of the disease has forced a small group of health care providers to accumulate a great amount of experience in a short period of time. Because of their social commitment, some of those health workers are still full of idealism.

A great effort is being made by the National AIDS Program. The National Plan for AIDS Prevention and Control is working hard to implement an effective health infrastructure that will provide the states with technical and financial support so that they can begin prevention and control of HIV infection. With so many health problems, it is a hard fight, and we are aware of the fact we will be forced to live with poverty for a long time to come, but there are no reasons why we should be forced to live with misery, great iniquities, inefficiency, corruption, opportunism, and social injustice. It will be crucial for Brazil to establish a balanced relationship between poverty and dignity, so that it may win its fight against AIDS.

Bibliography

Arruda, E. N. (1986). *Antibody serological evaluation for HIV in tribes of the Southeast of Pará State.*

Brazilian Institute of Statistics and Geography (IBGE). (1986). *Yearly Brazil Statistics* (pp. 126–159). São Paulo.

Brazilian Ministry of Health. (1987). *Control of donated blood in Brazil.* São Paulo.

Brazilian Ministry of Health. (1987). *Epidemiological bulletin.* São Paulo.

Brazilian Ministry of Health. (1987). *Estimate on hospital beds demand for full assistance for AIDS patients in Brazil in 1987.* São Paulo.

Brazilian Ministry of Health. (1987). *Hospital costs for AIDS patients.*

Brazilian Ministry of Health. (1987). *Monthly chronogram of resources allotment.* São Paulo.

Brazilian Ministry of Health. Statistics Department. (1987). *Report on national Blood and Its Products program in Brazil—PROSANGUE (National Blood Program) & HEMOCENTROS. (State Blood Donation Centers).* São Paulo.

Brazilian Ministry of Health. (1987). *Reported cases of notifiable diseases.* São Paulo.

Brazilian Ministry of Health. (1987). *Structure and projects for effective action on AIDS.* São Paulo.

Centers for Disease Control. (1985). Changing patterns of acquired immunodeficiency syndrome in hemophilia patients. *Morbidity and Mortality Weekly Report, 34,* 241–242.

Center for Infectious Diseases. (1987, November 2). *AIDS weekly surveillance report.*

Curran, J. W., Mawrence, D. N., Jaffe, H., Kaplan, J. E., Zyla, L. D., Chamberland, M., Weinstein, R., Lui, K. J., Shonberger, L. B., Spira, T. J., Alexander, W. J., Ammann, A., Solomons, S., Auerbach, D., Mildvan, D., Stoneburner, R., Jason, J. M., Haverkos,H. W., & Evatt, B. L. (1984). Acquired immunodeficiency syndrome associated with transfusions. *New England Journal of Medicine, 310,* 69–75.

Fernandes, M. E., & Del Bianco, R. (1987). *Clinical characteristics of 100 cases in Emilio Ribas Hospital, São Paulo.*

Fernandes, M. E., Ferrari, L., Schmall, M., Del Bianco, R., & Pereira, W. P. (1986). *Inter-hospital infections of AIDS patients interned in Emilio Ribas Hospital, São Paulo.*

Fernandes, Z. P. (1986). *Antibody research for HIV in donors and recepients of blood and its products* (pp. 39–1130).

Galvão, B., Moraes de Sá, C. A., & Furtado, M. F. (1985). *New proposals for research on acquired immunodeficiency syndrome.*

Information Center of State Secretary of Health. (1986). *Number of cases and coefficient of incidence of compulsory notifiable diseases.* São Paulo.

Information Center of State Secretary of Health. (1987). *AIDS cases reported in the State of São Paulo up to 10/31/87.* São Paulo.

Interministerial Commission for Planning and Coordination (CIPLAN). (1987). *Resolution no. 87.* São Paulo: Brazilian Ministry of Health.

Macedo, L. G. (1987). *Brazil's AIDS program.*

National Institute of Medical Care and Social Welfare (INAMPS). (1987). *Resolution no. 144.* São Paulo: Brazilian Ministry of Health.

National Institute of Medical Care and Social Welfare (INAMPS). (1987). *Resolution no. 162.* São Paulo: Brazilian Ministry of Health.

National Institute of Medical Care and Social Welfare (INAMPS). (1987). *Resolution no. 170.* São Paulo: Brazilian Ministry of Health.

Notifiable diseases summary of reported cases, by month. (1986). *Morbidity and Mortality Weekly Report, 34,* 782–786.

Pinotti, J. A. (1987). *Women: The victims of a distorted health policy.*

Seade, A. (1986). *State yearly statistics, São Paulo* (pp. 56, 61, 64, 73, 74, 77, 78, 79). São Paulo.

Selwyn, P. A. AIDS: What is now known. *Epidemiology Hospital Practice, 21,* 127–164.

State Secretary of Health. (1987). *100 years of government.* São Paulo.

State Secretary of Health. (1987). *Total care health program for women.* São Paulo.

State Secretary of Health. (1987). *Total care health program for children.* São Paulo.

Teixeira, P. R. (1985). *AIDS in Brazil.* São Paulo: National Reference Center for Acquired Immunodeficiency Syndrome in Brazil, State Secretary of Health.

11

The Unofficial Story of AIDS in Brazil

Maria Helena Matarazzo

Brazil, with a population of 140 million, is the largest country in Latin America. The majority of the population is concentrated in the southern part of the country. Full of all sorts of social contrasts, at the present moment Brazilians have to survive with a half-percent increase in inflation *every day* (December, 1987).

According to a report published by the World Bank in 1987, in a study of 50 countries, only in Brazil does 10% of the economically active population receive practically half of the national income (47.7% in 1985), while 50% of the economically active population has only 13% of this income. So, Brazil is currently the "leader" of socioeconomic differences. Furthermore, 40 million people are marginalized from the national economy, neither producing nor consuming.

Brazil has the largest external debt in the world (US$ 109 billion), and one of the lowest minimum salaries (around US$ 40.00 per month). Nine million people earn half a salary, and only 6% of the population earns more than US$ 800.00 per month. Poverty is everywhere.

The government is fighting against the crisis. In 10 years, the number of homes has grown 43.1%. Sewage treatment, water supplies, and electrical systems have been extended. This great effort has attained good results. In the period between 1975 and 1985, the water and electricity supply was doubled and the sewage system was tripled (Instituto Brazileiro de Geografia e Estatistica, 1987). However, with a population growth of 2.6 million per year, there is an inevitable unbalance. For instance, in São Paulo, with 10 million inhabitants, the housing problem is very serious: 1 million people live in slums and 2 million in tenements.

The Achilles' heel of Brazilian social structure is education. The annual report of the Institute for Economical and Social Research (IPEA, 1987) confirmed that the educational system, under the responsibility of the federal government, is in total collapse. The education level of Brazilian children is minimal. They remain in public schools less than 3 hours a day. In the poorest part of the country, the rural northwest, the average time of enrollment in school is around 1.5 years, and that in the most industrialized part, the urban southeast, 5.5 years. In the First-World countries, the average time in schools is more than 7 hours a day for 12 years.

Around 8 million children do not have the opportunity of learning because there are no schools for them. Officially 18 million people are considered illiterate, but unofficially we could say that approximately one third of the population is illiterate and another third is half illiterate.

The IPEA report also pointed out that college-level education is in a chaotic condition. At the end of 1985, there were more than 45,000 instructors at the federal universities, but only 44,000 students graduated. The average is one instructor per 7 students, while in the United States it is 1 per 23.

Between 1983 and 1985, there was an application decrease of 74,000 students in the federal universities, but new teachers were constantly hired, and in just one day more than 12,000 instructors were admitted by government designation. With so many teachers in the University system, we might expect a high level of research. However, the rate of publication is only one paper per 10 instructors per year!

The Unofficial Story of AIDS

Research has shown that in 1982 human immunodeficiency virus (HIV) was already circulating in Brazil. In the following year, representatives of the homosexual community requested the Secretary of Health of São Paulo State to provide services for AIDS patients. In September, 1983, Dr. Paulo Roberto Teixeira (presently a World Health Organization consultant) started the first program for prevention of AIDS and care of AIDS patients in Latin America (excluding Haiti).

As in the United States, the AIDS epidemic was first recognized in homosexual and bisexual males. Although male prostitution is a very old activity, it has considerably proliferated in São Paulo since the early 1970s (da Silva, 1986). This group consists of young, low-income men whose age ranges from 16 to 28 years. Male prostitution practices occur on streets, at public toilets, gay nightclubs, massage parlors, "love hotels," and saunas (São Paulo has 30 homosexual saunas). The average number of sexual partners is estimated to be 250 a year. A clear indication that homosexuals are using more condoms is that by 3 A.M., when saunas are swept, numbers of them are found lying on the floor.

Initial controversial news reports about homosexuals as an AIDS risk group reinforced the general public's notion that homosexuals are a social threat disseminating an incurable disease. In the city of São Paulo a true "hunting season" for homosexuals took place in the second quarter of 1987. The mayor, using the police force, forbade all gay students to attend classes at the Municipal School of Ballet; two violent murders occurred, and 10 cases of beatings were reported (Barros, 1987).

The homosexual community in different Brazilian cities organized groups (e.g., GAPA) to give support to AIDS patients and their families and also to conduct educational and AIDS prevention programs. In 2½ years, 1,000 people volunteered to help the São Paulo GAPA. However, our social misery and absolute lack of resources create great anxiety, and most volunteers lose courage and give up. The ones who are able to win over fear become warriors with no weapons.

São Paulo also has a very large number of transsexuals ("queens") compared with other big cities in the world. Since their clientele are basically married men, the risk of HIV dissemination to heterosexuals increases. In Rio de Janeiro, a study (Cortes, 1987) done in a well-known gay nightclub showed 60% of the clients were bisexuals and their female partners were totally unaware of this fact.

In the whole country, based on data from the United Nations International Children's Emergency Fund, there are 2 million girls between 10 and 15 years old working as prostitutes. Most prostitutes say that their clients refuse to wear condoms. The average cost of a condom is 25 cents, so people with low incomes cannot afford them.

The Condom Generation

In 1987, condom production reached 10 million units per month, and the consumption doubled from 1986 to 1987. Yet half of the condom boxes include no directions showing how to use them.

The popular name for condoms in Brazil is "tiny shirt," so when children and adolescents hear the educational campaign on radio and television, they do not really understand what it means. They think that if they have sexual intercourse wearing "T shirts," they will not get AIDS. Moreover, the majority of women do not know what condoms are for, and the AIDS hotline receives many inquiries from women asking about where they can buy them and how to use them.

Lack of information and sexual irresponsibility play an important role in the dissemination of HIV virus. For example, at Carnival (d'Alessandro, 1986) people dance, sing, and drink for almost 4 days in a row. During that time, anonymous sexual activities with many partners occurs in the nightclubs, in the parades, on the streets—just about any-where. The Carnival spirit of excitement, euphoria, and total freedom creates a perfect environment for HIV dissemination.

The Blood Market

Another form of HIV virus transmission uncontrolled up to now is blood transfusion. The number of blood banks that exists in Brazil is unknown; the government estimates that there are 2,100, but only 20% are under governmental supervision. The rest belong to private groups, many of them clandestine. At present, it is impossible to control the blood market in Brazil. The clandestine banks are attended by very poor people, who exchange their blood for food or money to buy drugs (L. Lobo, Com-munications Assistant and Special Campaign Coordinator, GLOBO TV Network, interview). For example, a donor who goes to a clandestine blood bank receives a coupon after donation. From there, the donor goes to a bakery where the coupon is exchanged for a sandwich, a glass of juice, and the "change" (his payment).

The buying and selling of blood was forbidden by law in the whole country in July 1980. However, in the state of Rio de Janeiro, during a 5-year period, a multinational pharmaceutical company did not adopt this official regulation in order to maintain their own commercial inter-ests, since this company exported a large amount of blood products. Only in August 1985 was the commercial purchase of blood finally for-bidden in Rio, but because there is no adequate enforcement of the law (in Rio de Janeiro there are only four inspectors controlling 164 blood banks), blood banks continue to buy and sell blood contaminated with HIV virus.

HIV dissemination also occurs as a result of ignorance. In an interview with J. Pasternack, MD, the clinical director of Public Servants Hospital (1987), I was informed that the clerk in charge of separating the blood after HIV testing is illiterate; thus, a red band is placed around the HIV-positive tubes to avoid confusion.

The Medication Mafia

There is great corruption in Brazil in the field of medicine (Raw, 1987). The pharmaceutical industries mainly produce medicines that are prof-itable and that only rich people can afford. Therefore, the Brazilian gov-ernment itself has to produce all basic medicines to supply low-income populations, and there is a constant shortage of basic products.

Self-medication is a Brazilian habit because most people cannot afford a doctor's visit, and because they have to wait 3 or 4 months for an appointment in a public health service facility. In the mean time, people go to a drugstore where a pharmacy clerk "prescribes" the medication.

Brazilian law permits multinational pharmaceutical industries to produce the same medication with many different names. The average number of items in stock in a United States drugstore is 2,000; in Brazil, it is 40,000. Frequently, pharmacy clerks change initial doctor's prescriptions to "similar" products because the more they sell of a particular brand, the higher the gratuity or kickback they will receive from the pharmaceutical industry.

Corruption also occurs at other levels of the medical profession. AIDS is allowing the appearance of a new kind of physician—the "opportunistic" doctor of the acquired syndrome. Some Brazilian physicians are administering bacille Calmette-Guérin vaccine, imported from France, to AIDS patients, and giving them aspirin as if it were azidodeoxythimidine. Also, in a recognized medical school in São Paulo, testicle biopsies are being performed on live patients only to publish academic papers (V. Petri, MD, Assistant Professor of Dermatology, Paulista School of Medicine, interviewed, September 1987).

What Is Being Done?

In December 1986, the National Program for Prevention and Control of AIDS defined education as its top priority. In 1987, a national compaign was started featuring a very famous soccer player and two well-known television actors to alert people about AIDS. The fourth largest TV network in the world—after NBC, CBS, and ABC—the GLOBO, showed 30-second spots during 1 month for a total of 12,000 insertions (L. Lobo, inteview). Never before had a Brazilian TV commercial network allocated so much time for an educational campaign.

In January 1988, the Health Secretary of São Paulo State will be distributing several pamphlets with basic AIDS information attached to household electric bills. The Ministry of Education is preparing booklets to be inserted in all science textbooks for 1988. Other public campaigns are planned.

It is very hard, and sometimes impossible, to obtain correct data and figures in the Third World. You follow your instincts, you make an intellectual or emotional guess, you talk to people: journalists, someone from the government, friends. You check and double check your numbers, and you are never sure. You constantly believe—not trusting, never knowing how close or how far you are from reality.

This chapter is the result of 400 hours of research, and the purpose is to present an idea of our reality. It is important to know the official story of AIDS, but the official speech can be sometimes quite dissociated from reality, so this is the negative picture of the same reality, a glimpse of the unofficial story of AIDS in Brazil.

References

Barros, R. (1987, November 15). *The Folha de São Paulo*, p. A-21.

Cortes, E. (1987). Unpublished study, UCLA School of Medicine.

d'Alessandro, S. M. (1986). *El Carnaval y el Brasilero [The carnival and the Brazilian]*, pp. 49–53. Unpublished doctoral thesis, University of Belgrano, Brazil.

da Silva, L. (1986). *Aids and homosexuality in São Paulo*. Unpublished master's thesis, Pontificia Universidade Catolica de São Paulo, Brazil.

Institute Brasileiro de Georgrafia e Estatistica [Brazilian Institute of Geography and Statistics]. (1987, September 20). *Annual report*, pp. A23–A26.

IPEA [Institute for Economical and Social Research]. (1987). *Annual report*.

Raw, I. (1987, August 19). *Veja*, p. 118.

World Bank. (1987). *Development of the world*.

12

Cross-Cultural Perspectives on AIDS:
A Commentary

Malcolm Potts

Human immunodeficiency virus (HIV) infection has reached all over the world, and the problem must be treated globally if the infection is to be contained. The pattern of AIDS differs geographically. Without doubt, the major spread in Africa is among individuals with numerous heterosexual partners, whereas in the United States high-risk behaviors have been male homosexuality and intravenous (IV) drug use.

Patterns of sexual orientation are difficult to measure in any society, but homosexual behavior is probably rarer in rural village communities than in cities. It is much more difficult for people to keep their sexual orientation anonymous when the population is subdivided into relatively small units; the chance of meeting suitable partners is also greatly reduced if there are only a few hundred people in a village, where everyone is controlled by the power of gossip.

Urbanization

AIDS is primarily an urban phenomenon. It has been propelled especially rapidly in Africa by the explosive growth of cities, the disruption of war (especially in Uganda and more recently in Sudan), and the migration of hungry people to cities as a result of famine (especially in Ethiopia and Somalia). As many people now live in the big cities of the Third World as lived in the whole of the world in 1950.

If, as seems likely, the rapid growth of cities in Africa has done more than anything else to create an environment in which HIV can be rapidly transmitted, then there must be a great concern that a similarly rapid

heterosexual transmission could occur in Latin America and Asia. Preceding chapters have emphasized how conventional labels of sexual
orientation can be misleading in America, and this is also true overseas.
For example, bisexual behavior has been reported in Brazil (Parker, 1987)
and anal intercourse, both heterosexual and homosexual, is relatively
common at least in the urban areas of Brazil, as indicated in Matarazzo's
and Fernandes's chapters (11 and 10, respectively) in this volume. Zulu
language has a word for "boy-wife," and Swahili has two words for
homosexual men.

Cities have become magnets for the excess population from the coun-
tryside. In leaving a traditional way of life and entering the work force
of a cash economy—or more commonly merely scraping a living on the
fringes of that economy—the current generation is facing a greater cultural change than any that has occurred in the past several thousand
years. Traditional practices of defining adulthood and restraining sexual
behavior are breaking down (Feldman & Taylor, 1987). The spread of
education is leading to a decline of polygyny, which, in turn, excludes
more and more women from the supporting lineage system at the village
level, driving such women into the towns. More men than women initially migrate to cities, and those men then seek sexual release with
prostitutes. As families and young women migrate to the cities, they
also have little chance of employment and are themselves often drawn
into prostitution (Jocano, 1975). In the West, city life thrives on a division
of labor, and there is a parallel ability for individuals to specialize in
certain sexual orientations.

Cultural Differences

Specific cultural practices may predispose to the spread of HIV infection,
although they are probably less important than the effects of poverty.
Wherever there is poverty, whether 19th-century London or 20th-cen-
tury São Paulo, there is sexual exploitation (Matarazzo, Chapter 11). In
the West, prostitutes often come from disturbed homes and many are
on drugs; in the Third World, they often come from stable, rural homes
and are rarely on drugs.

Sub-Saharan Africans may have more life-time sexual partners than
Asians or Latin Americans: the former have a life-style based on sub-
sistance agriculture, in which a woman and her children form an economic unit, whereas the life-style of the latter is based on plough-based
agriculture with male ownership of land (Caldwell, Caldwell, & Quig-
gan, 1989).

The strong pressure to have children found in many parts of Africa
may mitigate against the use of condoms and, even if a woman knows
she is at risk for HIV infection, she may still discard barrier methods of

contraception in order to get pregnant. Polygamy could account for clusters of HIV infection but need not necessarily disseminate HIV around the community. The practice of a brother caring for his sister-in-law if widowed and taking her as his own wife is common in many parts of Africa (Jocano, 1975). If the husband died from AIDS, the wife is also likely to be infected and, in turn, may pass it on to her husband's brother. In the Ivory Coast, a widow, after her period of mourning, is expected to have intercourse with a stranger. One topic that deserves more study is those areas in the AIDS belt of Africa where the prevalence is still low, such as parts of rural Zaire.

IV drug use, outside some of the countries of South America, such as Brazil, is less common in the Third World than in the West. Basically, the world trade in drugs is a "free market" and traffickers sell to those with the greatest disposable incomes. IV drug abuse is found along major drug trade routes, and many thousands of IV drug users are already HIV positive in Thailand. In contrast to the situation in the West, the use of inadequately sterilized needles in hospital practice, especially by traditional practitioners, has the potential for spreading HIV as has already occurred in the Soviet Union. Dirty needles have been suspected as a cause of transmission in Zaire (Quinn & Mann, 1986), and children with AIDS appear to have had more injections than those who are free of the disease, although it is difficult to separate cause and effect. On the whole there do not seem to be significant numbers of children or old people with AIDS whose disease cannot be explained as due to heterosexual or perinatal transmission.

Many of the larger cities of the developing world are served by extraordinarily busy obstetric hospitals, some of which, like the Mama Yamo hospital in Kinshasa, deliver 30,000 or more babies a year. In these situations, in which many women are laboring at the same time and the parturient mothers and their newborns are often two in a bed, the possibility may exist for HIV transmission through contaminated blood products and lochia, or through disposal of the placenta and through the stump of the infant's umbilical cord. It would be prudent to review current practices in obstetric hospitals, even in the absence of proven cases of HIV transmission.

More research on human sexuality is certainly needed in many Third-World countries, but it is also true that enough is known to set sound preventive policies.

Public Health Policies

The attributes that distinguish AIDS from all other diseases are also ones that deceive and bewilder decision makers. The interval between first infection and symptoms of AIDS can average up to 8 years, and this is often longer than many politicians are likely to be in office. Second, and more importantly, unlike most other infectious diseases, which spread passively in the air or through contaminated material, nearly all cases of HIV infection are spread by sexual intercourse or by blood transfusion or IV drug use. As a new disease, AIDS inevitably appears first in those groups who have the largest number of sexual partners or the most serious levels of IV drug use. The fact that AIDS was first recognized among gay men and junkies in America and among prostitutes and the urban poor in Africa makes it difficult for many in society to identify with the disease. The fact that AIDS largely passes from one individual to another as a result of volitional acts has suggested to some politicians and religious commentators that AIDS is punishment: What epidemiologists call variation about the median, the lay public all too easily sees as deviant perverts engaging in immoral activities. There is a strong consensus among public health experts about what needs to be done to slow the spread of HIV infection, but much less agreement among the public about what health policies should be implemented. For example, the Indian government has not taken any steps to slow the rapid spread of HIV among prostitutes in cities such as Bombay. If the gap between policies set by public health workers and health policies accepted by the public is to be closed, then the nature of the disease and our own sexuality must be much more clearly understood. In particular, contemporary society has almost as many explanations of human sexuality, from the biosocial to the fundamentalist religious, as it has sexual lifestyles.

Biosociology

As Voeller et al. said, sexuality is at the center of our biological evolution but it is too rarely studied (Introduction, this volume). The various denominations of psychology and sociology and religious interpretations of human sexual behavior all make assertions that are difficult or impossible to test experimentally, or to analyze in the framework of observational data on the reproductive behavior of other animals. I would like to suggest that the many insights that have come from biology in the past 2 decades provide useful insights into human sexuality (Symons, 1979) and, in turn, may provide a reason why scientific analysis has been so tardy.

We share 99% of our genes with chimpanzees; like us, they use tools, can be taught a language (Jolly, 1985), enjoy being hugged, and are divided into troops of cooperating males. What distinguishes humans from all other primates, including chimpanzees, is our sexual behavior:

Ovulation in women is concealed from the female and her mate

We copulate throughout the ovulatory cycle, pregnancy, and lactation

Sexual partners share the same nest at night

We share our food

A measurable proportion of our species focus their erotic activities on their own sex

We copulate in private

We rarely eat our placenta

We cover up our genitals

We have a sense of humor

Biosociology analyzes human sexuality in the framework of Darwinian evolution. Sperm are cheap and eggs are expensive. In viviparous mammals, the female must make an infinitely greater investment to reproduce than the male. All the large primates face the same issues: pregnancy takes many months to complete, breast-feeding goes on for several years, and the female can only reproduce herself on a few occasions during her fertile life. However, the male can inseminate many females during the interval that it takes one female to bring her offspring even half way through childhood.

Gorillas, chimpanzees, and humans are all sexually dimorphic in their anatomy, and there is no doubt that we have evolved from a stock in which males had multiple female partners. (Among the truly monogamous primates, such as the marmoset monkey or gibbons, males and females are approximately the same size because the males do not have to fight off other males in order to secure access to several females.)

Our relatively large body size among mammals and nonseasonal patterns of reproduction further differentiate the reproductive agendas open to either sex. It is easy to forget that in the struggle for survival, the worst competitor is the opposite sex of one's own species, and among the primates most closely related to humans, several different behavioral strategies have evolved to ensure a sufficient food supply. The orangutan lives at a low population density, and males and females only consort together at the time of mating. Gorillas live in troops of females with one silverback male who fights off other males so that the food supply in the territory that he and his harem of females occupies is not over exploited. Chimpanzees have an unusual social structure in which groups of related males form a cooperating troop and females from other troops migrate out of the troop of their birth at puberty to join a new group of related males. Several aspects of human social behavior, such as the role of men in politics (Tiger & Shapher, 1975) and the evidence from comparative anthropology that the female commonly joins her in-laws in marriage, may be shadows of a social structure that was once close to that of the chimpanzee.

Comparative anthropological studies show that most human societies are either polygamous or practice serial monogamy, and evolution appears to have capitalized on sexual dimorphism in behavior as well as in anatomy. The fact that ovulation is concealed in women probably helps strengthen the sexual bond with an individual man. Unlike most other animals, in humans the two sexes exploit rather different food supplies, probably accounting for our unique human drive to share food between sexes and between individuals.

Our biological drive toward multiple partners and numerous intercourses not associated with ovulation makes us an especially attractive species for the transmission of sexual diseases. Our vast increase in numbers, urbanization, and the division of labor and life-styles that go with the high degree of urbanization have further exaggerated this trend. All human societies cover up the genitals of adults, and it seems we do find it difficult to discuss sexuality, contraception, and sexually transmitted diseases objectively.

Human Sexuality and AIDS

I suggest that in following their biologically determined reproductive agendas, men and women do indeed behave in very different ways. Men look for numerous erotic experiences and many different partners and emphasize certain aspects of youth and beauty—as surrogates of fertility—in their mates. Women look for stable, long-term bonding with emphasis on human relationships, and choose their partners more by social status than by youth or beauty. It follows that heterosexual marriage is a compromise in which each sex adjusts its drives to the other's. Undoubtedly, cultures can emphasize the male or the female agenda in reproduction at the expense of the other. In many contemporary Third-World agricultural societies the male agenda often wins out. Western religions have leaned toward the side of the female agenda, to the extent that monogamy has become the social norm, but the "dual standard" (which represents the male reproductive agenda) has never been far beneath the surface.

Human sexual orientations appear to be very plastic. When an individual becomes oriented on his or her own sex, then the compromise that is central to marriage is no longer necessary. Male homosexuality represents free-wheeling male desire (Tripp, 1987) with emphasis on multiple partners and short-term erotic experiences, whereas lesbian relationships often put relatively little emphasis on sex and much more on supportive and loving relationships between the individuals involved. This is not to say that gay male relationships cannot last a long time or, as the AIDS epidemic has poignantly demonstrated, become a source of enormous mutual support and empathy, as dramatically documented in this volume by Drs. Mattison and McWhirter (Chapter 28). However, even in long-term male homosexual relationships, the partners often accommodate to outside erotic experiences.

If, as seems likely, HIV has jumped from some other primate into the human species, then it is reasonable to ask why this even took place in the 1960s and not 100 or 1,000 years ago. A plausible explanation is that the virus is very fragile and, although it may have infected human beings in the past (probably as a result of biting and scratching), it may have only caused a cluster of infections that went unnoticed. For the AIDS epidemic to catch fire in the human population, it required the dry tinder provided by heterosexuals with numerous partners in urban Africa, and homosexuals with numerous partners in industrialized countries.

Anderson and Johnson (Chapter 6) quoted data from the United Kingdom that heterosexual partners turn over (to use an appropriate term!) at approximately 0.9 a year, and new male homosexual partnerships are established at an average of 10.5 a year. When we also appreciate that there is great variation about the mean, then perhaps we are on the way to understanding something about the sexual transmission of AIDS.

One of the most extraordinary statistics to come out of the AIDS epidemic was that there were 17,645 deaths of gay men in the United States before the first case of AIDS in a lesbian couple (in 1984). Even given the greater difficulty of transmitting HIV infection in a female homosexual relationship, this observation tells us a great deal about the biology of male and female sexual drives.

The Key

In the long term, the challenge that faces us is not only how to understand human sexual behavior in relation to transmission of AIDS but, just as important, how to understand the sexual attitudes of those who set public health policies, so that those who are closest to the mainstream of behavior, and tend to control most of the resources, can accommodate in a respectful and sympathetic way those whose behavior is far removed from the mean.

Enough is known about AIDS to set realistic public health policies to prevent spread. Earlier contributions have shown that relatively modest changes in sexual behavior and a commonsense use of condoms can slow the spread of HIV infection. A great deal can and should be done to control AIDS. The problem is that policymakers are often afraid or embarrassed to assist groups with high-risk behaviors, such as IV drug users, gay men, and prostitutes. A similar issue arose in family planning in the 1970s. In generic terms, the most politically acceptable interventions are those that are least controversial, and therefore, by definition, the least effective. National AIDS committees, or administrators in the

U.S. Agency for International Development find it easier to pay to clean up a nation's blood supply—which is laudable but does not prevent the major source of infection—than to educate and provide condoms to prostitutes.

This problem may be particularly acute in the United States, with its fundamentalist tribal groups and where many of the voting public are scientifically illiterate—polls show more Americans believe in extraterrestrial visitors than in Darwinian evolution!

References

Caldwell, J. C., Caldwell, P., & Quiggin, P. (1989). The social content of AIDS in sub-Saharan Africa. *Population and Development Review, 15*(2), 185–234.

Feldman, D. A., & Taylor, C. C. (1987). *Blood, sex and AIDS in Rwanda.* Unpublished manuscript.

Jocano, F. L. (1975). *Slums as a way of life.* Quezon City: University of the Philippines Press.

Jolly, A. (1985). *The evolution of primate behaviour.* New York: MacMillan.

Parker, R. (1987). Acquired immunodeficiency syndrome in urban Brazil. *Medical Anthropology Quarterly, 1*(2), 155–175.

Quinn, J., & Mann, J. (1986). AIDS in Africa: An epidemiologic paradigm. *Science, 234*, 955–963.

Symons, D. (1979). *The evolution of human sexuality.* New York: University Press.

Tiger, L., & Shapher, J. (1975). *Women in the kibbutz.* New York: Harcourt-Brace.

Tripp, C. A. (1987). *The homosexual matrix.* New York: New American Library.

V
PERSPECTIVES FROM THE MEDIA

13

AIDS: Media Coverage and Responsibility

Marlene Cimons

One of the first things taught in Journalism 101 is not to get involved. As reporters, we are supposed to function as observers and chroniclers of the news, not participants. That is a standard I believe in . . . most of the time.

But I think AIDS is different, and has presented us with a dilemma of sorts. I think newspapers, magazines, and broadcast news programs, with their tremendous reach, have a duty to participate—especially when lives are at stake. I have said this over and over again at many meetings such as the Kinsey Conference on AIDS and Sex, during many media panels discussing AIDS coverage. I think the media have a public education responsibility. And I think we have failed miserably to fulfill it.

I began covering AIDS nearly 3 years ago as part of the federal health policy beat. I have long had an interest in health and medicine. I believe health is a subject that has been vastly undercovered by the media, although I've never quite understood the reason for this, since I am also convinced that these stories are among the best read in the newspaper. As time went on, and as the AIDS story began to grow, I found myself writing more and more about AIDS—and less and less about anything else. AIDS is now almost a full-time job.

As we all know, AIDS has become much more than a medical and science story—it is a story that encompasses social, political, ethical, and legal issues, and involves one of the most basic of human emotions: fear. It touches us on all levels, but specifically in our feelings about sexuality and death, where we are especially vulnerable.

And it has raised many complicated issues for journalists.

Getting the Message Out—and Getting It Right

Reporters are accustomed to hearing and writing about the symptoms of a disease—*Pneumocystis carinii* pneumonia, for example, or Kaposi's sarcoma—and what causes it and how to treat it. Or we report the statistics of a disease: how many people have it, and how many have died.

But this epidemic is different from any we have ever known. It involves sex. It involves gay men. It involves intravenous drug users. It involves life-styles that to many people are unknown at best—and anathema at worst. It think for a long time the media were uncomfortable dealing with those aspects of the epidemic and, as a result, performed a great disservice to the public.

We have been criticized for taking far too long to take this story seriously, and I think the criticism is justified and I share the blame. This epidemic involved a limited patient population and people could not relate to it. Gay rights groups and others complained early on that we were dismissing AIDS as a "gay man's disease," and I regret to say they were probably right. It's like the newspaper that fails to report the hundreds of murders of black people in the inner city, but runs a prominent page 1 story when a white Yuppie is stabbed to death on a city rooftop.

I think several things finally made us pay attention: transfusion and hemophiliac cases, Rock Hudson, and the schoolchildren. Transfusions, because you did not have to be "different" to get AIDS. Rock Hudson, because suddenly somebody we all knew had AIDS. We could see a "before" and an "after." And the schoolchildren—well, they were children. Everyone relates to children. Suddenly, AIDS was NOT just a gay man's disease anymore.

To compound the problems, I think the media also contributed to the epidemic—and to the public hysteria—by our failure to communicate graphically and specifically exactly how this disease is transmitted. We have been so paralyzed by our hang-ups over what is in "good" taste that we simply lost sight of the fact that people were dying.

For months, we dealt in euphemisms about not exchanging bodily fluids. I don't even know what that means. What body fluids? How do you not exchange them? How many people thought we were talking about sweat? I was more fortunate than some of my journalistic colleagues; I was able to get the words "semen," "blood," and "anal intercourse" into the *Los Angeles Times*. But that's as far as I was able to get—and it wasn't always easy.

One of my colleagues in Los Angeles, in a story he wrote about AIDS, attempted to describe, at the bottom of his story, behavior that was considered possibly risky and very risky: anal intercourse without a condom, swallowing semen, "fisting," and "rimming." He didn't attempt to explain or define those terms because, as he says, "the people who need to know don't have to be told what they mean. If I had explained them, it never would have gotten into the newspaper."

Well it didn't get into the newspaper anyway. The bottom of his story was sliced off before it ever got into print. Yet, when people die of cyanide-laced Tylenol, we are very quick to publish the lot numbers and store locations of every poisoned bottle—we feel a definite responsibility to protect the public. Why should AIDS be any different?

We've gotten better—much, much better. But it's taken us far too long to get there. We are still not reaching into the very communities hardest hit by AIDS, populations that desperately need the information. And I'm not sure what that approach should be. What worries me even more is that, as the danger to the heterosexual population is reassessed (and perhaps minimized again), editors and communicators will again dismiss AIDS as a gay man's disease shared only by junkies and their sexual partners, which would be a tragedy of monumental proportions.

We have also been guilty of communicating the wrong messages, although perhaps motivated by well-meaning intentions. *Newsweek* magazine, some months ago, ran a cover story on an AIDS doctor. I thought it was powerful stuff—one of the finest pieces of journalism I've ever read. It made almost all the right statements about this disease—all but one. While the physician and every member of his staff and family was identified, every single AIDS patient described in the article was disguised.

I think everyone has the right to privacy. People are certainly entitled to have an illness and not have the world know it. Yet, I still think that when we hide names or disguise people who have AIDS, we contribute to an environment that perpetuates an unfortunate notion: that having AIDS in something to be ashamed of. There is a stigma associated with this disease, and I think this kind of anonymity just makes it worse.

In 1985, I began a series of stories following an AIDS patient and his physician in a study of azidodeoxythimidine, the experimental AIDS treatment drug, at Massachusetts General Hospital in Boston. I envisioned these articles as ongoing, hopefully over a period of years. My intention was to humanize the disease—to enable the public to get to the heart of the emotional experience of both the patient and his physician. I wanted the readers to feel they know these men personally and get a real sense of what they were going through.

They were eager to do it. The physician, Chip (Dr. Robert T.) Schooley, was enthusiastic without reservation, but the patient—whose name is Jeff Mullican—was a little wary. He was concerned about being identified, not so much for himself, but for his family. He asked if I could change his name just a little—put a "g" where the "c" was, making it Mulligan, perhaps.

I refused. I was very emphatic about it, too. He could change his mind and back out—but if he decided to do it, he would have to be honest about who he was, and what he was, and what was happening to him.

He agreed to participate with no ground rules, which I thought was a very gutsy decision. And the feedback has been tremendous—universally positive. I have written seven articles so far, all of which were extremely well received.[1] Jeff is very pleased, and quite happy now that he decided to be so open and accessible. Jeff and Chip and I feel that, together, we are making a very important contribution.

Other Thoughts on Confidentiality

I have been asked whether I would have printed the nature of Roy Cohn's illness while he was still alive. I think, if I had been able to document that he had AIDS, I would have made every effort to write a story, although I'm not sure whether my editors would have permitted it.

I would have fought to run the story because Cohn was using his dying in a very public way—and he was lying about what was killing him. During his disbarment proceedings in New York, he pleaded for compassion because he had a terminal illness, which he claimed was liver cancer. He was asked publicly on several occasions whether he was suffering from AIDS. He said no. He was a public figure who was dying publicly—but not honestly.

The day after he died, I found it interesting to watch the media dance around the cause of his death. Most were very careful to use the precise words issued in an official statement by the National Institutes of Health: the primary cause of death was "cardiopulmonary arrest." The secondary causes were: "dementia" and "underlying HTLV-III infection." "Underlying HTLV-III infection" means only one thing to me. And that's what the *Times* said in his obituary—that Roy Cohn died from "complications stemming from AIDS."

Former Washington Redskins football player Jerry Smith, however, is a different story. I got a call from our sports editor last year. He wanted my advice. He told me that his reporters had found out from several iron-clad sources that Jerry Smith had AIDS. "We have confirmation all over the place," he said.

But he had decided not to run the story. He told me that he had asked

---

[1] Jeff Mullican died on December 24, 1987. The eighth and last article in the series was published on December 29, 1987.

himself what would be gained by running that story—and the answer was that he would gain nothing but sensationalism. He asked himself whether his readers had a burning desire to know that Jerry Smith had AIDS. The answer was no. Was Smith still in the news? Again, the answer was no. And could Jerry Smith be hurt by this story? Yes.

He decided that the only way to do a story like this was to have Jerry Smith tell it. He asked me if he had made the right decision. I told him yes. Absolutely.

As it turned out, Smith himself decided to talk—but to the *Washington Post*. So, in the competitive newspaper wars, we lost a story. But so what? The story was not ours to tell. I really believe that. Additionally, I think Smith performed an enormous service by speaking about his illness publicly—but it had to be his choice to do so.

But each case is different. Obituaries, for example, have been the subject of spirited debate. Should AIDS be listed as the cause of death? My answer to that is yes—if it can be documented. Obituaries are news stories, and the cause of death is part of a news story. I am occasionally irritated by obituaries in which the obvious cause of death is AIDS and the newspaper knows it but can't print it, and the buzz words give it away anyway. So-and-so died of viral encephalitis, according to his family, or according to his doctor. Invariably, these are obituaries of young men in their 30s or 40s who are survived only by their parents and siblings.

Reporting how the disease is contracted, however, is another issue— a private matter unless the situation is one in which the public is deceived or hurt. Conservative Terry Dolan, for example, spent his entire career supporting candidates who worked against gay rights causes. To me, the fact that he was gay and died of AIDS was a critical piece of information in reporting his death.

Rep. Stewart McKinney is another example—perhaps the best example of this. When he died last May, his office announced he had AIDS. Very courageous. But his office did not stop there. They said he contracted the virus from a blood transfusion several years earlier. It was well known within the Washington gay community, however, that McKinney, married and the father of five, had had homosexual relationships. Although it was possible he contracted the disease from a transfusion, it was statistically much more likely he got it from homosexual behavior. His staff could have remained silent on the source of his infection. Instead, they unnecessarily frightened thousands of blood transfusion recipients with their announcement.

A Few More Words About Words

Much has been made of the language of AIDS. I have tried, as I have written my stories, to be aware of how I use the language. Early on, I was criticized by someone for using the term "general population" when what I meant was "heterosexual population." The criticism was deserved, and I will never make that mistake again.

The word "victim" is another example. Initially, I thought it a very appropriate word to describe someone with AIDS: an innocent person, struck by a potentially fatal disease he or she did not deserve—and powerless to do anything about it. That sounds like a victim to me. My feelings have changed since then, mostly because the term is so offensive to the people I write about. I respect their feelings. Therefore, I don't use the word.

But it has been very hard to communicate this to headline writers. It has been many, many years since I have written a headline, but I do recall the frustration of making words fit. "Victim," I guess, must have exactly the right number of letters. I'm sympathetic to that kind of problem. Yet, I have managed to write thousands of words about Jeff Mullican and Chip Schooley without using the word "victim" once. I thought the headline writers could surely find another word to describe my protagonist. But the headlines, over and over again, call him a victim.

Furthermore, I fail to understand the resistance I encounter every time I suggest we should try to find an alternative word—such as "patient"—because the term "victim" offends people. I don't think it hurts us to be responsive to the people we cover, especially since we are dealing here with a national tragedy. But the arbiters of style are stubborn. They insist that "victim" is appropriate. I'm sure the same thing happened when blacks began complaining about the use of the term "Negro."

"Epidemic" was another word I had trouble with in the beginning. There was the time when an editor—after 12,000 people had become ill—questioned my use of the term "epidemic" to describe what was happening.

"Couldn't we call it an outbreak," he said. "I don't know anyone who has this disease—do you?"

At the time, I didn't. But I do now. And within the next 5 years, almost every American will know someone who has AIDS. No one quibbles with my use of the word "epidemic" anymore.

Final Comments: Fear and Media Responsibility

There is one final word I want to bring up again: fear. I mentioned it earlier and I'd like to mention in once more. There is a great deal of it out there, and most of it is misplaced.

People are entitled to be afraid of a fatal disease. AIDS is a terrifying disease. But people can't decide why they are afraid. It is human nature to be afraid of an unknown, but I think we have helped make people afraid of the wrong things. And, in doing so, we have caused them not to want to believe the right things anymore.

I think our federal health agencies have contributed to the problem. Let me give you an example. The government's Centers for Disease Control, which has a major public education responsibility, 3 years ago released guidelines for schoolchildren with AIDS. They were released Labor Day weekend, only a scant few days before children were scheduled to return to school. The timing provided little—really no—opportunity for teachers and parents and school boards to assimilate the data, and think about how to plan for kids with AIDS in the classroom.

Also, when material was presented, it was often couched in uncertain terms. Public Health Service officials were reticent in their presentation, or they didn't speak the English language at all. At press conferences, scientists would declare there was "overwhelming evidence" that AIDS could not be transmitted by casual contact. Then, almost in the same breath, they would add—as scientists often do to protect themselves—that "of course, we can't give you any guarantees."

This is a very scary disease. Parents who must decide whether or not to send their child to school with a child who has AIDS do not hear the "overwhelming evidence." What they hear is: "We can't give you guarantees."

They WANT guarantees. We all want guarantees. But there are no guarantees in life. When an airline advertises its services, it never says: "We can't guarantee our planes won't crash." But airplanes crash. These same parents who are keeping their children home or setting up separate schools—because you can't given them any guarantees—would not hesitate to put those children on an airplane. They probably face a greater risk of losing their children in an airplane accident than of having the children get AIDS in the classroom.

I hope we can continue to correct the damage. We're getting better and better. But when there are incidents such as the burning of the Ray family house in Arcadia, Florida, I know we're not there yet. We in the media have to recognize that there is more we can and must do. We have to continue to find better, more effective ways of getting critical information across.

We have a responsibility NOT to print distortions and misinformation. We also have a duty to correct that misinformation when it is spoken or otherwise communicated in the public arena in situations we cover.

We have to say the things that have to be said—and we have to say them over and over again. Words do not kill people, AIDS does. Therefore, if words can save lives, they need to be said.

14

Media Coverage of AIDS

Steven Findlay

The media always play an important role in public health issues, but their role in the AIDS epidemic has been even more integral, visible, and palpable than usual. Television news specials and magazine cover stories have been a reference point not only for the public but for many leading AIDS researchers and public health officials as to what is known and understood about the disease and what the critical AIDS issue of the day is. It was an essay by a gay activist in the *New York Native* in 1983 that sparked a reassessment of sexual ethics and behavior in the gay community. And reporting by Randy Shilts for the *San Francisco Chronicle* played a big part in the eventual closing of gay bath houses in San Francisco, the first public health action that signaled the extent of the crisis. Shilts later went on to write the first provocative book on the AIDS epidemic, *And the Band Played On*.

But it took some time for the national media to catch on. From 1980 to 1983, lack of media attention significantly contributed to a delay in perception of AIDS as a major health crisis, thus delaying public health action. I fault journalists in that period, including myself, for failing to grasp the urgency. But I reject the charge, often made in the gay community, that this had anything to do with prejudice against gay men. Looking back, I think ignorance and uncertainty were the real culprits— ignorance of sexually transmitted disease and gay life-styles and the extent to which the two mixed to produce a potentially explosive epidemic, and uncertainty over whether this was a phenomenon that truly was going to be more than a blip on the scene.

A further problem was the lack of a tradition of investigative journalism in medicine. Medical reporters usually do not go much beyond what the researchers, doctors, and health officials they cover tell them. It is rare that any real digging takes place. With AIDS, if some had, perhaps the story would have moved to center stage earlier than it did. After AIDS, science and medical journalists are more alert to strange medical phenomena. A case in point is an outbreak of a rare disease that occurred in several dozen people in the United States in late 1989 and early 1990. Tamar Stieber, a reporter for the *Albuquerque Journal*, started asking questions of several women in New Mexico who had developed a crippling disorder calld eosinophilia-myalgia syndrome. She discovered all the women had taken a popular over-the-counter supplement called L-tryptophan. Investigations by the Food and Drug Administration eventually proved that a contaminant of some batches of the supplement had indeed caused the illness. Steiber won a 1990 Pulitzer Prize for articles exposing and probing the outbreak.

Beginning in late 1983, the media "found" AIDS and propelled it to the level of a major national story. In a survey of 1,000 adults for *U.S. News and World Report* in late 1987, for example, the spread of AIDS was the number one national concern, ahead of the budget deficit and nuclear arms control. AIDS had become almost a daily page 1 story in many of the nation's leading newspapers and popped up regularly on the evening news programs. In contrast to the period 1980 to 1983, the media since 1983 not only have been on top of AIDS developments, both medical and social, but have played an activist role in getting out the prevention message. Reporters have become health educators, often taking the initiative—above and beyond any official public health statements—in informing people about the practical steps they can take to protect themselves against human immunodeficiency virus (HIV) infection. Indeed, conservative elements in the Reagan and Bush administrations and in Congress have continually stalled development of the kind of explicit educational material that might help some groups at high risk for infection. To a certain extent, the media filled the gap. With the explicit nature of the information being given, AIDS has ushered in a new era of the media as a conduit of public health information. I think, by and large, in this respect, we have done a good job.

The same could not be said, at least until recently, about coverage of the political and social dimensions of the epidemic. Only since the beginning of 1987 has the press become more probing of this aspect of AIDS. Here again, with a few exceptions, most reporters did not bother to look deeper into the growing debate about testing, legal issues, or government policies, until it emerged fully into the public limelight. However, since early 1987, the public policy aspects of AIDS have been well covered, notably in the *New York Times*, a newspaper that was widely criticized for ignoring the epidemic in the early days.

There have been other problems as well in the media's coverage of AIDS. It has occasionally been sensationalistic, fanning the flames of fear. AIDS became a good "panic" story. As much as many responsible journalists were trying to provide antipanic messages in their reporting, the temptation to highlight the fear reaction often overcame their sense of responsibility. This was especially true in local coverage of the school AIDS issue—should children with AIDS be allowed in classrooms? Do they pose a threat to other children? The media often, wittingly and unwittingly, pandered to the panic mongers in its coverage of this issue. The national attention accorded the death of Ryan White in April 1990 reminded many observers of how far we had come in accepting AIDS and HIV-infected people, how the fear had died down and a deep compassion had, for much of the population, taken its place. But it also reminded many of how far there is still to go to end persistent discrimination against those with HIV infection. Even as thousands mourned the death of the brave teenager, some insurance company somewhere was surely denying health coverage to some HIV-infected person. And those are stories, too, that the media should not stop covering.

Another issue also deserves special mention as an area in which coverage was confusing at best, and truly irresponsible at worst. That is the issue of whether the disease is going to spread rapidly into the broad heterosexual population. The low point in this coverage occurred with the July 1985 issue of *Life* magazine, whose cover blared "Now, No One Is Safe from AIDS." Not soon after *Life* ran its cover story, another Time, Inc. publication, the science magazine *Discover* (since sold to Family Media, Inc.), had a lengthy cover story with the basic message that unless you engage in anal intercourse, you need not worry about contracting the virus, at least not much. The *Discover* story went on at length to suggest that the heterosexual spread of AIDS in Africa was likely linked to other factors not present in the United States. Not to be outdone, *Time* magazine itself and its competitor *Newsweek* weighed in on the topic in 1986, suggesting the fear of wider spread was justified and that the future of the epidemic looked ominous. *U.S. News and World Report* (for whom I was not yet working) did not run a major story on this until January 1987. It, too, took the tack that spread to the larger population was likely unless behavioral changes occurred. Working for the newspaper *USA Today* at the time, I had less of an opportunity to take on the issue as a whole. But I admit to seizing upon, for spot stories, any research that seemed to indicate AIDS was spreading outside the high-risk groups. The paper also ran several "cover" stories on women and children with AIDS.

In fairness, leading researchers and public health authorities also took sides on this issue, and a debate has raged for several years in the absence of any solid data on the true prevalence of the infection. But that does not exonerate the media. The confusing coverage seemed to stem from reporters' and editors' understandable, but lamentable, desire to get an answer where none yet existed. For many reaons, it was not possible to even make an educated guess as to the AIDS threat to the broader population until very recently. Yet, the early cases outside the high-risk groups, in women and babies born to them, for example, seemed license to come to the conclusion that HIV would eventually spread like wildfire throughout the population. There were also powerful underlying forces at work here. For some AIDS researchers and journalists, as horrible as this disease was and is, it had also become very exciting. For researchers, it meant more research dollars, more papers published, possible prizes, a challenging crisis and biological problem to work on. For the journalists, AIDS meant more space in their publications and more recognition. Make no mistake about it, people in both camps were victims of the well-known psychological phenomenon that makes the adrenalin pump when you come upon a car accident and dictates the usual grisly murder and mayhem stories on the local news; part of their minds wanted to believe that the crisis would worsen, that the virus would spread to "the rest of us," although not to them personally.

This issue continues, too. Evidence has now begun to grow that HIV is not spreading rapidly into the mainstream heterosexual community, and more and more stories of late are taking that angle. Actually, the truth is, it is still too early to know the future course of this epidemic. But stories have begun to pop up that take the new data too far. *Cosmopolitan* magazine ran an essay by a New York doctor in its January 1988 issue that had as its basic message: if you're heterosexual and only engage in "ordinary sexual intercourse" with people who are not gay and do not use intravenous drugs, then you've nothing to worry about. AIDS experts lambasted the article and ABC's "Nightline" did a show on it.

The press fell down significantly in one other aspect of AIDS coverage, that involving the disproportionate percentage of black and Hispanic people who had AIDS or were infected. Although the Centers for Disease Control was reporting the numbers that made this plain as early as 1984, it was not a significant story until late 1986 and early 1987. Why? The chief reason is that most of the minorities affected were intravenous drug users. We were talking not about middle-class blacks or Hispanics, but down-and-out druggies, mostly in New York and other large cities. So there was no rush to focus on the angle that, in many inner-city areas, AIDS was a having a devastating impact. In truth, many of us in the media were more comfortable being sympathetic to the plight of well-educated gay men than we were to poor, inner-city drug abusers.

What lessons can be drawn from the successes and lapses in the media's coverage of AIDS? Foremost, in today's informational society, the media are an indispensable purveyor of health and medical information. Those of us in the media should perhaps take that role even more seriously than we do. AIDS has taught us that we can make a difference. The AIDS epidemic should also remind us once again that it is dangerous to second-guess science. As frustrating as it is to have to wait for answers, that is the nature of the beast. The temptation to extrapolate from insufficient data can seem harmless when it actually has serious negative consequences.

15

AIDS in Medical News

Marsha F. Goldsmith

Reporting on the pandemic of AIDS for *JAMA*, the *Journal of the American Medical Association*, is quite a bit different from covering the news for the public at large. The first *JAMA* Medical News story on AIDS appeared in September 1982, when the medical community was said to be "still baffled" by the 600 reported cases in this country, with more than 200 deaths, of the then newly named disease.

But the physicians for whom the Medical News reporter was writing about the startling new syndrome understood what she meant by an underlying defect in cellular immunity, peripheral blood lymphocytes, opportunistic infection, and so on. From the beginning, there was no need, as there was for lay publications, to resort to the euphemistic "avoidance of exchange of bodily fluids" and such. Although the early stories of necessity took a speculative approach, in unraveling the mystery of a new disease the writer could and did call a sperm a sperm.

It is surprising now to look back and see stories headlined, "Preventing AIDS Transmission: Should Blood Donors Be Screened?" and "'Contaminated' Plasma: No Automatic Recall." I believe those stories, and scientific papers subsequently published in the *Journal*, were important in alerting the medical community to the potential danger in blood products—and in compelling the removal of that hazard.

Similar examples abound. Medical News is the six-page section at the front of *JAMA*, the only part not written routinely by scientists. The six of us on the staff who travel to medical specialty meetings around the country and make endless phone calls to busy investigators attempt to give *Journal* readers an idea of ongoing developments in medicine. Time after time, Medical News stories have come first, followed by more detailed scientific explication.

Where AIDS is concerned, this task has taken on special urgency because the appetite for every bit of news as it is announced is so voracious and the need to communicate sometimes very complex information quickly and clearly—so as to enlighten without unduly frightening even sophisticated physicians—is so great. Sometimes, however, one gets a "kill the messenger" reception.

A case in point is the matter of increasing heterosexual transmission of HIV. When I wrote up Dr. Robert Redfield's presentation of this subject at the First International Conference on AIDS in 1985, some of our reviewers—Medical News stories are peer reviewed, like the rest of the *Journal*—demurred, not wanting to believe the findings. Redfield had convinced me, and I convinced them, so my story ran right away. Redfield's report appeared in *JAMA* 4 months later.

Physicians are afraid of AIDS, and rightly so. Although the behavior of the human immunodeficiency virus may be scientifically fascinating to contemplate, the thing kills you. So another aspect of the AIDS story we have been concerned with is how doctors ought to act in the face of this pandemic—and how they do act.

From the vantage point of Medical News, I have learned much about the way human beings individually and institutionally come to grips with unforeseen situations. I never expected to hear surgeons debate the ethics of operating on HIV antibody–positive patients, yet this became commonplace. And while surgeons and emergency department staff may be at the greatest risk because they have the most exposure to blood, one also hears from many others about the fear of accidentally contracting AIDS. For example, at a Bristol-Myers symposium, an investigator in a field far removed from AIDS research told me he finds it increasingly difficult to get laboratory technicians to do work involving blood samples.

Early on, I wrote about the problem of antibody tests and confidentiality. Gay spokesmen had expressed their concern at a meeting of the American Association of Physicians for Human Rights, and while some medical people wanted to downplay the need to deal with the question, *JAMA* editors agreed it was important. The doctor-patient relationship is a continuing theme of AIDS stories.

At the American Medical Association (AMA) AIDS education begins at home. Not only have materials been prepared in-house and distributed to the entire staff, but the AMA has prepared numerous pamphlets, books, and videotapes for practicing physicians to use in discussing AIDS with their patients or in giving talks in their communities. It also sponsors a number of education programs on AIDS prevention. Information that originated in Medical News stories has been incorporated into these materials.

The impact Medical News has is brought home to me repeatedly when doctors request reprints of my articles. I was surprised at the number of responses to a story that listed addresses and phone numbers for AIDS information and support groups—some public health, but most of them gay organizations—that ran in 1985. Many physicians simply had not known such information was available. And, in the spring of 1987, I think I received the most reprint requests for any story I ever wrote when *JAMA* ran my complete-with-user's manual contribution to "condom sense." The fact that this previously unmentionable information appeared in a respectable medical journal evidently made it okay to give it to patients.

I have been writing about AIDS for 6 years, and I could not have done it without unfailing cooperation from the scientists and clinicians themselves. It may not sound surprising that nobody says no to *JAMA*, but I have found the patience, good humor, and all-around helpfulness of hundreds of busy people to be outstanding.

And a good thing it is, too, because we are in this for the long haul. The best therapeutic agents tried so far only postpone the inevitable for a while, and add enormously to the cost of care. The word "breakthrough" is taboo at Medical News, but a breakthrough is what is needed. Only a so-far elusive understanding of the lethal link between the human immunodeficiency virus and the human immune system will enable us one day to "write '30'" to the AIDS story.

16

Thoughts on AIDS

Joseph F. Lovett

James Baldwin, the great American writer, said that he believed his duty was to "bear witness . . . to be a disturber of the peace . . . to wake the people sleeping." In a way, that has been the goal of most people involved in the AIDS epidemic: the patients, their clinicians, the researchers, and those of us who report on the epidemic. We all want "to wake the people sleeping" so that no one else becomes infected with this virus. But it's so difficult to wake the people sleeping. It's so damned frustrating.

I have been doing stories on AIDS for the ABC News magazine "20/20" since May of 1983. Had Larry Kramer, one of the founders of the Gay Men's Health Crisis and the author of "The Normal Heart," had his way, I would have been doing stories a lot earlier.

Larry lived across the street from me in New York, and we knew each other socially. I went to a benefit he had organized to raise money for what was then known as GRID—Gay-Related Immune Deficiency. There Larry asked me why "20/20" hadn't done a story on the disease.

I told him that I wasn't sure that GRID was a "national" story, but that I would bring the idea up at work. I gave him the names of some local New York television news people who I thought would have a better chance of getting the story done.

Larry took the names, but continued to hound me and everyone else he knew in the national press to get the story told. He knew 10, then 20, then 40 people with the disease. Many had died. He wanted people to know. He wanted something done. He wanted it to stop.

Basically, Larry singlehandledly brought the press' attention to the AIDS crisis.

"For Christ's sake," he would yell into the phone. "Why aren't you guys doing something?" he would ask, clearly mystified at how so many deaths were virtually being ignored by the press.

His tenacity was at times unbearable. But I never doubted his message. Clearly something horrible was happening. Happening quietly. And silence would allow it to continue.

The New York Native, New York City's gay newspaper, regularly published articles by Lawrence Mass, James D'Eramo, and Ann Giudice Fettner about AIDS. The articles were clear, humane, and disturbing. They showed the steady growth of the epidemic and the "knock on the door" mentality of so many people in the gay community. Yet in the general press, the few articles published were scientific in nature. Few described the emotional impact the disease had on its victims, their families, and their friends.

But print can't do that as well as television can. Television, at its best, brings people to other people. We can deliver facts, but we are at our best when a person can deliver the essence of those facts to an audience.

Within a few months, Larry and the statistics had me convinced that everyone needed to know what was happening. We had to "bear witness," "to wake the people sleeping."

My experiences trying to get our executive staff to approve a story on AIDS early in the epidemic were similar to those of other reporters. We were all met by the same responses:

"I really haven't heard too much about it."
"It's a very limited group of people who are affected, isn't it?"
"I don't know if our audience would be interested."
"I don't think it's right for our audience."

This should not be a surprise. There is a decided lack of empathy in the power structure of corporate America—including the news media—for the groups initially recognized as being affected by AIDS. In fact, this power structure includes very little black representation, and next to no gay male representation. And if a deadly infectious disease ever hits lesbians, they will all be wiped out before anyone in the power structure decides to take notice. That's how invisible they are.

It wasn't until AIDS hit the blood supply that the power structure finally had to take notice. In March 1983, 2 months after the blood industry's "joint statement" on the blood supply, I convinced our executive producer, A. V. Westin, to let me shoot one day of an International AIDS conference, held at New York University Medical Center. Drs. Linda Laubenstein and Alvin Friedman-Kien had organized the conference.

I edited that day's shooting into a 20-minute clip showing the dramatic concern of doctors from all over the world. In particular, the Europeans were amazed that more wasn't being done here in America where there were so many cases. I showed the clip to Geraldo Rivera, who at that time was a correspondent and a senior producer of the show. The footage spoke for itself, and within 2 months we had the first of 15 stories "20/20" has done on AIDS to date.

At the time our first story aired, the picture people had of AIDS was a thin young man lying in a hospital bed. Sometimes he had a few purple spots, sometimes not. We changed that. We introduced a young man of 26 named Ken Ramsaur. Our camera opened on a photo of Ken taken 8 months earlier, when his hair was full and dark and his thick moustache emphasized his dark, thoughtful eyes. The camera panned from that image to a wizened creature propped in a wheelchair, a man 40 pounds lighter, so totally disfigured from Kaposi's sarcoma that the handsome full eyes had been reduced to slits. His muscled arms were reduced to bone.

Ken wanted to tell his story. He wanted people to know. He wanted to wake the people sleeping. We were accused of exploiting . . . sensationalizing . . . capitalizing on the misfortunes of others. But we told a story that Ken wanted told. We told a story that no one wanted to hear. We told it in as real terms as we could. We tried to wake the people sleeping.

We showed communities of people who had scores of friends diagnosed as having AIDS every month. We showed how the Centers for Disease Control (CDC) was not getting an accurate count of the cases. We showed that federal leadership did not appreciate that at the time the number of cases was doubling every 6 months. And we showed how prevailing attitudes toward homosexuality had given this deadly epidemic a tremendous head start.

We heard that people paid attention. We heard that people watched the show in bars all the way from New York's Christopher Street to San Francisco's Castro Street, and that the mood was different when people went home that night.

Two years later, in 1985, when Jerry Falwell was helping matters by explaining how AIDS was God's punishment, we did a story called "AIDS in the Heartland." It was about Amy Sloan, a 24-year-old born-again Christian from Lafayette, Indiana. Amy found out that she had AIDS and was pregnant in the same week. She had gotten AIDS through a blood transfusion she received in 1982. If you remember, in 1982 the public was told that AIDS was an urban, gay disease: If you weren't both, it wasn't your concern.

Amy helped change the face of AIDS. The fact that this heterosexual woman living in Indiana had contracted the disease demonstrated that

AIDS was everyone's concern. Her husband, parents, physician, child, and friends showed how hugs and kisses and sharing a meal helped the person with AIDS and did not hurt his or her friends and family members.

In that same story we introduced a couple called "Jack" and "Diane." They had been seeing each other for about a year when Diane was diagnosed as having AIDS-related complex (ARC). Jack went to be tested for human immunodeficiency virus (HIV) and turned up positive. It came out that Jack was married to Mary. He also maintained a sexual relationship with Jim. Jim was married to Wendy. Diane had not been exclusive to Jack either. This web of relationships showed how the virus can spread.

Jack had stopped having sexual relations with his male lover, but was continuing to have sexual intercourse with his wife. No, he hadn't told her of his diagnosis. And, no, he wasn't using a condom.

"Couldn't it be said that you are condemning her by not using one?" Geraldo asked.

"If it's possible she isn't already infected, I guess you could say so," Jack replied quietly.

In that same story, Geraldo did a "stand up" in a drug store in Lafayette, Indiana, explaining how the use of a condom can offer protection against the AIDS virus. He recommended their use on a regular basis several months before the Surgeon General spoke out.

The next week, Barbara Walters held a baby with AIDS in her arms and explained to our audience that they couldn't get AIDS by holding, touching, or wiping a tear from a child with AIDS.

But do people listen to the messages of one or two prominent television journalists? I don't think so. Not when what they are being told means they have to change their behavior and their value systems so radically. Denial is so much easier. Denial was as strong in the gay community when AIDS first hit as it is among heterosexuals today. Why is that denial so strong?

AIDS is such a horrible concept . . . the idea of being condemned to death by a moment of pleasure . . . that it is only natural to try to keep it as far away as possible. Everyone wants to keep AIDS at arm's length, including some of the most informed members of our society—our scientific researchers. They tell us, "AIDS is different in the United States than in Africa. You can't extrapolate from African figures, where the cases are almost equally divided between men and women with most of them being heterosexual."

"Why can't we extrapolate from the African experience?" I've asked.

"Because so many Africans have genital ulcers!" I've been told. "This allows the virus easier access."

Yet, 25% of the American adult population have genital herpes. Forty million people! Surely that indicates some similarity to the African experience. Or is that too horrible an idea? If we can't (or should I say "won't") learn from the African or Haitian experience in terms of the changing epidemiology of AIDS, why bother to study any history or sociology at all? We must learn from others.

More denial exists. To protect the myth of the unthreatened heterosexual, government spokesmen have said that heterosexually spread cases make up the same percentage of the case load that they always have. This is not reassuring. The actual numbers are growing. Heterosexually spread AIDS cases are doubling every 6.6 months. Homosexually spread AIDS cases are doubling every 13 months. How is it that AIDS is "spreading" among gays, but not among straights? Why do we continue to deny something so obvious?

Dr. Donald Francis of the CDC once observed that in the gay community it took knowing three people with AIDS to fully appreciate the reality of the disease. But if we wait until every middle-class heterosexual personally knows a few people with the disease, where will we be?

No one who is not already infected with HIV need become infected tonight. No one. This is true as long as people are convinced of the necessity of taking responsibility for their personal health, as long as they are educated as to how to keep the virus from entering their systems. It's so simple, and so important, that if we do not commit to getting the message of condoms and nonoxynol-9, and unshared needles, across to the public, we are basically contributing to genocide in this country. It's easy to avoid AIDS. It's so simple. So why are we having such a difficult time getting the message across?

At the beginning of the epidemic, people were told that AIDS was a disease that affected homosexual men, bisexual men, intravenous drug users, hemophiliacs, and Haitians. They were told, "No need to worry." Then they were told, "No need to worry, *yet*." Now the Surgeon General is recommending condoms while the editorial pages of the papers say "Most people are at low risk." No wonder the public is confused.

In the beginning, the "experts"—the government health workers and the clincians—saw the possibility of panic among the "general population." A very serious mistake was made at that time: The experts chose to *allay* fears rather than to *address* fears. They said, "You don't need tc be afraid." Period. They could have said, "This is a very serious disease. This is how you get it. This is how you can avoid getting it." Had they addressed people's fears and given them the tools to deal rationally with those fears, they would not be faced with such a confused public today.

It's no wonder people are confused. We, the press, continue to confuse them by quoting you, the scientists. Let's be frank. The terminology of this epidemic is ridiculous. Let's stop using terms like "low risk" and "high risk" unless we are talking about Russian roulette. We do not advocate playing Russian roulette with a gun with six chambers. That's too high a risk. Would we recommend playing Russian roulette with a gun with 1,000 chambers? Of course not; it only takes one bullet to kill you. There is no "low-risk" Russian roulette. There is no "low risk" of HIV transmission when there are condoms and nonoxynol-9 available. "Risk" is confusing to the public.

Would it be possible to dump the definition for AIDS? Could we get rid of "HIV" infection? Could we say that infection with the "AIDS virus" can be deadly because the longer the virus is in your system, the more damage it can do? Wouldn't that be clearer and more to the point than arcane terms like "AIDS," "ARC," and "HIV," which need a definition every single time they appear in print or on the air? We have to demystify this disease so that people can see it for what it is. AIDS is the result of coming into contact with a virus. That contact can be avoided. Everyone must know that.

Clinicians can help journalists help the public. Let your patients talk to us. So often doctors have told me that to tell a patient that "20/20" is looking for someone to tell his or her story would be breaking confidentiality. Yet when we are looking for kidney transplant candidates or arthritis sufferers, there never seems to be such a problem.

When physicians start to get over their own hang-ups with the disease and allow their patients the option of talking with journalists, more people will know people with AIDS as just people. Unless we can show real people with real faces who have this disease, the stigmatization that afflicts patients will continue.

Friends and colleagues of mine often ask, "Don't you hate doing these stories? Aren't they depressing?" I don't hate doing the stories. I hate the disease. I don't find doing the stories depressing. I find the people I meet uplifting.

But, more than that, these stories give me the illusion of having some power against this disease. Working on AIDS stories makes me feel, in my own naive way, that I am having an effect in some way, that I am fighting back. But it's very frustrating as we watch these numbers grow and grow. It's sad to know that the brave people who have spoken with us in their living rooms are now dead. It's even sadder to think that more will be needlessly infected.

Television can help make the threat of AIDS a reality by introducing people with AIDS to others. People like Kenny, who, disfigured by Kaposi's sarcoma, wanted his story told. Or Amy, the young mother in her 20s who got AIDS from a blood transfusion in Indiana in 1982. Or Bob, Mark, and Bill, who had to go to Paris for any hope of treatment. Or Tom, who started to treat himself, impatient with the slow government trials. Or Mair and Betty, two sisters who were both infected by the same man. Or Debbie, an intravenous drug user who didn't think she was at risk because she had never shared a needle with a gay man. All of these people have told their stories to the press to help others understand how important it is not to put yourself in contact with the AIDS virus.

Will the press continue to be the main purveyor of AIDS education in this country? I don't think so. We cannot depend upon the press for our information because the press is a fickle lover. According to your newspapers and evening news, there is no more war in Ireland, the hostages are no longer held in Lebanon, and Mrs. Aquino has everything under control.

Summer months are filled with AIDS stories because of the international conferences. Will they keep up in the winter? Today we are hearing about AIDS, but will we tomorrow? The business of the news is that which is new. Will AIDS become commonplace? If it does, will it be treated like the weather and the traffic, with regular reporting? Or will it become the occasional news item upon the hospitalization of a celebrity or when one more vaccine trial is begun?

As individuals we must do everything we can to alert the uninfected so that they can remain that way. We must elect officials who understand that the agony this disease has spawned is only the seed of what is to come upon our nation unless the spread is stopped. We must encourage corporate heads to wage war on AIDS as they have waged war on illiteracy and drug and alcohol addiction. They must be prepared for the effects of the disease on their employees and their corporate strength, both financially and emotionally.

We must teach people that fears of mosquitoes and schoolchildren and neighbors with AIDS are misplaced. As we look with fear on these "will-o'-the-wisps," we cannot see the quagmire that will trap us: our unwillingness to take responsibility for our own protection. Laws that make it a crime for a person with the AIDS virus to have sex with another person keep people from realizing that both members of a sexual relationship must be responsible—both to themselves and to their partner.

We must talk about AIDS openly and frequently among our friends, our families, and our colleagues so that no one will be unaware of the risk of unprotected sex or shared needles . . . so that no one who is not infected with the virus today will get it tonight because of ignorance.

We must "wake the people sleeping."

VI
MODES OF
TRANSMISSION

17

Heterosexual Transmission of Human Immunodeficiency Virus from Intravenous Drug Users: Regular Partnerships and Prostitution

Don C. Des Jarlais, Samuel R. Friedman, Douglas Goldsmith, and William Hopkins

Intravenous (IV) drug users are the largest group of heterosexuals to have developed AIDS in the United States and Europe. They have also become the predominant source for transmission of human immuno-deficiency virus (HIV) to heterosexuals who do not inject drugs and of perinatal transmission (Des Jarlais & Friedman, 1987). To a great extent, the future course of the AIDS epidemic in the United States and Europe will depend upon the rate of spread of HIV from IV drug users to non–drug-injecting heterosexual partners. The worst case scenario is that IV drug users will form a large reservoir of HIV infection with little sexual risk reduction among themselves and their non–drug-using sexual partners, and eventually create the potential for self-sustaining heterosexual transmission.

In this chapter, we will review our current research on the potential for heterosexual transmission from IV drug users, including potential transmission from IV drug–using prostitutes. Prior to AIDS, there was some research on the sexual behavior of IV drug users, including those engaged in prostitution. The extent of the research, however, was not sufficient to quantify the behavioral data needed in models of hetero-sexual transmission, and there was almost a total absence of data on how the sexual behaviors of IV drug users might be modified to reduce sexually transmitted diseases. We want to emphasize that much of the research that we will be reviewing is ongoing and that the findings must in many cases be considered preliminary. Observations made without references are based on our current ethnographic studies in New York City.

Basic Demographics

There are two demographic factors that greatly influence heterosexual behavior among IV drug users. First, IV drug use in the United States is heavily concentrated in minority populations that have traditionally suffered from a wide variety of health, social, and economic problems. Eighty percent of the cases of AIDS among IV drug users have been blacks or Latinos, as have been 90% of the heterosexual transmission cases (Centers for Disease Control, personal communication, 1987). There is great diversity among the black and Latino communities in the United States, as has been emphasized by many others in this volume, and we do not want to ignore the significance of this diversity for sexual behaviors that can affect the transmission of AIDS. Nevertheless, minority communities that have high rates of IV drug use generally have very high rates of unemployment among their youth. This unemployment, along with other factors, has served to undermine the stability of the traditional married father-mother-child family unit (Wilson, 1987).

The second basic demographic factor is that the great majority—approximately 75%—of IV drug users in the United States are males. High male-to-female sex ratios within a group are usually associated with males having sexual relationships outside of the group and with males utilizing prostitutes as a sexual outlet. Among IV drug users, the great majority (approximately three quarters) of males have their primary sexual relationships with women who do not inject drugs, and the number of females who do not inject drugs but are regular sexual partners of male IV drug users is at least half as large as the number of IV drug users (Des Jarlais, Chamberland, Yancovitz, Weinberg & Friedman, 1984). (A portion of female IV drug users also have regular male sexual partners who do not inject drugs, but the currently available data are too scarce and inconsistent to estimate the percentage.)

Effects of Drug Use on Sexual Activity

The basic biology of human beings seems to include a strong potential for experiencing sexual activity as pleasurable. IV drug use subcultures in the United States strongly value intense drug-induced pleasure, and sexual pleasures are valued within the same framework. When IV heroin users try to explain the quality of injecting very good heroin to persons who have never injected, the most commonly made analogy is that the heroin rush is "like having an orgasm all through your body."

Noninjecting sexual partners of IV drug users also value sexual pleasure, and expect the IV drug user to perform sexually. This performance is often in exchange for providing food, shelter, or small amounts of money (see Johnson et al., 1985, for examples of these exchanges). It is important to note that neither party defines such exchanges as "prostitution." These exchanges are seen as part of a committed relationship (discussed below) and, in the words of one of our street researchers, it is a "c'mon, honey" expectation that the male will provide the female with sexual pleasure.

The physiological effects of drugs on libido and sexual functioning are complex, and comprehensively reviewing them is beyond the scope of this paper. It is important, however, to discuss some of the interactions between drug use, sexual pleasure, and sexual performance. For the purposes of this discussion, an important distinction is between the "functional" IV drug user and the "low" IV drug user (these terms come from our ongoing ethnographic studies in New York).

"Functional" IV drug users are capable of earning more money than is needed for their drug use, although the activities for earning money may be illicit. They avoid most problems with the law, may keep drug use hidden from the public at large and even from significant others, and try to "look sharp." For these IV drug users, if sex while "straight" is good, then sex while stoned is likely to be even better. The fact of enjoyment of "stoned" sex is not dependent upon the pharmacological effects of any specific drug, but upon the psychological appreciation of the differences between the straight and the various stoned states. Alcohol effects may be perceived as relaxing, marijuana effects may be perceived as expanding time perception, cocaine effects as a generalized sexual arousal, and heroin (for males) as delaying ejaculation and prolonging intercourse. Thus, despite the wide variety of effects from both injected and noninjected drugs, drug subcultures have been able to emphasize some aspects of each drug to enhance sexual pleasure.

The additive qualities of sex and being stoned will therefore lead to a tendency to have sex while stoned whenever the sexual opportunity is readily available. This is linked to considerable sexual activity outside of committed relationships. It may also lead to a disinhibition effect against safer sex practices when stoned, although data on this effect among IV drug users are not yet available.

For the "low" IV drug users, however, too much drugs overwhelms their ability to cope with life. These drug users are likely to be physiologically addicted to heroin, cocaine, alcohol, or some combination of these. They are often desperate for money to obtain drugs, taking very large risks to obtain both. Their distressed state is believed to be reflected in their physical appearance—unkempt, scruffy, with torn and dirty clothes, and often with open sores or abscesses. For heavily addicted, debilitated drug users, the drug use replaces sexual pleasures. The combination of the intensive drug use and the physical deterioration leads to a loss of libido and, for many males, a loss of the ability to perform sexually.

IV Drug Use and Long-Term Relationships

While there is a strong hedonistic component in the approach to sex among functional IV drug users that leads to many casual relationships, there is also a strong tendency to form long-lasting sexual relationships. Contrary to common perceptions, the life of an IV drug user is not rapid alteration between being stoned and being in withdrawal, nor a continuous decline from seeking sex for pleasure to being too strung out to enjoy sex. For many days and months there will be no drug injections, and for many other days and months the drug use will be modulated so that the user spends most of his or her time feeling normal. While drug injection may be a recurring predominant theme in the person's life, it is not likely to be the only interest. They will have concerns for companionship, commitment, and having a family, even though drug use will often interfere with fulfilling these needs.

The great majority of IV drug users develop relatively stable relationships with a single partner. In one study we conducted, male IV drug users entering treatment were asked about their "most recent primary sexual relationship" (Des Jarlais et al., 1984). Over 90% stated that they had a stable sexual relationship, and the average duration of these relationships was over 5 years. These relationships between IV drug users, or more frequently between a male IV drug user and a female who has never injected drugs, are not trouble free—they suffer from economic difficulties as well as the considerable difficulties associated with IV drug use. The two partners often do not live together, and mutual monogamy is rare. A common pattern is for the male IV drug user to spend several nights per week with the noninjecting female partner, with her providing food and shelter and he occasionally bringing gifts. Despite these problematic issues, the relationships typically persist for long periods of time.

These relationships also frequently involve children. In one study of IV drug users in treatment, they averaged two children apiece (Deren, 1985). In addition, one quarter of the subjects reported that they intended to have additional children, and another one quarter were "not sure" if they wanted to have additional children. While IV drug users often intend to have children, and often are not certain whether they want additional children, deliberate planning for childbearing is rather rare. Much more common is a simple lack of any contraception.

Despite common perceptions to the contrary, the interpersonal commitment aspect of sexual relations is a strong part of sexual activity among IV drug users, and between IV drug users and persons who do not inject drugs. It is these long-term relationships that have the greatest potential for heterosexual transmission of HIV.

Table 17-1
Cumulative Number of Cases of AIDS
in New York City[a]

Year	Heterosexual IV Drug User	Heterosexual Partners	Ratio
1980	6	0	—
1981	26	1	25.0:1
1982	136	9	15.1:1
1983	390	25	15.6:1
1984	881	55	16.0:1
1985	1682	117	14.4:1
1986	2719	214	12.7:1
1987	3997	318	12.6:1

[a] These cases represent "domestic" heterosexual transmission of HIV. They do not include immigrants from countries where heterosexual transmission is believed to be "the dominant source of HIV transmission," since those cases may represent transmission that occurred prior to immigration.

AIDS Among the Sexual Partners of IV Drug Users

Table 17-1 shows the cumulative numbers of cases of AIDS among heterosexual IV drug users and among "heterosexual contacts" in New York City from 1980 through 1987 (New York City Department of Health, personal communication, 1987). The "heterosexual contact" cases are those in which the index cases report no known risk factor for HIV infection other than having heterosexual activity with a person with AIDS or with a person known to be at high risk for AIDS (Table 17-1).

In 88% of the 318 heterosexual contact cases, an IV drug user was reported as the likely source of HIV infection. Despite the presumably large number of HIV-infected bisexual men in New York, IV drug users account for almost all of the domestic heterosexual transmission cases. All but six of these heterosexual contact cases have been male-to-female transmission. The six cases with female-to-male transmission have all been from a female IV drug user.

Long-term sexual relationships between IV drug users, and those between IV drug users and persons who do not inject drugs, have traditionally been subject to problems, many but not all of which were associated with the drug use itself. Despite these problems these relationships usually lasted for relatively long periods of time and often produced children. With the AIDS epidemic, heterosexual and perinatal transmission of an often fatal virus has been added to these problems.

In agreement with common perceptions, prostitution occurs frequently among IV drug users, although the actual behavior patterns may vary greatly from the media image of the prostitute driven by her addiction. Both male and female IV drug users may be involved in the exchange of sexual activity for money. Male IV drug users may participate in these sexual exchanges with other men while retaining a self-definition as being "straight" (heterosexual in orientation). Maintaining this self-definition is usually dependent upon accepting money for the sexual activity—so that the IV drug user can define it as exploitation of the customer—and sometimes by restricting the particular sexual activities, usually to oral intercourse with the IV drug user as the insertive partner.

Female IV drug users will frequently engage in prostitution in order to obtain money for drugs. The common stereotype is of the female who becomes addicted and then must turn to prostitution in order to support her IV drug habit. This does occur, but is not the only relationship between prostitution and IV drug use. Female prostitutes will also often use drugs as a way of spending profits from their trade or as a form of self-medication for their stressful life-style. Drug use is part of the environment in which much prostitution occurs. Thus a prostitute who does not inject drugs will come to know persons who do, and be offered opportunities to inject. Drug use does provide a way of spending money that provides intense and immediate gratification. It does not require the stable personal situation that would make delayed gratification consumption a rational economic decision.

A final aspect of the commercial sexual activity among IV drug users that should be noted is the frequent use of prostitutes by male IV drug users. The great majority of IV drug users are males, so they tend to have their stable sexual relationships with females who do not inject drugs. Male IV drug users will often desire casual sex for pleasure, without the complications of a long-term relationship; and the long-term relationship may also be in a state of trouble in which sexual activity is not occurring. In these instances, a male IV drug user will frequently turn to a prostitute for sex. If the female is also a drug user the exchange may be sex for drugs rather than sex for money. In these circumstances, there is the possibility that the exchange will be negotiated and defined as simple sharing of pleasures among friends rather than as prostitution.

AIDS Among the Customers of IV Drug–Using Prostitutes

New York City would seem to have a very large potential for transmission of HIV from prostitutes to their customers. The virus has been present among IV drug users since at least 1977 (Des Jarlais et al., 1987).

Studies of street prostitutes indicate that about half of them have histories of injecting drugs (Des Jarlais et al., 1987; Goldstein, 1979). (Workers in other forms of prostitution are likely to use illicit drugs, but are much less likely to inject drugs [Goldstein, 1979].) The studies of HIV infection among IV drug–using prostitutes in New York City indicate that at least half of them had been infected with HIV by 1985 (Des Jarlais et al., 1987; Schoenbaum et al., 1986; Wallace, Christonikos, Marlink, Guroff, & Weiss, 1986).

Despite this large potential for transmission of HIV from IV drug–using prostitutes to their customers, the amount of transmission has been relatively little. Studies of HIV infection among clients of sexually transmitted disease clinics do not find use of prostitutes to be a statistically significant risk factor for HIV infection (Chaisson, Fleisher, Petrus, & Miller, 1987; Marmor et al., 1987; Rabkin, Thomas, Jaffe & Schultz, 1986). Of the first 12,180 cases of AIDS in New York City, only 22 occurred in men who reported contact with prostitutes as their only known risk behavior (Des Jarlais et al., 1987). (Because the man is not likely to know whether or not the prostitute injects drugs, these cases are recorded as "no identified risk" in the city.) This compares to 777 cases of AIDS among female IV drug users by the same date. A best estimate is that a third of these female IV drug users (259) have practiced prostitution (Ginzberg et al., 1986). Thus, the feared scenario of many IV drug–using, HIV-infected prostitutes transmitting the virus to large numbers of customers does not appear to reflect reality.

One important reason for the currently low rate of transmission from IV drug–using prostitutes to their customers in New York City may be the frequent practice of safer sex within these exchanges. Over 90% of interviewed street prostitutes report that they "regularly" have their customers use condoms, and at least charge an extra fee if the customer does not want to use a condom (Des Jarlais et al., 1987). This use of condoms preceded concern about AIDS; it originated from concern about the economic consequences of contracting other sexually transmitted diseases. In addition, oral sex is by far the most common activity among street prostitutes and their customers in New York. Oral sex offers many practical advantages: it is relatively fast, does not require disrobing, and can be done in a truck cab, a car, or a doorway. Oral sex also appears to pose a very low risk of HIV transmission to the insertive partner, as Dr. Detels and Dr. Stevens report in their chapters (1 and 2, respectively) in this volume (also Detels et al., 1983).

While potential transmission of HIV from IV drug–using prostitutes will require further monitoring, the New York City experience to date suggests that this transmission is very infrequent, at least if there is a high practice of safer sex techniques by the prostitutes and their customers.

AIDS-Related Changes in Sexual Behavior Among IV Drug Users

Concern about heterosexual transmission of AIDS has been increasing among IV drug users in New York City over the last several years. IV drug users were concerned about AIDS as early as 1983, learning about it primarily through the mass media and the oral communication networks of the IV drug use subcultures. The concern about AIDS has grown with the still-increasing number of cases of AIDS among heterosexual IV drug users and among their non–drug-injecting heterosexual partners. Knowledge of the heterosexual transmission of HIV has led to some sexual behavior changes among IV drug users.

One of the more important methods of spreading information about AIDS transmission, and of symbolizing the need for behavior change, has been the free distribution of condoms by drug abuse treatment programs in the city. In many methadone maintenance treatment program facilities, boxes of condoms are placed near the door, and anyone leaving the facility can simply take as many as he or she wants. A substantial number of the condoms are taken from the methadone programs, and many of these are sold in the streets. The methadone treatment staff are aware that many of the condoms are being resold, and do not object to this.

IV drug users in the city are now relatively well supplied with condoms. Many of the male IV drug users now carry condoms. The actual use of the condoms, however, depends greatly on the meaning of the sexual activity. The greatest changes have been reported in the areas of sex as pleasure and sex as commerce. Drug users have been reducing their numbers of casual sexual partners and increasing their use of condoms with casual partners. Male IV drug users have also changed somewhat in their use of prostitutes. They are more likely to use condoms when with a prostitute, more likely to select a "regular" prostitute for patronage rather than seeing many different prostitutes, and more likely to insist that the prostitute they use look "clean," well dressed, and in apparent good health.

It is when sexual activity implies long-term commitment that we have observed the least amount of AIDS-related behavior change. It appears very difficult to practice safer sex within such a relationship. These long-term relationships have at least an implication of mutual monogamy. Use of condoms implies that one or both partners are not being faithful. These relationships are also characterized by some degree of mutual interdependence and conflict over IV drug use and other issues. The threat of also contracting AIDS may be too much to bear for many of these relationships, with the breakup of the relationship having significant costs for both partners.

Additional Special Issues for Future Concern

IV drug users are already the predominant source of heterosexual transmission of HIV in the United States and Europe. Up to now this transmission has been a gradual increase rather than the explosive spread that occurred through homosexual activity or through the sharing of drug injection equipment during the early part of the HIV epidemic. The most likely future spread of HIV from IV drug users will be a continuation of this gradual spread. There are several specific factors, however, that might lead to a more rapid heterosexual spread of HIV from IV drug users. The first factor is the potential that currently seropositive IV drug users may become more infectious as their HIV-related immunosuppression increases. There are preliminary data indicating that HIV seropositive hemophiliacs become more infectious as their immunosuppression increases (Goedert, Eyster, & Biggar, 1987). Since there are already a large number of asymptomatic HIV-seropositive IV drug users in the United States and Europe, with difficulties in adopting consistent safer sex practices by IV drug users, there is a clear potential for greater heterosexual transmission if infectiousness should increase with increasing immunosuppression.

A second factor for concern is a potential relationship between the use of "crack" and heterosexual transmission of HIV. Crack is a potent form of cocaine that can be vaporized and then inhaled. It produces a very intense but short-lived euphoria—roughly equivalent to injecting cocaine—and can produce rapid dependence. Although the smoking of crack in itself does not lead to transmission of HIV, crack houses frequently become sites for prostitution; the males smoking crack may interpret the crack high as increasing their desire for sexual activity and female crack users may be quite willing to exchange sex for additional crack. It is unknown whether crack house prostitution is more or less conducive to practicing safer sex; recently however, there has been an increase in heterosexual cases of syphilis in many large cities in the United States (Centers for Disease Control, personal communication, 1987) that may be related to crack house prostitution. (We are currently conducting research on this issue.) There is also the potential that syphilis might increase the infectivity of persons who have been infected with HIV (Curran et al., 1985), which could amplify the transmission of HIV.

A final concern about future heterosexual spread of HIV from IV drug users regards the sexual abuse/exploitation of young females by older males. With the weakening of the traditional family structure in many of the ethnic minority, low socioeconomic status, high drug use neighborhoods, a man living with a woman and children is likely to be a current boyfriend of the woman and not the biological father of the children. In this situation, if there is a young adolescent daughter in the family, the mother's boyfriend may initiate the daughter into sexual activity. We do not yet have enough data on how often this occurs, whether force and/or seduction is used, or the later consequences for the daughter. Information on this type of behavior is often hidden behind guilt, embarrassment, psychic pain, confusion of sexual identity, and anger. We have, however, found examples of these events from therapy groups with IV drug users, and recently from AIDS prevention efforts.

The AIDS prevention example is worth presenting. One of our ex-addict street outreach workers was working with a group of female IV drug–using prostitutes. He was encouraging them to practice safter sex with their customers and to come in for counseling and possible HIV antibody testing. One of the members agreed to come in for counseling on the condition that the outreach worker arrange to have a social worker visit her mother and younger sister in an effort to stop the sexual abuse of her sister. It was her own sexual relationship with this man that led the female IV drug user to leave home. In further discussion with this group of 12 IV drug–using prostitutes of which this woman was a member, 8 of the women reported that they had also been sexually abused by a mother's boyfriend. They stated that, in their opinion, if they had been exposed to HIV, it was through this sexual abuse rather than by their own IV drug use or activities as prostitutes.

We want to emphasize that we do not yet have sufficient data on the frequency of this type of sexual behavior or any causal data on whether it leads to IV drug use and/or prostitution. We realize that collecting data on this topic will not be easy. The highly private nature of the events make them difficult to discuss, and their emotionally charged nature requires that any investigator be able to provide some form of assistance for the emotional distress that may be provoked by discussing these types of events.

Despite the great uncertainties regarding this type of sexual abuse/ exploitation, we believe that it has considerable potential relevance to the prevention of the spread of AIDS through heterosexual transmission. Clearly this type of situation is almost the opposite of the standard situation of two consenting adults rationally discussing AIDS, with the woman refusing to have sex unless the man agrees to use a condom. The standard AIDS educational materials may not be applicable to sexual situations in which the woman has little or no power, and the behaviors are kept highly secret.

Summary

IV drug users have become the predominant source for heterosexual transmission of HIV in the United States. The heterosexual behavior of IV drug users is complex, and often does not conform to popular stereotypes of drug users. Functional IV drug users value sexual pleasure and will frequently engage in sexual activity both while using drugs and while not using drugs. Intense drug use—to the point of addiction and being "strung out"—will usually lead to a reduction in sexual activity. Both male and female IV drug users will engage in prostitution to obtain funds for drugs. Contrary to popular fears of prostitution-linked HIV transmission, there has been relatively little such transmission in the areas where it would be most likely to have occurred. This may be because of safer sex practices, such as condom use and oral sex.

IV drug users also form relatively permanent sexual relationships that typically involve a male IV drug user and a female who does not inject drugs. It is within these long-term relationships that heterosexual transmission of HIV is most likely. IV drug users frequently have children because they actively desire children and, to a lesser extent, because they do not consistently practice contraception.

Concerns about AIDS have led many IV drug users to change their sexual practices, including a reduction in the number of partners and increased use of condoms. Most of the AIDS risk reduction appears to be with casual partners, however, and not within the long-term relationships where heterosexual transmission is most likely.

The rate of heterosexual HIV transmission to date has been slow in comparison to the very rapid transmission that occurred with male homosexual activity and through the sharing of drug injection equipment. Factors that might lead to a rapid increase in heterosexual transmission in the future include infectiousness increasing with progressive immunosuppression, crack use associated with frequent unsafe sex, and prior exposure to syphilis.

Acknowledgment

Preparation of this chapter was supported by grant RO1 DA03574 from the National Institute on Drug Abuse.

References

Chaisson, M. A., Fleisher, E., Petrus, D., & Miller, B. (1987, June). *Epidemiologic characteristics of women with AIDS in New York City* [abstract]. The Third International Conference on AIDS, Washington, DC.

Curran, J. W., Morgan, W. M., Hardy, A. M., Jaffe, H. W., Darrow, W. W., & Dowdle, W. R. (1985). The epidemiology of AIDS: Current status and future prospects. *Science, 229,* 1352–1357.

Deren, S. (1985). *A description of methadone maintenance patients and their children.* New York: New York State Division of Substance Abuse Services.

Des Jarlais, D. C., Chamberland, M. E., Yancovitz, S. R., Weinberg, P., & Friedman, S. R. (1984). Heterosexual partners: A large risk group for AIDS. *Lancet, 1,* 1346–1347.

Des Jarlais, D. C., & Friedman, S. R. (1987). HIV infection among intravenous drug users: Epidemiology and risk reduction [review]. *AIDS, 1,* 67–76.

Des Jarlais, D. C., Wish, E., Friedman, S. R., Stoneburner, R., Yancovitz, S., Mildvan, D., El-Sadr, W., Brady, E., & Cuadrado, M. (1987). Intravenous drug use and heterosexual transmission of immunodeficiency virus: Current trends in New York City. *New York State Journal of Medicine, 87,* 283–285.

Detels, R., Fahey, J. I., Schwartz, K., Greene, R. S., Visscher, B. R., & Gottlieb, M. S. (1983). Relation between sexual practices and T-cell subsets in homosexually active men. *Lancet, 1,* 609–611.

Ginzberg, H. M., French, J., Jackson, J., Hartsock, P. I., MacDonald, P. I., & Weiss, S. H. (1986). Health education and knowledge assessment of HTLV-III diseases among intravenous drug users. *Health Education Quarterly, 13*(4), 373–382.

Goedert, J. J., Eyster, M. E., & Biggar, R. J. (1987, June). *Heterosexual transmission of human immunodeficiency virus (HIV): Association with severe T4 cell depletion in male hemophiliacs* [abstract]. The Third International Conference on AIDS, Washington DC.

Goldstein, P. J. (1979). *Prostitution and drugs.* Lexington, MA: Lexington Books.

Johnson, B. D., Goldstein, P. J., Preble, E., Schmeidler, J., Spunt, B., & Miller, T. (1985). *Taking care of business: The economics of crime by heroin abusers.* Lexington, MA: D.C. Health and Company.

Marmor, M., Sanchez, M., Krasinski, K., Cohen, H., Bartelme, S., Weis, L. R., et al. (1987, June). *Risk factors for human immunodeficiency virus (HIV) infection among heterosexuals in New York City* [abstract]. The Third International Conference on AIDS, Washington DC.

Rabkin, C. S., Thomas, P. A., Jaffe, H. W., & Schultz, S. (1986). Prevalence of antibody to HTLV-III in a population attending a sexually transmitted disease clinic. *Journal of the American Medical Association, 255,* 2167–2172.

Schoenbaum, E. E., Selwyn, P. A., Klein, R. S., Rogers, M. F., Freeman, K., Friedland, G. H., et al. (1986, June). *Prevalence of and risk factors associated with HTLV-III/LAV antibodies among intravenous drug abusers in methadone patients in New York City* [abstract]. The Second International Conference on AIDS, Paris.

Wallace, J. I., Christonikos, A. B., Marlink, R., Guroff, M. R., & Weiss, S. (1986, June). *HTLV-III/LAV exposure in New York City prostitutes* [abstract]. The Second International Conference on AIDS, Paris.

Wilson, W. J. (1987). *The truly disadvantaged.* Chicago: University of Chicago Press.

18

Oral Sex: A Critical Overview

William Simon, Dianne M. Kraft, and Howard B. Kaplan

In modern sex research, there has been a nearly exclusive concern for coitus. Thus, until recently, Herdt (1984) could properly observe that outside of reproductively significant behaviors "the question of who gives and receives sperm is not one that much concerns Westerners today" (p. 167). However, the role of semen as a mode of transport for AIDS has now created substantial interest in just this question (Kaplan et al., 1987). The initial goal of this chapter, then, will be to provide a descriptive overview of the prevalence of oral-genital sexual contacts. We will attempt to draw, however tentatively, appropriate conclusions about the possible meanings of the patterns that emerge. Some of these may speak rather directly to the broadest issues of how we think about the sexual, which has potential relevance as to how the AIDS crisis will be interwoven into the fabric of sexual consciousness and behavior. The final goal will be to begin—but by no means exhaust—an examination of the emergent patterns and appropriate conclusions in terms of oral-genital contacts both as a source of risk and as a potential alternative to forms of sexual activity that pose higher or less manageable risks.

Biological Perspective

Alfred Kinsey, the first to provide contemporary behavioral sciences with large-scale, empirically based data on human sexual behavior, was, as Robinson (1976) observed, among the most important figures in the achievement of the "modernization of sex." The modernization of sex can also be conceptualized as the "naturalization of sex" or the attempt to bring human sexuality within the conceptual jurisdiction of the biological sciences (Simon, 1989). One consequence of this attempt to naturalize the sexual was the scanning of the behaviors of other mammalian species in order to find legitimating analogies between those and human sexual behaviors.

It is in this vein that Kinsey sought and found many such analogies, but these were essentially instances that provided only the most limited sense of possible distributions of such behaviors. As a result Kinsey's very consideration of oral-genital behaviors was colored by his assumption of its inherent "naturalness"; this is evident in one of his most sweeping generalizations:

> No legislation or social taboos have been able to eliminate them [masturbation, oral-genital behavior, homosexuality] from the history of the human animal. (Kinsey, Pomeroy, & Martin, 1948, p. 578)

Similarly, in commenting upon the fact that oral-genital behaviors appeared to be strongly correlated in a positive manner with levels of attained education, he noted:

> Once again, it is the upper [educational] level which first reverted, through a considerable sophistication, to *behavior which is biologically natural and basic.* (Kinsey et al., 1948, p. 369; our emphasis)

Our own provisional examination of available literature suggests that this position at best represents an excessive and potentially misleading overgeneralization. This literature suggests that oral-genital behaviors play, at best, a marginal role in the sexual lives of other mammals, including the higher primates (Goodall, 1971). Among the latter, oral-genital activity is rarely present as source of sexual excitement or as an alternative to other behaviors associated with reproduction or dominance. Thus, when using mammalian behavior as criteria, it cannot be said that oral-genital activity is either "basic" or "natural" in the sense that it might be strongly potentiated by elemental drives or sensitive tissues that respond as quasigenital areas (Dewsbury, 1981).

The Anthropological Perspective

Oral-genital activity decidedly tends to be a human pastime. However, more surprising than its presence is how relatively infrequently it appears in either heterosexual or homosexual guises. In some small number of cultures, oral-genital acts appear as a ritual or "nutritional" requirement and then the experience is exclusively "homosexual." It usually requires age and role criteria that are rigorously applied to the participants. It rarely appears to compete with or act as an alternative to heterosexual requirements, but is seen by those cultures as a developmental stage or requirement. As stage-specific behavior it is expected to be experienced and then relinquished, and this is what occurs with only the rarest of exceptions.

The Sambians of New Guinea represent one cultural setting in which fellatio has been extensively reported and where it appears to have a great deal of significance (Herdt, 1981, 1984; Stoller, 1985). Sambians practice obligatory fellatio between prepubescent male children (recipients) and postpubescent unmarried males (donors) and between husbands and wives, particularly as the initial form of marital intercourse. And while emotions clearly generate sexual arousal, they appear to be emotions that "fetishize" semen and not the penis as such; semen attests to one's effective masculinity, just as its absence (in prepubescent males and all females) confirms the "dangers" and "inferiority" of the female.

Even where forms of behavior that might be labeled "homosexual" occur, oral-genital activity neither is found universally nor, where it is observed, is the preferred mode of sexual interaction or the preferred mode of realizing sexual pleasure or orgasm. Gebhard (1971) observed: "Anal intercourse is the usual technique employed by male homosexuals in preliterate societies" (p. 215). Davenport's (1965) description of a Melanesian group reinforces this view: as paraphrased by Karlen, "Sodomy is their only homosexual activity; all men deny fellatio" (1971). Even more infrequently do we find references to or cultural scenarios suggesting the presence of either heterosexual or homosexual cunnilingus.

A number of cultures contain representations of the sexually explicit or, more appropriately, erotically explicit, such as the *Kama Sutra*, *The Perfumed Garden*, or the tradition of the Japanese "Pillow-Book"; these, however, were hardly marriage manuals for the vast majority of individuals in their societies. These were more likely to be as narrowly available as were the idioms and postures of "courtly love" in feudal Europe. In *The Perfumed Garden*, no reference is made to oral-genital activity. Oral-genital activity seems most commonly to have been a part of cultures in which ritualized bathing also appears, as in India, China, and Japan (Tannahill, 1980). Among the Chinese it characterized the privileged of the society and was reinforced by beliefs that made oral sex an occasion for exchanging vital forces ("yin" and "yang") (Bullough, 1976).

This too-brief examination of the cross-cultural perspective suggests the possibility that oral-genital activity occurs in too few cultures to suggest that it is a near-constant within the human experience. At the same time, however, it appears with sufficient frequency to place it comfortably within the normal range for the evolution of uses of the sexual by humans.

Western History

History is at the mercy of its sources as well as the bigotries that fill the empty spaces between our recognized historical facts with what we currently accept as our current ahistorical, natural truth. A not uncommon error is to assume that people have been engaging in the same sex acts across time and are doing so for essentially the same reasons or in pursuit of the identical experience. Indeed, one legacy of the "naturalization of the sexual" was that its move to the organismic level for explanation may have led to an overgeneralization of what has been little more than a culturally and historically limited construction of human sexuality.

Homosexual relationships among men were accepted by the ancient Greeks and Romans, but there is little evidence that oral-genital activity played a conspicuous role. If anything, there was a strong antipathy to it centering largely upon the view that those performing it were of a degraded or inferior status. In Dover's examination of homosexual behavior in ancient Greece (1978), there are only three references to oral sex. He notes that its representation in "vase paintings" (Dover's major source of representations of sexual behavior) is largely limited to satyrs (p. 99). Elsewhere it is described as an act performed by inferiors.

Among the Romans, who had a relatively permissive erotic culture, particularly with reference to same-gender sexual contact, much the same is true: Oral sex was a highly devalued form of sexual activity that did not seem to occur with great frequency. When it did occur it was not treated by itself as an extreme offense or one that justified extraordinary punishments. Boswell (1980) observed:

> Although neither oral nor anal intercourse was specifically condemned in the Jewish or Christian Scriptures, the former was the object of considerable contempt among the citizens of the ancient world. (p. 145)

The later elevation of oral sex to the status of a more serious violation obviously fed upon this unenviable reputation.

There are very few references to oral-genital activity in the literature and records of the Middle Ages. Interestingly enough, what one does find in the legal records is a great deal of mention of other sexually "deviant behaviors," of which sodomy appears as one of major concern. "Sodomy" was a term used to cover more than one activity, but the majority of times it has been used in the writing of both the medieval and Renaissance periods it refers to anal penetration or frottage (Goodich, 1979; Bullough & Bullough, 1977).

It is significant that within a tradition that enumerated and classified sexual violations with great detail, as in the above, oral sex sometimes did not find a place or did so in ways that suggested that it was far from common. Bullough (1976) observed:

> . . . although one of the manuscript sources of Finnian does have three paragraphs dealing with oral-genital contacts, [it is] the first time these are mentioned. Most scholars feel that this is a later interpolation. Why such references are missing in the earlier pennitals is difficult to explain, unless the writers' experience was somewhat limited, and whenever they found some new variation, they added to it. (pp. 358–359)

If, as we suspect, both the representations and occurrence of fellatio were relatively rare up until the modern era, representations of cunnilingus may have been still more rare. When it is mentioned, as in the fabliaux of the Middle Ages, included among which is Chaucer's *Canterbury Tales*, oral sex is mentioned as a humiliation:

> Absolon in *The Reeve's Tale* is humiliated when he is duped into unknowingly kissing the lady's crotch. In "Beranger Longbottom" ("Beranger au Lonc Cul"), the cowardly knight—like Absolon— is humiliated by being forced (this time knowingly) into kissing his wife's crotch. (Bullough & Brundage, 1982, p. 164)

There is little to suggest that oral-genital activity was widely engaged in before the 19th or 20th centuries in Western Europe. When it is mentioned it is with relatively mild disapproval, being treated as violative, as were virtually all forms of noncoital interaction (Bullough, 1976, p. 547).

The Modern Era

By the end of the 19th century, several dynamics are in place that may not have been operative previously. In the Middle Ages, erotic or lustful interest between husband and wife was condemned (Flandrin, 1985). However, by the 19th century the expectation of an eroticized romantic love as part of marriage became an ascendant cultural theme, setting the stage for the shifting patterns of sexual activity seen in modern and postmodern times (Gardella, 1985).

A second major 19th-century phenomenon was the increasing tendency of labeling an individual by the sexual partner or activity he or she chooses. Accompanying this was an elaboration of both scientific and popular categories of sexual types that served as an anchorage around which specific behaviors could be normalized and extended, creating, as it were, such entities as the "typical" homosexual (Robinson,

1976; Weeks, 1985). A commitment to a specific sexual identity could lead the individual to an elaboration of specific behaviors more readily than specific behaviors directly could lead to given identities. Foucault (1985) described this process as follows:

> The term itself [sexuality] did not appear until the beginning of the nineteenth century, a fact that should be neither underestimated nor over interpreted. It does point to something other than a simple recasting of vocabulary, but obviously it does not mark the sudden emergence of that to which "sexuality" refers. The use of the word was established in connection with other phenomena: the development of diverse fields of knowledge . . . the establishment of a set of rules and norms—which found support in religious, judicial, pedagogical, and medical institutions; and changes in the way individuals were led to assign meaning and value to their conduct, their duties, their pleasures, their feelings and sensations, their dreams. In short, it was a matter of seeing how an "experience" came to be constituted in modern Western societies, an experience that caused individuals to recognize themselves as subjects of a "sexuality," which was accessible to very diverse fields of knowledge and linked to a system of rules and constraints. (pp. 3–4)

Contemporary Social Science

At the outset, it must be noted that those who have attempted the bookkeeping of the sexual within modern times have had the problem of generating data based upon adequate and unbiased samples, a problem that few have effectively resolved. This is true for research on heterosexual as well as homosexual behavior. There are no data currently available that could provide a precise estimate of the prevalence of oralgenital behavior. However, making the best of an unfortunate situation, there are a number of studies—all flawed in critical dimensions of sampling—that allow for cautious interpretations. In effect, gross distinctions may be drawn in terms of the direction of change (increasing or decreasing), the magnitude of involvement (rare, or common but infrequent, or common and frequent), and differentials describing gender and sexual preferences. Greater confidence in such estimates is possible where findings can be interpreted in terms of related shifts in other aspects of human sexual behavior.

The Heterosexual Experience

The major shift in the 20th-century sexual landscape was a growing elaboration of images of the sexual as a positive experience, as distinct from earlier themes of the sexual as only a dangerous pleasure or as a matter of obligation and reproductive responsibility. Indeed, enacted within the context of an appropriate relationship, it became a positive experience that could signify or give testimony to the individual's psychosexual maturity and sociosexual competence.

What followed was as much reflected in, as created by, a succession of marriage manuals that tutored in an increasingly elaborated description of sexual techniques. These techniques, with decreasing ambiguity, made tests of competence out of the ease with which varied techniques were practiced and mutually enjoyed. The basic target population for this new sexual ideology was the emerging urban middle class. One might say that, in a sense, it was this group that represented the earliest staging area for what would be the succession of dramatic changes in patterns of sexual behavior, as well as in prevailing erotic ideologies and public and semipublic representations of the sexual.

In 20th-century American society, the increasing incidence of oral sex can be noted in the original Kinsey data (Gagnon & Simon, 1987). Thus, of those men with 13 or more years of education, only about a third (34.5%) reported engaging in premarital fellatio and slightly less than half that (15.5%) reported premarital cunnilingus. This difference between "receiving" and "giving" may reflect part of the dominant imagery of oral-genital activity as it is represented in much of 19th-century pornography/erotica. As Gagnon and Simon observed:

> Part of the difference could be a bias toward under-reporting cunnilingus by males in this time period [70% born between 1915–29; most of the remaining born before 1915]; oral genital contact might have been easier to report when it was received than when provided. A more substantial explanation for this difference is that in this historical period many males might have received a substantial portion of their premarital fellation experiences from prostitute contacts. Sex with prostitutes during young adulthood in this period was fairly common, and fellation was then, as now, a significant element in the prostitute script. In addition, sexual encounters with women defined as "bad girls" were often the occasion for experimentation with forbidden techniques. (p. 6)

The institutionalized values that proscribed almost all forms of premarital sex during the first quarter of the present century gave rise to a split between the erotic and sentimental; love and lust were forced, as it were, into an uneasy collaboration. In part this was widely recognized as the "whore and madonna syndrome," and that continued as one of the dominant themes in the erotic dramas of the rest of the century (see Freud, 1963/1912). If the erotic moment was commonly a moment of great potential anxiety, a willingness to perform oral-genital acts could be read as a confirmation of that person's degraded position or lascivious character, attributions that could eroticize anxiety and blunt the claims of guilt.

Oral sex, whether or not experienced premaritally, could clearly be seen as having earned a conspicuous place in the male sexual agenda. Its introduction into marital or proto-marital relationships followed what must have been its partial destigmatization. When it was performed in the context of a marital relationship, fellatio increased by about 25% (being reported by 42.7% of this same category of men—those with at least some college education) and the performance of cunnilingus increased by about three times (45.3%).

This overall increase in the behavior and convergence of the two forms of oral sex is confirmed when data on the marital oral-genital activity of females of the same educational background and same essential cohorts are examined: slightly more than half (52%) reported having performed fellatio, slightly more (58%) reported having received cunnilingus. This contrasts with the premarital experiences of such women, during which less than a fifth reported engaging in oral sex. In other words, increasing numbers of "respectable" females were both doing it and allowing it to be done (Gebhard & Johnson, 1979).

It is important to note that the increase in a willingness to receive (permit) cunnilingus and to perform fellatio occurs in the marital relationship, although much of the erotic significance of oral-genital activity derives from its initial address in the realm of the nonmarital. The domestication of oral sex, clothed in the eroticism of its initial cultural introduction, could be assimilated into conventional marital scripts by invoking the intensity of feelings associated with it, as well as having it serve as reciprocal confirmation of one's liberated or sophisticated status.

This change did not occur in isolation, but rather has to be seen in numerous life-style changes over the present century that have had an impact upon virtually every aspect of family life. Not the least of these were dramatic modifications in concepts of the very architecture of the urban middle-class family and the role of the sexual within it.

These data are based upon what was to become the American middle class (i.e., those who had attended at least some postsecondary school). It is worth noting that among these, those who received some postcollege education were significantly more likely to have engaged in premarital oral sex than were those whose education stopped at 16 years or less. And while adequate data on those with an education of high school or less are not available, it is not an unlikely inference that experience with reciprocal oral sex among such groups would have been substantially less frequent than among those with some college education or more (Rubin, 1976).

By the later 1970s and early 1980s, the American middle class had almost fully conventionalized oral-genital activity among married couples as well as among those who can be described as "cohabiting"— the ease with which nonmarital cohabitation has been accepted may itself represent a major cultural shift with profound implications for far more than the future of sex (Blumstein & Schwartz, 1983). The proportion of such heterosexual couples who rarely or never engaged in either fellatio or cunnilingus (28% and 26%, respectively) was slightly smaller than the proportion among whom such behaviors occurred every time or usually (29% and 32%, respectively). Blumstein and Schwartz also found a positive relationship between the frequency of oral-genital activity and their respondents' expressed satisfaction with both their sex lives and their overall relationships. They also observed a negative relationship between the frequency of oral-genital activity and the frequency of intercourse:

> Once intercourse becomes one of several sexual choices available to a couple, it is no longer a "required" part of every sexual session. For example, couples who have more oral sex have less intercourse (p. 229)

Perhaps the most significant change during recent decades in American sexual patterns with reference to heterosexual conduct was the virtual normalizing of premarital intercourse. Along with increasing numbers of individuals engaging in premarital intercourse, the age at which such activity occurs also dropped. Thus, while about half of all women included in the initial Kinsey study (Gebhard & Johnson, 1979, p. 267) reported having experienced premarital intercourse, currently about half of all young women completing high school report having had an initial coital experience. What for the cohorts studied by Kinsey was premarital in a very literal sense (half of women having had premarital intercourse having done so with only one partner, the one they ultimately married), the larger part of current premarital activity might more aptly be described as pre-premarital sex—that is, women engaging in coitus with individuals they were not planning to marry.

Moreover, along with this shift, it is clear that as the behavior has become more legitimate there is a decline of a negative stigmatization of females who are sexually "active" and, as an aspect of this, these women are more likely than ever before to receive social support for this activity by female peers (Kallen & Stephenson, 1982).

Given the tendency for oral sex to follow the beginnings of significant sociosexual activity, it is not unexpected that experience with oral sex— either fellatio or cunnilingus—should begin to occur more frequently and at younger ages. Describing their 1967 study of a national sample of college students, Gagnon and Simon (1987) observed:

> About 30% of the men and about 25% of the women in the study had ever experienced oral genital contacts. These figures represent a substantial change from the reports of premarital oral genital contact by the youngest birth cohorts (1920–1929) in the Kinsey populations. For the men there is twice the proportion reporting cunnilingus and a somewhat smaller proportion reporting fellatio, and for the women about twice the incidence for both fellatio and cunnilingus. The similarities of the rates reported by males and females and the decline in the male reports of fellatio suggests a reduction in "excess" experience of fellatio from commercial contacts. (p. 15)

Sexual experience, including coitus, appears for the vast majority to serve as a precondition or context for explorations of oral sex. Only about 5% of both males and females who have not experienced coitus report having experienced fellatio or cunnilingus. Moreover, the majority of these virgins report having engaged in oral sex only rarely; as the frequency of coital activity increases so does oral sex.

It is worth noting that among those college students who had not as yet engaged in oral sex, but who had at least experienced some genital contact, there is a poignant symmetry in the reasons cited for not having done so. The most frequent reason given by the young men was "partner not want" (49%), and its counterpart among the young women was "too shy." This, in turn, suggests that for many—if not most—cunnilingus or fellatio already exists as a possibility: a possibility layered with a history of meanings and associations.

Two distinct factors, at this point, appear to shape heterosexual oral-genital activity. The first of these is experience and comfort with conventional interpersonal scripts for sexual interaction (Simon & Gagnon, 1987). The second is the develoment of erotic meanings associated with oral sex; these can be both negative meanings (as degraded, dirty, or particularly sinful acts) and positive meanings (a display of liberated sophistication or an expression of intimacy and passion). Traditionally, a range of largely negative meanings ("cocksucking" and "cuntlapping") predominated. For contemporary youth, a more positive vocabulary for the behavior exists (e.g., "giving good head"). As a consequence, participation in oral sex can be expected to occur among a larger proportion of youth and young adults.

One study undertaken in 1973 (Delamater and MacCorquodale, 1979) of a population similar to that in the 1967 study [described above] suggests that there was a sharp increase in both coitus and oral sex in the intervening six years. In a sample ranging in age from 18 to 23, the proportion of coitally experienced young people was about 75% (for all males and non-student females) and 60% for student females which represent increases of about 50% among males (from 48%) and 100% among females (from 30%) from the 1967 study. Oral sex was also substantially more common in this population which was on the average about age 20 at the time. Thus about 60% of the student males and 57% of the student females had had oral sex . . . with the average age at first oral sex about 18.1 for both genders. Once again . . . oral sex occurred largely in the context of a coital but not necessarily premarital script. A further important point is that both males and females in the student samples report oral sex beginning at earlier ages than among non-students suggesting the continuing significance of these techniques among more educated groups even during this teen-age period.

A more recent study of high school students (Newcomer and Udry, 1985) reporting on data collected in 1982 suggested, on the basis of the age of subjects at interview (16.3), that oral sex may have become very common in . . . the pre-premarital period. Thus 53% of the boys and 42% of the girls interviewed had had oral sex. . . . Indeed, a slightly larger proportion of the sample had had oral sex than had had coitus. This is the first carefully done study in which the "technical virgin" represented a substantial portion of the sample. (Gagnon & Simon, 1987, pp. 22–23)

In a still more recent study of adolescents ages 13 through 18 (Coles & Stokes, 1985), similar findings are reported. They found that about a third of these teenagers had already engaged in coitus (about 50% of those 18) and a fifth had already engaged in oral sex. Forty-one percent of the 17- and 18-year-old females acknowledged performing fellatio, while about a third of the same-age males reported performing cunnilingus—the difference reflecting the persistence of traditional age difference in dating patterns.

Coles and Stokes also reported (p. 59) a small, but significant, number of teenagers who experienced oral sex without as yet having experienced coitus. Fellatio was performed by 59% of those who were coitally experienced, and by 16% of those without coital experience. Their suggested explanation, which on the surface has an aspect of plausibility about it, is that the deferring of coitus tended to be more an expression of concerns for avoiding pregnancy than an investment in the preservation of virginity.

It is difficult to avoid the conclusion that oral-genital contact, while

not necessarily a universal phenomenon, has been widely conventionalized in contemporary heterosexual practice. Observing a number of
"studies" undertaken during the 1970s by popular magazines, as well
as noting the dramatic increase in the attention paid to oral sex in texts,
sex manuals, and popular culture, Ehrenreich, Hess, and Jacobs (1986)
could note:

> Oral sex gained prominence in the seventies in part because of the
> increasing awareness of gay and lesbian sexual practices but also
> because it seemed to offer a solution to the crisis of heterosexual
> sex. Unlike intercourse, oral sex isolated the most sexual respon-
> sive organs—the clitoris and the penis—and dispensed with the
> relatively inert vagina. And unlike intercourse, oral sex seemed to
> offer the possibility of making heterosexual sex more reciprocal and
> egalitarian: either partner could do it, and either could, presum-
> ably, enjoy it. (p. 83)

The Experience of Gay Men

If research-based data on heterosexual oral-genital activity tend to be
both sporadic and thin, data on the oral-genital activity of gay men are
no more impressive and may, in fact, be far less reliable. As indicated
above, historically fellatio has, at best, occupied a shadowy position. Its
emergence as a modal or representative homosexual act may have followed the convergence of several developments.

The first and most important of these was the very emergence of (with
the introduction and conventionalizing of the very word) "homosexual"
as predominantly describing a kind of person rather than a category of
acts. A second aspect was the heightened policing of homosexuality as
the modern world appeared to increase the policing of almost all forms
of the sexual (Donzelot, 1979). This, in turn, encouraged forms of sexual
encounters that minimized risks of disclosure, forms of sexual contact
that allowed for a minimum of disrobing and a maximum of anonymity
(Pollak, 1985). Fellatio was and remains an adaptive way of responding
to these pressures. The growth of public images of the homosexual and
the corresponding development of a homosexual community, with language and scripted rituals of its own, also served to highlight the oral
option.

The third factor was the increasing concern for bodily cleanliness.
While there is some erotic enrichment of intimate genital contact because
of the association with eliminatory functions for both homosexuals and
heterosexuals alike, the aesthetics of such contacts were obviously enhanced by expanding commitments to cleansing the body and the wearing of clean clothing. As Kinsey et al. observed: "The human is excep-

tional . . . when it abstains from oral activities because of learned social proprieties, moral restraints, or exaggerated ideas of sanitation" (1948, p. 588).

Undoubtedly fellatio is a common form of sexual interaction among modern gay men. However, in itself, there is little to suggest that desire for fellatio acts as a major motivation for homosexual encounters. Data on men with extensive homosexual histories gathered by Kinsey and his associates, for example, indicated that receiving fellatio served as the initial form of homosexual contact for only 10% of nondelinquent whites and for only 20% of nondelinquet nonwhites. More strikingly, only 2 to 3% of all nondelinquents reported performing fellatio, with an additional 6% indicating that mutual fellatio described their initial homosexual acts (Gebhard & Johnson, 1979, p. 496). Indeed, the Kinsey data suggested that by age 18 only half of gay men had performed or received fellatio.

In addition, some data are available that suggest that oral sex may not be the technique of preference for all gay men. A relatively early, largely ethnographic study of gay men in England by Schofield (1965) indicated that frottage was the preferred or most common form of sexual interaction. In terms of the Kinsey data, about 15% of white gay men and 18% of nonwhite gay men reported receiving oral sex either never, rarely (1 to 3 times), or little, and slightly more than 20% of whites and almost half of nonwhites reported performing oral sex never, rarely, or little.

More than three decades later, Bell and Weinberg (1978) reported that 5% of white gay men and 11% of black gay men reported never performing oral-genital acts; 6% and 4%, respectively, indicated they had never received oral sex. However, in these same data, white gay men mentioned oral sex as their favorite sexual activity, with 2% preferring performing oral sex, 14% favoring mutual oral sex, and 27% favoring receiving oral sex. Over half of the respondents reported both performing and receiving oral sex once a week or more during the year preceding the interview. Similarly, Blumstein and Schwartz (1983) reported that for half their sample of "coupled" gay men, oral sex occurred "usually" or "every time."

Clearly, although homosexual populations are varied in their preferred sexual techniques, oral-genital activity remains the most widely preferred and most widely practiced. However, as suggested above, there is little reason to assume that homosexuality is an accommodation of a desire for, or a fixation upon, oral sex, or that when oral sex is critical to the desire, it is rarely indifferent to the gender of the participating oral cavity. Rather, much like the heterosexual activity considered above, oral-genital contacts tend to appear as part of a movement into and experience with prevailing sexual scripts that provide guide-lines for introducing the behavior and managing the performance.

The Lesbian Experience

In some regards, the experiences of lesbians resemble those of gay men. Among the major parallels are, first, the rarity of oral sex as an initial form of lesbian sexual contact. No nondelinquent lesbian included in subgroups of the Kinsey research reported performing oral sex as an initial experience; less than 5% of white lesbians and less than 10% of nonwhite lesbians reported have received oral sex in their first lesbian experience (Gebhard & Johnson, 1979, p. 496). Even more strikingly, only a small minority of white lesbians were the recipients of oral sex (13%) and only 16% performed oral sex prior to their 18th birthday. Lesbians tend to reflect the general patterns of sexual experience reported by females in general, and the differences between them are consistent with differences in the staging of sexual experience for the two categories (Gagnon & Simon, 1973).

A second similarity is that oral-genital techniques appear to be both the most favored and the most common sexual techniques among lesbians. As reported by Bell and Weinberg (1978), oral sex in its varied modalities was clearly the most preferred (6% of white lesbians preferred performing oral sex, 20% receiving oral sex, and 20% mutual oral sex; among black lesbians the corresponding figures were 0%, 29%, and 24%). Blumstein and Schwartz (1983) reported that while 39% of the lesbians in their study engaged in oral sex "every time" or "usually," 38% did so "some times," and a not inconsiderable 23% did it "rarely" or "never." (It should be noted, however, that this latter group was limited to lesbians who were currently "coupled.") Bell and Weinberg (1978) reported figures consistent with these data; a fifth of the lesbians interviewed reported they had not performed oral sex during the preceding year and comparable proportions reported not having received it, with more than an additional tenth saying that neither had occurred (p. 329).

Thus the patterns of oral sex as practiced by lesbians are not significantly different from those of other women and men, other than a lowered frequency that may reflect little more than the general tendency for lesbians to report a lowered frequency of sexual activity (Blumstein & Schwartz, 1983).

Discussion

There is little reason to accept the idea of an elemental drive or impulse to engage in oral-genital activity; by the same token, there is little reason to assume a general predisposition not to. The appearance of oral sex as erotically significant behavior is associated with the appearance of populations for whom the value of the sexual became distinct from reproductive issues. Erotically significant behavior can be defined as (a) behavior that is treated as significant by the surrounding social world, as indicated by the special emphasis given to either its inclusion in or exclusion from available sexual scripts; and (b) on the individual level, by the fact that the idea is itself capable of generating sexual excitement.

The shift that occasions the continuing diffusion of oral-genital activity as part of recent Western sexual history must be understood in terms of an emerging transformation of the larger role of the sexual. Following Foucault's (1978) suggestion, it seems that from the 17th century on, sexuality took on a heightened significance for both institutions and individuals as a result of institutionalized practices that encouraged not merely a repression of the sexual, but an intensified search for the very possibility of the sexual in almost all behaviors. The "attempted" repression of the sexual actually proved to be its elevation. One consequence was that this very search for the sexual tended to encourage the investment of all manner of behaviors, gestures, and costumes with heightened erotic powers. Consistent with this is the observation of Masters and Johnson (1979) that even under laboratory conditions men performing cunnilingus, while not reaching orgasm, generally maintained erections.

The investment of nongenital body parts with sexual meanings is neither a process with a single dimension nor a process that can be understood without reference to the previous and continuing uses and meanings of that body part. Included in this process to varying degrees are the following:

1. The very presence of strong societal efforts to provide effective interdiction increases the significance of the behaviors being restricted; they signify, they become exemplary, accusatory, a test, characteristics associated with the sexual in general. An example would be the ritual uses of cunnilingus by "outlaw biker" groups, and the special emphasis given by them to performing it upon a menstruating female or a black woman.

2. The opposing uses of the two body parts, the mouth (ingestive) and the genitalia (eliminatory), reinforces and multiplies the metaphoric uses of the opposition—in particular its metaphoric representation of the multiple postures and uses of power and control or the creation of a climate consistent with a sense of the sexual as "dirty" business.

3. The "taint" of the preceding, echoes of which persist despite the "refurbishing" of oral-genital contacts in contemporary human sexuality texts and manuals, makes it expressive of special intensities, special claims for intimacy.

4. Given the enlarged requirements of competence in sexual per-
    formances, particularly as manifested in the "production" of
    orgasm, which continues as perhaps one of the more enduring
    legacies of recent decades, oral-genital contacts serve to recruit
    additional numbers to this practice by providing more control
    in the manipulation of the other's genitalia than that afforded
    generally by various coital positions.
5. Added to the latter, oral-genital contacts become attractive as
    additions to the individual's sexual repertoire in contexts in
    which definitions of sexual competence tend to include the qual-
    ity of performance as well as frequency.

In general, as can be seen from the above, oral-genital contacts within
both heterosexual and homosexual contexts appear to be firmly rooted
in contemporary public sexual culture or, if you will, available cultural
scenarios. While probably still exhibiting differential levels of incidence
and frequency within various social classes, it is reasonable to assume
that oral-genital contacts will continue to see substantial increases. Indeed, it is not unlikely that an associated trend toward viewing oral sex
not as an aspect of foreplay in anticipation of coitus, but as an alternative
to coitus, will also continue to increase. It might be hypothesized that
historians of the future may view the present period as one in which
the inherited and dominant cultural scenarios describing and instructing
coitus as the most pleasuring and culminating aspect ("the real thing")
of the sexual encounter began to fade.

It is also clear that while a small number of individuals may have
fetishized oral contact as a nearly exclusive, if not obsessive preference,
for most it will be a learned skill that for many will be sufficiently attractive to be self-reinforcing. As an acquired preference, it may, in fact,
be amenable to influences that might either encourage or discourage its
appearance. However, like much that is associated with the sexual, oral
sex will retain a capacity to elicit responses that can effectively resist the
appeals of reason or prudence. Moreover, the complex of emotions associated with oral-genital contacts are such that attempts at influencing
preferences for them will have to be as varied as the socioemotional uses
to which they are put. For the relatively small group that has fetishized
oral contacts, particularly where, as with many forms of fetishism, preference moves toward the exclusive, not much is really known, although
recourse to such simplisitically opaque explanations as "oral fixation"
afford little insight.

Implications for Public Policy Regarding AIDS

The implications of oral sex for public policy regarding AIDS are dependent upon two considerations that are beyond the scope of this chapter; these are the degree of risk of transmission associated with fellatio

and cunnilingus and the degree of risk, if any, that is acceptable. The degree of risk associated with fellatio, which includes the ingestion of semen, may be different from that associated with cunnilingus, which, aside from the ingestion of vaginal fluids, may, under some circumstances of heterosexual contact, also include the ingestion of semen. Moreover, the level of risk attached to fellatio may be mitigated by the effective use of condoms. There may be little by way of comparable protection in the case of cunnilingus.

Clearly, at present, there is great confusion about the role of oral sex in the transmission of AIDS. To the degree that oral sex of either kind involves either no risk or a tolerable level of risk, then it provides a possible alternative to higher risk forms of sexual contact. To the degree, however, that it presents an unacceptable level of risk, developing strategies for persuading individuals to modify their specific practices may have to consider the many different motives for and the commitment to such practices.

This, in fact, may be the single most important contribution to facilitating the formulation of public policy with reference to AIDS that the present examination might provide: The recognition that behind the seeming uniformities of all sexual behavior there inevitably is a plurality of cultural and developmental factors that either encourage or discourage specific forms of erotic/sexual behavior either unconditionally or within specific contexts. Moreover, the fact that such sexual behaviors are motivated by the manifestly variable conditions of social life, instead of by the more culturally transcendent regularities of human biology, does not by definition make them either more superficial in their claims upon the behavior of individuals or necessarily less resistant to modification. In other words, the ability of the sexual to repress or distort the voice of reason, its ability to construct compelling reasons of its own, is not lessened by being rooted in the transitory complexities and heterogeneities of "postmodern societies"; in fact, it may be just this cultural conditioning that creates the potential for an increased enpowering of what are experienced as erotic desires.

Acknowledgments

This work was supported in part by research grants RO1 DA 04310 and RO1 DA 02497, as well as by a Research Scientist Award (KO5 DA 001905) to the third author, from the National Institute on Drug Abuse.

References

Bell, A., & Weinberg, M. (1978). *Homosexualities: A study of diversity among men and women*. New York: Simon and Schuster.
Blumstein, P., & Schwartz, P. (1983). *American couples*. New York: Morrow.

Boswell, J. (1980). *Christianity, social tolerance and homosexuality: Gay people in western Europe from the beginning of the Christian era to the fourteenth century.* Chicago: University of Chicago Press.

Bullough, V. L. (1976). *Sexual variance in society and history.* Chicago: University of Chicago Press.

Bullough, V. L., & Brundage, J. (1982). *Sexual practices & the medieval church.* Buffalo, NY: Prometheus Books.

Bullough, V. L., & Bullough, B. (1977). *Sin, sickness, and society: A history of sexual attitudes.* New York: Garland Publishing Co.

Coles, R., & Stokes, G. (1985). *Sex and the American teenager.* New York: Harper.

Davenport, W. (1965). Sexual patterns and their regulation in a society of the Southwest Pacific. In F. A. Beach (Ed.), *Sex and behavior.* New York: W. W. Norton & Co.

Delamater, J., & MacCorquodale, P. (1979). *Premarital sexuality: Attitudes, relationships, behavior.* Madison: University of Wisconsin.

Dewsbury, D. A. (1981). *Mammalian sexual behavior: Foundations for contemporary research.* Stroudsberg, PA: Hutchinson Ross.

Donzelot, J. (1979). *The policing families: Welfare versus the state.* London: Hutchinson.

Dover, K. J. (1978). *Greek homosexuality.* Cambridge, MA: Harvard University Press.

Ehrenreich, B., Hess, E., & Jacobs, G. (1986). *Re-making love: The feminization of sex.* Garden City, NY: Anchor Books.

Flandrin, J. (1985). Sex in married life in the early Middle Ages: The Church's teaching and behavioural reality. In P. Aries & A. Bejin (Eds.), *Western sexuality: Practices and precept in past and present times.* London: Blackwell.

Foucault, M. (1978). *The history of sexuality: Vol. 1. An introduction.* New York: Pantheon.

Foucault, M. (1985). *The history of sexuality: Vol. 2. The uses of pleasure.* New York: Pantheon.

Freud, S. (1963). On the universal tendency to debasement in the sphere of love. In J. Strachey (Ed. and Trans.), *The standard edition of the complete psychological works of Sigmund Freud* (Vol. 11, pp. 177–190). London: Hogarth Press. (Original work published 1912)

Gagnon, J. H., & Simon, W. (1973). *Sexual conduct: The social sources of human sexuality.* Chicago: Aldine.

Gagnon, J. H., & Simon, W. (1987). The sexual scripting of oral genital contacts. *Archives of Sexual Behavior, 16,* 1–25.

Gardella, P. (1985). *Innocent ecstasy.* New York: Oxford University Press.

Gebhard, P. H. (1971). Human sexual behavior: A summary statement. In D. Marshall & R. Suggs (Eds.), *Human sexual behavior: Variations in the ethnographic spectrum.* Englewood Cliffs, NJ: Prentice-Hall.

Gebhard, P. H., & Johnson, A. B. (1979). *The Kinsey data: Marginal tabulations of the 1948–1963 interviews.* Philadelphia: W. B. Saunders Company.

Goodall, J. (1971). *In the shadow of man.* Boston: Houghton Mifflin Company.

Goodich, M. (1979). *The unmentionable vice: Homosexuality in the later medieval period.* Santa Barbara, CA: Ross-Erikson, Publishers.

Herdt, G. H. (1981). Guardians of the Flutes: Idioms of Masculinity. New York: McGraw-Hill.

Herdt, G. H. (Ed.). (1984). *Ritualized homosexuality in Melanesia.* Los Angeles: University of California Press.

Kallen, D., & Stephenson, J. J. (1982). Talking about sex revisited. *Journal of Youth and Adolescence, 11,* 11–23.

Kaplan, H., Johnson, R., Bailey, C., & Simon, W. (1987). The sociological study of AIDS: A critical review of the literature and suggested research agenda. *Journal of Health and Social Behavior, 28,* 144–187.

Karlen, A. (1971). *Sexuality and homosexuality: A new view.* New York: Norton.

Kinsey, A. C., Pomeroy, W. P., & Martin, C. (1948). *Sexual behavior in the human male.* Philadelphia: W. B. Saunders Company.

Masters, W. H., & Johnson, B. E. (1979). *Homosexuality in perspective.* Boston: Little, Brown.

Newcomer, S. F., & Udry, J. R. (1985). Oral sex in an adolescent population. *Archives of Sexual Behavior, 14,* 41–46.

Pollak, M. (1985). Male homosexuality—or happiness in the ghetto. In P. Aries & A. Bejin (Eds.), *Western sexuality: Practices and precept in past and present times.* London: Blackwell.

Robinson, P. (1976). *The modernization of sex.* New York: Harper & Row.

Rubin, L. B. (1976). *Worlds of pain.* New York: Basic Books.

Schofield, M. (1965). *Sociological aspects of homosexuality.* London: Longmans.

Simon, W. (1989). the post-modernization of sex. *Journal of Psychology and Human Sexuality, 2,* 9–38.

Simon, W., & Gagnon, J. H. (1987). The scripting approach. In J. Geer & W. Donough (Eds.), *Theories of human sexuality.* New York: Plenum.

Stoller, R. J. (1985). Theories of origins of male homosexuality: A cross-cultural look and psychoanalytic research on homosexuality. In *Observing the erotic imagination.* New Haven, CT: Yale University Press.

Tannahill, R. (1980). *Sex in history.* New York: Stein and Day.

Waterman, C., & Chiuzzi, E. (1982). The role of orgasm in male and female sexual enjoyment. *Journal of Sexual Research, 18,* 146–159.

Weeks, J. (1985). *Sexuality and its discontents: Meanings, myths, and modern sexualities.* London: Routledge & Kegan Paul.

19

Heterosexual Anal Intercourse: An AIDS Risk Factor

Bruce Voeller

The first clinical descriptions of AIDS were published in 1981 (Friedman-Kien et al., 1981; Gottlieb, Schanker, Fan, Saxon, & Weissman, 1981). Almost concurrently, Drew and his colleagues were finding that a virus in the herpesvirus family, cytomegalovirus (CMV), known to exist in high titer in semen (Lang, Kummer, & Hartley, 1974), was *sexually* transmitted among homosexual men (Drew, 1982; Katzneslson & Drew, 1984; Mintz, Drew, Miner, & Braff, 1983). Indeed,

> . . . only passive [receptive] anal intercourse correlated either with the initial presence of antibody to CMV, or with seroconversion to CMV during the course of the study . . . these data suggest that exposure of the anorectal mucosa to CMV-infected semen may constitute a major mode of infection with CMV in homosexual men (Katzneslson & Drew, 1984, p. 157)

a perspective confirmed by others (Collier et al., 1987).

In 1983 Voeller suggested that AIDS, too, might be spread through anal intercourse, and that *hetero*sexuals also engage in anal intercourse and might be at increased risk, both in this country and places such as Brazil (e.g., areas with Latin cultures) (Voeller, 1983c, 1988).

The emphasis of the Drew group upon the unidirectional transmission of semen-borne CMV to the "receptive" partner in anal intercourse was a prophetic lodestone orienting attention to the role anal coitus would gradually be recognized to hold in the extensive spread of AIDS among homosexual men in industrialized Western nations. While the potential

276

for contracting the AIDS virus (human immunodeficiency virus; HIV) through oral-oral sex, oral-genital sex (including the swallowing of semen), and oral-anal sex exists, the hazard has proved low in comparison with receptive anal intercourse. Indeed, the major AIDS epidemiological studies, including those of Stevens and of Detels presented in this volume (see Chapters 2 and 1, respectively), strikingly demonstrate that among homosexual males *receptive* anal intercourse is quite distinctly the highest risk sexual practice for contracting HIV (Boyko et al., 1986; Darrow et al., 1987; Kinsgley et al., 1987; Osmond et al., 1988; Polk et al., 1987; Winkelstein et al., 1987).

Perhaps because so few physicians are aware of the extent of heterosexual anal intercourse, some AIDS authorities, such as Peterman and Curran (1986), and Guinan and Hardy (1987), all at the Centers for Disease Control, have discounted the significance of heterosexual anal intercourse in AIDS. Peterman and Curran, for example, stated that, "nearly all infected heterosexuals . . . have no history of receptive anal intercourse" (1986, p. 2222), a statement definitionally true for Kinsey Scale "0" males, but demonstrably incorrect for females. Similarly, Wofsy stated, "It has been suggested that rectal intercourse enhances the probability of a woman becoming infected, but this has not been proved" (1987). Taking exception, Bolling and Voeller warned that "we underestimate the practice among heterosexuals at some peril" (Bolling & Voeller, 1987, p. 474); Lorian (1988) has recently expressed similar concern.

Padian and her associates studied women whose sexual partners were HIV-infected males. They found that Northern California women who engaged in both vaginal and anal intercourse placed themselves at double or triple the risk of HIV infection experienced by women who limited coitus to vaginal intercourse (Padian, Marquis, et al., 1987). Similar results have been reported by Sion and colleagues in Brazil, where 75 female partners of HIV-infected bisexual men were studied. Eighteen of 33 women (54.5%) who engaged in anal as well as vaginal intercourse were infected; 8 of 42 women (19%) who indicated they engaged only in vaginal intercourse were infected (Sion et al., 1988).

In a similar European study recently reported by De Vincenzi, 29 of 104 female partners of HIV-infected men were infected. Of these infected women, 58.6% had engaged in anal coitus; 25.2% did not acknowledge doing so (De Vincenzi, 1988). Among 65 European couples from six countries, the odds ratio for male to female anal sex was 4.8 (De Vincenzi & Ancelle-Park, 1989). In Italy, the crude risk ratio was 2.5 for 80 (22%) women acknowledging anal coitus, compared with 288 women engaging in vaginal, but not anal, sex (D'Arminio-Monforte et al., 1989).

In the Bronx, New York, Steigbigel and coworkers (Steigbigel et al., 1987, 1988) studied 114 steady heterosexual partners of men with AIDS or AIDS-related complex (ARC), 41 (36%) of whom had engaged in anal intercourse, and stated:

> The number of episodes of anal intercourse was the only significant independent predictor of HIV seropositivity ($p < .01$) in a multiple logistic regression analysis of various sex practices. Of the 73 females who did not have anal intercourse 26 (36%) were sero (+); of the 41 females who had anal sex 25 (61%) were sero (+) ($p = .01$). (Steigbigel et al., 1987).

Despite the demonstrations that *receptive anal intercourse* is the sexual practice carrying far the highest risk for HIV infection among both homosexual men *and* heterosexual women, and despite the magnitude of the world AIDS crisis, several deficiencies in information bearing on anal intercourse emerge to foil attempts to predict the scope of the spread of AIDS, and thence to address strategies to contain that spread in a sound fashion.

First, apart from the half-century-old "Kinsey" data base, there is but limited information as to how many American males have occasional to frequent same-sex contacts, which might place them at particularly elevated risk of contracting the AIDS virus and of transmitting it to their female sex partners (McWhirter, Sanders, & Reinisch, 1990). We have even less notion what *portion* of same-sex participants engage in *anal* coitus. Clearly not all homosexual or bisexual men do so, some preferring or restricting sexual expression to oral sex, mutual masturbation, or other sexual practices. What proportion does practice anal coitus, and to what extent? How does this vary among representative ethnic, regional, and other groups?

Data are both scarce and usually based upon unreliable interview methods and upon samplings flawed through use solely of readily reached populations of *self-identified* gay males—populations that constitute a special and limited representation of males who participate in same-gender sex, and that are overrich in middle-class whites. Scarce as the existing, flawed information for American society is, incidence figures similar to the Kinsey data, and pertinent to other parts of the world or even to American subpopulations, are essentially nonexistent.

Second, the proportion of *bisexual* men who are *anal receptive* with male partners, maximizing their risk of contracting HIV and then transmitting it to their female sexual partners, is unknown. As Dr. C. A. Tripp has suggested at this conference, bisexual men who "export" vaginal or anal practices with their female partners (i.e., be insertive participants) may tend to "import" this act with male partners—that is, be *receptive* coital participants with their male sexual partners. So far as HIV transmission is concerned, the term "bisexual" designates a highly diverse population of men—varied in sexual backgrounds, needs, self-perceptions and outlets. Some eroticize both women and men, others maintain relationships with women to protect themselves from being perceived as or labelled homosexual. The pertinence of Tripp's perceptive "export-import" concept should be assessed for each of the various patterns subsumed under the catchall label "bisexual."

Bisexual men pose possibly greater risk to women than their heterosexual counterparts not only because of their higher HIV seroprevalence, but because of their greater tendency to engage in anal intercourse with their female partners. Nancy Padian and her associates (Padian, Marquis, et al., 1987) found a strikingly higher prevalence of anal intercourse among the female partners of bisexual men, compared with those of exclusively heterosexual men. Of 29 women in their study who participated in anal intercourse, 75% did so with bisexual men. Similarly, Reinisch, Sanders, and Ziemba-Davis (1990) found that 23% of 81 women with heterosexual partners engaged in anal intercourse, compared with 45% of 38 women with bisexual male partners. Such observations are underscored by Bateson and Goldsby (1988), and Day, Des Londe, Houston-Hamilton, and Steiger (1989).

Third, by definition, women engaging in heterosexual anal intercourse are *receptive* partners in the act, thence at grave risk of contracting HIV from infected bisexual men, hemophiliacs, and intravenous drug users, among others. Despite this grave risk, the traditional and continuing medical and scientific silence about heterosexual anal intercourse is so deafening that it begs to be addressed. Either such intercourse is so rare that physicians and researchers have little reason to comment upon it, or the topic is so taboo that few care or dare to, even in the face of one of the worst infectious scourges in modern medicine.

Fear about anal intercourse is bolstered by a widespread viewpoint that it is accompanied by severe trauma to the rectal and anal walls, commonly including bleeding. These notions find slim backing in fact, rape and force excepted. So far as willing, consensual anal intercourse is concerned, the notions of trauma are so widespread that they serve as the foundation for the most widely held medical hypothesis of the high HIV infectivity linked with rectal coitus in comparison with other sex acts—entrance of the virus into the rectal blood or lymphatic systems through the bleeding portal and through the infectable lymphocytes that are mobilized to the wound sites. The notions of bloody trauma are contrary to the observation of most persons participating in simple penile-anal intercourse (exempting use of dildos, engaging in fisting, marathon sex, etc.) and the observation of health care providers counseling those who engage in anal sex. Most researchers and health care providers subscribing to the idea of severe anorectal trauma linked with penetration seem to overlook the fact that each day most humans pass stool specimens comparable in circumference and diameter to the average human penis. Healthy people do so without pain, discomfort, bleeding, or tissue trauma.

Finally, it should be noted that Western males widely consider being "sodomized" to be the ultimate degradation that can befall them—at once a devastating physical injury and an irreparable psychological blow to masculinity.

Too often scientists and physicians forget that they share most of the cultural values and myths about sex. Their ethnocentrism has unquestionably colored their perceptions about anal intercourse and their willingness to approach it in their research and their clinical practices.

The purpose of the present chaper is to present and evaluate various data that suggest that the extent of heterosexual anal intercourse is greater than most physicians and researchers appreciate, as has been proposed by Bolling and Voeller (Bolling, 1976, 1977; Voeller, 1983c, 1988; also see Abramson & Heardt, 1990). Such information can

1. Focus future research into sexual practices and their medical and social consequences.
2. Play a role in assessing the directions and scope of the spread of AIDS *into* heterosexual populations and *within* heterosexual populations, as well as in recognizing the similarities and differences of heterosexual and homosexual dispersal of HIV.
3. Assist in the design of more effective AIDS and sexually transmitted disease prevention programs—including the educational messages to target individuals at risk, and the meaningful allocation of prevention resources.
4. Lead to wiser, more compassionate care for the medical and psychological needs of women and men engaging in anal coitus.
5. Affect the understanding of legislators and courts in labeling sex acts as aberrant or deviant.
6. Lead to new therapies for ailments such as hemorrhoids, as discussed by a few undaunted physicians.
7. Facilitate more realistic assessment of the value of condoms and spermicides in reducing the sexual spread of HIV.

The Perception of Rarity

Chapter 3 in this volume (by Reinisch et al.) demonstrates the ubiquity of the practice of heterosexual anal intercourse throughout recorded history and around the world—even if cross-cultural anthropology, art, and literary history do not permit conclusions about incidence or prevalence, or demographic distribution, in other places or times. In our own part of the world, heterosexual anal sexuality has been documented through modern history (Fauconney, 1902; Gebhard & Johnson, 1979; Morin, 1986; Sinistrari de Ameno, 1879; Voeller, 1983c, 1988). However, anal intercourse has been regarded by most physicians and by popular sex authorities as characterizing homosexual males' behavior, and as being a rarity among heterosexual women and men.

To the extent that university and medical school texts have touched upon the issue, they have generally reinforced the common belief in the scarcity of heterosexual anal intercourse. McCary (1973), for example, in the second edition of his textbook, *Human Sexuality*, stated:

> Sexual analism is seldom practiced in heterosexual contacts, except for occasional experimentation; but up to 50% of male homosexuals in the U.S. use this technique as a preferred method of sexual intercourse. (p. 368)

and:

> Sodomy is not widely practiced, however; only about 3% of husbands engage in it with their wives. (p. 368)

Marino and Mancini wrote:

> Anal sexual eroticism is a fact of modern life. While it is a minor variant in heterosexual intercourse, it is a major part of the male homosexual relationship. (1978, p. 513)

However, a Consumers Union survey of sex took exception and observed that, "Stimulation of the anus during sex is often [erroneously] considered an exclusively homosexual activity" (p. 363), and reported that even the older men and women in their survey acknowledged enjoying anal stimulation (Brecher, 1984).

Occasional mentions of anal sexuality in general medical research journals have usually cited medical curiosities such as the removal of foreign object from the rectum of a patient (Haft, 1973, 1976), or have had tabloid titles such as "Removal of Exotic Foreign Objects from the Abdominal Orifices" (Benjamin, 1969), underscoring the physicians' sense of the deviance and uncommonness of anal sexuality. The other contexts in which heterosexual anal intercourse come to the attention of the physician are cases of anorectal sexually transmitted diseases (STDs) and medicolegal cases, including within marriage ("Sodomy and," 1969).

Alfred Kinsey and his colleagues implicitly and explicitly contributed to the prevailing viewpoint on heterosexual anal intercourse. The perception is *implicit* in their near silence on the topic in their pioneering works, *Sexual Behavior in the Human Male* (Kinsey, Pomeroy, & Martin,

1948) and *Sexual Behavior in the Human Female* (Kinsey, Pomeroy, Martin, & Gebhard, 1953). While three pages in the male volume and five in the female are listed in the index under "anal eroticism," none provides more than brief mention of the act, nor are incidence figures presented—a striking contrast to the myriad tables quantifying a great range of other erotic and sexual behaviors. In fact, the Kinsey group *explicitly* contribute to the perception of rarity in the only mention of heterosexual anal intercourse indexed under "anal eroticism" in the male volume:

> There is some anal play in some of the marital histories, usually as an additional source of stimulation during vaginal coitus; and there is an occasional instance of anal coitus. However, anal activity in the heterosexual is not frequent enough to make it possible to determine the incidence of individuals who are specifically responsive to such stimulation. (Kinsey et al., 1948, p. 579)

Similarly, Ford and Beach (1951), in their classic survey of the anthropological and ethobiological literature on sexuality, cite several pages on anal coitus, *none bearing on females*. Also, as commented by Witz, Shpitz, Zager, Eliashiv, and Dinbar (1984), while there are quite a few, if scattered, anecdotal medical and emergency room reports of foreign bodies in the rectum of a man, "There are no reported cases of rectal autoeroticism with resulting rectal entrapment in females" (p. 332).

The perception that heterosexual anal sex is rare is perpetuated and reinforced by physicians' and researchers' failure to ask subjects about anal sex. This oversight derives in part from the inhibition most professionals feel in approaching patients or subjects with questions about "unusual" sexual practices. Marino and Mancini (1978) noted that, "a common question [put to them by medical colleagues] is, 'How do you ask a patient if they are having anal sex?'" (p. 513). Perhaps the most comprehensive discussions of the taboo on mentioning anal sex are presented in Morin's *Anal Pleasure & Health: A Guide for Men and Women* (1986) and by Voeller (1988).

Texas gynecologist David Bolling formally surveyed his medical colleagues in gynecology and psychiatry and wrote, "The respondents acknowledged asking no questions relating to anal intercourse in their routine evaluation, routine sexual counseling, or intensive sexual therapy" (Bolling, 1976, p. 567). Indeed, in discussing heterosexual anal intercourse with many hundreds of physicians—chiefly in obstetrics and gynecology—most seem embarrassed or even shocked that such practices occur, or that they, as physicians, ought to intrude into them as part of patient care.

Even psychotherapists are not exempt. Roman and associates' study

Table 19-1
Therapists' Responses to Sex Acts: Percentage Acceptable
for Oneself and Others[a]

	Oneself			Others		
	Yes	Possibly	No	Yes	Possibly	No
Cunnilingus	91	3	6	98	1	1
Fellatio	89	4	7	98	1	1
Anal intercourse	43	28	29	85	13	2

[a] Adapted from Roman et al. (1978).

of 124 staff psychotherapists at Albert Einstein College of Medicine in New York contrasted psychotherapists' acceptance of various sexual acts for themselves versus for others (Roman, Charles, & Karasu, 1978). As can be deduced from Table 19-1, most of the psychiatrists surveyed were comfortable with the concept of oral sex, nearly as much so for themselves as for others. Far the most striking difference between acceptability for oneself and for others (e.g., one's patients or research subjects) existed for anal coitus. Inasmuch as therapists are more highly trained to suspend personal judgment than are other physicians or health providers, the attenuated level of "acceptability" reported for the Einstein therapists probably represents a significantly higher level of tolerance than would be anticipated from other professionals.

Inasmuch as physicians are also members of our culture, and shaped by it, they are reluctant to press questions they think may embarrass their patients—patients who *are* ill at ease with the topic. Physician Martin E. Plaut, in his book, *The Doctor's Guide to You and Your Colon: A Candid Guide to Our #1 Hidden Health Complaint*, wrote: "Bowel concerns are private, and often so embarrassing to mention that people wanting answers to common complaints hesitate to speak of them even in a doctor's office" (1982).

Similar avoidance of discussion of heterosexual anal intercourse in polite society is the rule in Latin America, even though the act may be yet more common there than in the United States (Parker, 1987; Leal de Santa Inez, 1983; Voeller, 1988). In a survey of some 5,000 households, over 40% of those interviewed in rural Brazil considered anal intercourse a normal part of sexuality, and over 50% of urban citizens in Rio de Janeiro did (Leal de Santa Inez, 1983). The extent of this practice may find its widespread pre-Columbian roots in native Amazonian and Inca culture. Hoyle (1965), in his classic study of erotic elements in Peruvian Mochica art, documents that, "more than 95% [of pottery vessels] represent coitus *per anum*." In fact, anal intercourse appears to have been commonplace throughout much of Meso-America, judged from the pre-

Columbian records documented by such researchers as Francisco Guerra (Guerra, 1971).

The bearing of interviewing methodology upon evidence of heterosexual anal intercourse is also worth considering. Even among researchers who are comfortable asking sexual questions, many have not had specific training in taking valid sex histories. They make an understandable error: they presume that subjects who talk about *some* aspects of their sexuality will freely discuss *all* aspects. Even when dealing with overtly homosexual subjects or with gynecological and obstetric patients who acknowledge their sexuality, entire areas of sexual behavior are often withheld from the physician or researcher. What may seem to the unwary interviewer to be an opened door may be the merest crack from the perspective of the informant.

As noted by Drs. Mays and Cochran, Nichols, and Wyatt in their chapters in this volume (Chapters 5, 26, and 4, respectively), these problems can be greatly increased *even within our own country and culture* when the patient or subject and the interviewer do not share the same gender, racial, ethnic, linguistic, and other characteristics. Their observations underscore Kinsey's strong reliance on the face-to-face interview (compared with the written questionnaire)—Kinsey "recognized very early that many people . . . lack an adequate sexual vocabulary" (Christenson, 1971, p. 101).

The problem of retrospective recall also colors the dependability of sexual data, as recently reviewed by Abramson (1990) and by McLaws, Oldenburg, Ross, and Cooper (1990).

Incidence Data

General Change in Sexual Practices

As discussed in detail in this volume by Reinisch et al. (Chapter 3) and by Simon et al. (Chapter 18), varied evidence suggests that diversity of sexual practices among heterosexual couples in our culture is far more widespread than generally appreciated. Moreover, as the educational level has steadily increased in the United States, as contraceptive methods have been revolutionized, and as the women's and gay movements have grown, sexuality has become more openly discussed and the numbers of persons participating in varied sexual practices has grown.

With respect to *oral* sex, Kinsey et al. (1953, Table 100, p. 399) indicated that *within marriage* (as distinct from sex with prostitutes or singles, for example) the extents of both cunnilingus and fellatio have increased decade after decade. The Kinsey group found a steady increase in the range of women's sexual exploration linked to the decade of their birth. Moreover, the detailed discussions of oral sex in a wide array of texts on human sexuality (Katchadourian & Lune, 1975; McCary, 1978; Rosen & Rosen, 1981) and in sex manuals (Comfort, 1972), as well as in the results of surveys (Hite, 1976; Hunt, 1974), support this view. One would thus be surprised if, as part of that progressive increase in general sexual experimentation, heterosexual anal sex had not also increased over the past decades.

Anecdotal Data

Anecdotal commentaries on changing attitudes and behavior surrounding anal coitus can be found as early as the 1950s from physicians such as G. D. Jensen (1975), who wrote:

> . . . it [anal intercourse] is now considered a "normal" sexual variation. Not too many years ago it was regarded as a perversion, but this is no longer true in most quarters. (p. 115)

Neubardt (1974) wrote:

> Since early 1970 I have been investigating the role of anal intercourse in bacterial vaginitis in my own practice. I have examined many women who indulge in this activity, and have not been impressed with an unusual incidence of irritation or infection. (pp. 74–75)

Masters, Johnson, and Kolodny (1986) stated:

> Anal intercourse is often thought to be primarily an act of male homosexuals. However, numerically speaking, far more heterosexual couples engage in this activity than homosexuals, and many homosexual men have not had experience with this type of sexual behavior. (p. 53)

The STD literature has scattered references to urethral or bladder infections of men attributed to heterosexual anal intercourse (e.g., Meares, 1975). In addition, several researchers have noted that anal stimulation of women (and of heterosexual men), leads to sexual orgasm (Agnew, 1982). Masters and Johnson observed that during anal intercourse in seven heterosexual women, on 11 out of 14 occasions females reached orgasm—three times including multiple orgasms (Masters & Johnson, 1979). Such piecemeal observations are instructive, but allow no conclusion about the extent of anal intercourse.

Quantitative Survey Data

While the original Kinsey volumes left the topic of heterosexual anal intercourse barely broached, more recent additional "Kinsey" data published by Gebhard and Johnson confirm that anal intercourse was recorded by Kinsey and his colleagues during the period 1938 through 1963, both for married couples and for those engaging in premarital coitus, underscoring the Kinsey group's puzzling earlier silence on the matter (Gebhard & Johnson, 1979). They in fact had collected data from 3,618 men and women. Premarital percentages for both college and non-college white males who had attempted anal intercourse ran about 9; the percentage for black college men was 5. In very interesting contrast, the data for women were much higher: white college women, 26%; white noncollege women, 30%; black college women, 11%. Married informants acknowledged less exploration of anal intercourse with their spouses, although 11% of both married white women and men reported attempting anal intercourse (Gebhard & Johnson, 1979). The Kinsey group also published data on the histories of heterosexual anal intercourse by sex offenders with those they assaulted, as well as with the sex offenders' wives (Gebhard, Gagnon, Pomeroy, & Christenson, 1965). McCary (1978) claimed that only about 3% of husbands and wives practice anal sex.

Cornthwaite, Savage, and Willcox (1974) interviewed 105 women named as sexual contacts by men being treated for gonorrhea in a London clinic. "During the three months before referral 19 women (18.1%) had indulged in rectal coitus with one or more partners" (p. 306). Padian, Marquis, et al. in Northern California reported that, "Twenty-nine (31%) of 95 women enrolled in this [AIDS] study practiced anal intercourse" (1987, p. 789). In the Bronx, New York, Steigbigel et al. (1987, 1988) studied 114 steady sexual partners of men with AIDS or ARC; most were Hispanic. Forty-one of these women (36%) engaged in anal intercourse.

In a study of Nevada prostitutes, Padian and colleagues compared incarcerated women and women at legal brothels. Thirty percent of those at brothels ($n = 53$) engaged in anal intercourse; 44% of the incarcerated prostitutes ($n = 185$) did so (Padian, Carlson, Browning, Nelson, Grimes, & Marquis, 1987). Gill et al. (1986), in Los Angeles, indicated that of 113 prostitutes they interviewed, 30% had a history of rectal intercourse. Two of the three prostitutes who were HIV infected "had a history of rectal intercourse and sexual contact with bisexual men." Judith Cohen and her colleagues surveyed over 400 women in the sex industry (prostitutes) and other sexually active women (Cohen, Hauer, Poole, & Wofsy, 1987; Cohen et al., 1986). "Anal intercourse was reported by 51% of SI [sex industry] women and 58% of others" (Cohen et al., 1986).

Interviews with 46 Philadelphia streetwalkers (15 white, 22 black, 8 Hispanic, and 1 oriental) were conducted by Savitz and Rosen (1988). In their private lives, with lovers (as distinct from customers), 73.8% of these women acknowledged enjoyment of anal intercourse some of the time to all of the time; 58% indicated enjoyment most or all of the time.

Table 19-2
Percentage of Girls Practicing Anal Coitus[a]

	Age (years)		
	13–15	16–18	19–21
Number	23	56	32
Percentage	8.7	25.0	37.5

[a] Adapted from Jaffe et al. (1988).

Barton and associates interviewed 50 London prostitutes, of whom, "9 [18%] engaged in regular anal intercourse" (Barton, Underhill, Gilchrist, Jeffries, & Harris, 1985, p. 1424).

Leslie Jaffe, in New York City, investigated the sexual behaviors of 111 sexually active inner city adolescent girls (44.0% Hispanic, 46.8% black). She and her colleagues found that

> A total of 28 girls (25.2%) acknowledged having had anal inter-course in addition to vaginal. Of these, 19 (67.8%) reported having had anal sex within the preceding 3 months. The practice of anal intercourse was significantly associated with increasing age . . . (Jaffe, Seehaus, Wagner, & Leadbeater, 1988, p. 1006)

as shown in Table 19-2. There was no significant difference between blacks and Hispanics in the incidence of anal sex (Jaffe et al., 1988). R. R. Bell and coworkers analyzed written interviews of 2,262 married women from across the United States. Eight percent indicated experience with anal intercourse (R. R. Bell, Turner, & Rosen, 1975).

In an investigation into the etiology of anal cancer, Daling et al. (1987) contrasted 90 women with anal cancer with a control group of 102 women with colon cancer. They found that 16.9% of the former group acknowledged having engaged in anal intercourse, whereas 10.8% of the control group had done so.

Much larger surveys of heterosexual anal intercourse were conducted through mail questionnaires targeted to audiences including magazine subscribers to *Redbook* and *Cosmopolitan*, as well as to women's groups, such as National Organization of Women chapters. Shere Hite, who extensively explored female sexuality through such written question-naires, touches on anal intercourse. Several versions of the questionnaire she used each contain one item such as:

> Do you enjoy rectal contact? What kind? Do you enjoy penetration? How often are you requested to do this and how often do you do it? (Hite, 1976)

Of 1,394 women who responded as to whether or not they enjoy rectal contact (by touching, penile penetration, or finger penetration), 441 (32%) answered yes; an additional 136 (10%) said "sometimes," for a total of 42%. Of those 441 women (42%) who enjoyed rectal contact, 146 (26%) enjoyed penetration by the penis (10.5% out of the total population of 1,394 women) (Hite, 1976).

While such studies as Hite's indicate that a population of females does engage in anal intercourse, the inherent population biases of her survey methods preclude evaluation of the generality of her findings. Also, written questionnaires, while providing some persons a sense of ano-nymity, lock in language and cultural features that may be misunder-stood by, or may be incomprehensible to, the informant. *In the hands of a skilled interviewer*, the face-to-face interview allows far greater flexibility, and of course, the developing rapport and trust Kinsey (and as discussed here, Bolling) found so critical in getting people to disclose practices and fantasies they closely guarded.

As noted, retrospective studies, enquiring into subjects' remembrance or *estimation* of the number of occasions on which they engaged in par-ticular sexual acts, are fraught with problems surrounding the accuracy of recollection, exaggeration and underestimation, biases related to ex-pectations from sex (e.g., the remarkable difference in "Kinsey" data between reportage by male and female partners in the frequency of various kinds of sex acts), and the like. Some of these problems have been perceptively and accurately reviewed by Abramson (1990) and McLaws et al. (1990). Such problems color not only the trust that may be placed in written questionnaire surveys, but also personal interviews. The problem of the validity of reportage is an ill-measured variable of great importance in sex research.

Subject to similar interview limitations (but with demographic traits closely paralleling those collected by the U.S. Bureau of the Census), a 1974 *Redbook* magazine survey of some 100,000 women who responded to a questionnaire tallied 43% (the majority married) who had tried anal sex, 19% who engaged in it from time to time, and 2% who did so often (Levin & Levin, 1975; Tavris & Sadd, 1977). A more recent *Redbook* survey of 26,000 readers (no longer just married women) reported that 40% had tried anal intercourse and 12% enjoyed it (Rubenstein & Tavris, 1987). In a *Cosmopolitan* magazine survey of 106,000 women, 13 to 15% of re-spondents indicated that they "regularly" engaged in anal intercourse (Wolfe, 1980, 1981). The *Playboy* survey of 100,000 readers (80% male), reported that 47% of men had tried anal intercourse and 61% of women had. Thirteen percent of married couples engaged in anal intercourse more than once each month (Peterson, 1983).

Athanasiou, Shaver, and Tavris (1970) reviewed questionnaires re-turned by more than 20,000 *Psychology Today* readers, of whom 10,600 respondents were female. The authors divide the population into four groups based on personality traits. Regrettably, results relating to anal intercourse are presented only in the percentages of each group who, "Have never tried anal intercourse, unlikely that they will," (70, 48, 56, and 47%). Even if one assumes that the remaining percentage of each group (30, 52, 44, and 53%) *did* engage in anal intercourse, the number of persons in each group is not indicated. Thus it is not possible to calculate the overall percentage of the sample that engaged in anal intercourse. Nor does the presentation separate out the percentages pertaining to women from those for men.

In an underappreciated study, Hunt (1974) reported results from questionnaire interviews of some 2,000 people, half of the group female and half male. The population was carefully selected to be representative of people from around the nation in terms of a wide range of socioeconomic, age, and racial characteristics. Of married heterosexual couples, 25% of those under age 35 indicated that they had participated in anal intercourse during the past year. Fourteen percent of those between ages 35 and 45 had, perhaps indicating a temporal evolution of the extent (or at least acknowledgment of) anal intercourse, paralleling the Kinsey oral sex data referred to above.

In a recent Kinsey Institute study, Reinisch, Sanders, and Ziemba-Davis (1990) took sex histories of 262 self-described lesbian women, a startling 119 of whom had sex with men since 1980. Thirty percent (36/119 women) had engaged in penile-anal intercourse since 1980. The Kinsey group also conducted a larger study of heterosexual students at the University of Indiana. At this Midwestern institution, 426 women (23.7%) and 177 men (27%) acknowledged having engaged in anal intercourse (Reinisch, Hill, Sanders, & Ziemba-Davis, 1990).

Interview of 100 women consecutively attending a New Jersey medical school AIDS testing clinic included enquiry into anal intercourse. Twenty-one (21%) of the women reported participating in anal intercourse. Of the 100 women, 82 were white and 18 were women of color (Connolly, Brennan, Winters, & Gocke, 1989). In another AIDS clinic study, chiefly of black and Hispanic heterosexual couples in Brooklyn, New York (Nichols, Paroski, Nieves-Rosa, et al., 1989), "A staggering 53% of the total sample engaged in heterosexual anal-intercourse at least sometimes . . . the incidence of anal sex is much higher than is usually presumed to be true for heterosexuals."

European data from AIDS-related populations parallel those from the United States. In Italy, 80 of 368 (22%) of the female partners of HIV-seropositive men indicated engaging in anal intercourse (D'Arminio-Monforte et al., 1989). Twenty-six (37%) of 70 European women in a European Community Multicentre Study acknowledged anal coitus; 27% of men in the same study did (De Vincenzi & Ancelle-Park, 1989) and in a separate investigation, 35 of 104 women (35%) (De Vincenzi, 1988).

Another important study of anal sex was conducted in the 1970s via clinical interviews of women patients. David Bolling, M.D., a professor of obstetrics and gynecology at the University of Texas in San Antonio, interviewed 526 women sequentially presenting in his medical practice. He reported that about 25% of these patients acknowledged having engaged in anal intercourse with their male partners. Even more striking, 8% of the women reported engaging in anal intercourse regularly for pleasure. The women were racially diverse—292 (56%) Hispanic, 42 (8%) black, 192 (36%) white—and represented a broad range of educational and socioeconomic levels (Bolling, 1977). More recently (1986 to 1987) Bolling has surveyed his current clinic populations again. In interviews of 1,007 healthy women patients consecutively presenting in his gynecology practice, 723 (72%) acknowledged having tried anal intercourse, and 238 (23%) women engaged regularly in anal intercourse for pleasure (Bolling, 1987).

Bolling also noted that at first enquiry, patients very commonly denied engaging in anal intercourse; only at the second or third interview did they acknowledge and discuss this aspect of their sexuality. As noted above, such reticence in selected areas of behavior is to be expected. This may have played a role in some AIDS studies that have failed to identify anal sexuality in subjects—perhaps correctly, perhaps not. For instance, Fischl, Dickinson, Flanagan, and Fletcher (1986) reported that 40 South Florida prostitutes denied receptive anal intercourse, in puzzling contrast to the several other prostitute studies referred to above.

Separating Anal Coitus from "Rear-Entry" Vaginal Coitus and "Anal Stimulation"

In surveying the art and the archeological materials used in their chapter in this volume, Reinisch et al. carefully discriminated between *bona fide* heterosexual anal intercourse and rear-entry vaginal intercourse (Chapter 3). In assessing the anecdotal and survey data from the numerous reports reviewed in the present chapter, the potential confusion of informants about this distinction must be considered as a possible confounding variable. Language clear to researchers may be meaningless or misleading to informants.

Several pieces of evidence moderate against such confusion accounting for many of the cited reports of heterosexual anal intercourse. In Bolling's large studies of racially and ethnically diverse women, he habitually interviewed women who were positioned in gynecological examination stirrups where they could see their own genitals and anus in a mirror, and where the anatomy under discussion was pointed to or touched, minimizing possible confusion. Also, large-scale questionnaires returned by tens of thousands of magazine subscribers rather pointedly segregate or group questions about anal intercourse separately from those about vaginal intercourse and "dog-style" or "rear-entry" intercourse, again minimizing the likelihood of confusion.

Meticulous interviews, such as those already described by the past and present Kinsey staff, by Padian and her coworkers, and by many other investigators, clearly distinguish various sex acts for their subjects, making confusion quite improbable. These studies would not likely have demonstrated a severalfold graver risk of HIV-infection through anal intercourse, compared with vaginal, if their subjects had confused the act with rear-entry vaginal coitus.

Moreover, researchers repeatedly comment upon the relief many women express at the opportunity to discuss the possibility of disease or injury when a penis is inserted into a cavity other than the vagina, or whether condoms or spermicides can or may be used there, too. Many want to know if fecal matter adhering to the penis is dangerous when the male withdraws and then inserts his penis into the vagina.

Some women indicate they enjoy penile-anal intercourse while their genitals are being digitally manipulated. Others state they practice anal sex to avoid pregnancy, menstrual blood, or injury to the hymen prior to marriage.

Finally, data presented below about rectal infection with a range of STDs (particularly where no concomitant genital infection exists) are dependent upon the results of bacterial cultures taken by a physician or nurse from each orifice—thence not dependent upon reportage by the woman patient or informant, even though she often acknowledges anal sex when asked about positive rectal cultures.

Inevitably, some incautious interviewers will collect flawed results, of course. However, the broad concordance between levels of anal intercourse reported where the distinction between rear-entry and anal sex was rather deliberately and carefully made, and the studies in which anal sex was enquired about but not explicitly separated from rear-entry sex, makes confusion between the two sex acts seem a limited problem, if a serious one at all.

As noted by Voeller (1988) and Abramson (1988), more problematic are studies that seem to assume that anal stimulation is equivalent to anal coitus, or fail to make a clear distinction, such as those of Kronhausen and Kronhausen (1960) and the Consumers Union (Brecher, 1984). An otherwise very interesting and valuable survey of 4,244 subjects over age 50, conducted by the Consumers Union, asked women which sexual activities they had tried since age 50, including, "Having your anus stimulated during sex" (Brecher, 1984). The report observed that,

Stimulation of the anus during sex is often considered an exclusively homosexual activity. Our data show that assumption to be a mistake: 16 percent of our heterosexual men and women report that, since age 50, they have had their anus stimulated during sex. Of those who have tried it, 86 percent of the men and 67 percent of women say they like it. (p. 363).

Regrettably, the wordings of the question and of the results leave open whether the data refer to penile-anal penetration, digital stimulation, tongue or lip stimulation, or some other form. Indeed, the question immediately following the one about anal stimulation in the interview enquired about, "Stimulating your clitoris with your fingers during intercourse." Such juxtaposition enhances the possibility that respondents might include experience with digital or other stimulation in their response. The interpretation placed on the question by interviewees was not clarified by their comments: "despite the fact that most of them say they enjoyed it, we failed to find a single discussion of anal sex on our Comment Pages . . . unlike masturbation, cunnilingus, and fellatio" (Brecher, 1984, p. 363).

Wyatt and her associates (Wyatt, Peters, & Guthrie, 1988) interviewed Los Angeles women selected by random-digit telephone recruitment. A subset of 120 white women was asked if they had *ever* engaged in anal sex; 43% said yes. Regrettably, the published record of this research does not make clear whether anal sex meant penile penetration, or if the term was intended (or perceived by subjects) to include other forms of anal stimulation.

Rectal Sexually Transmitted Diseases

Additional data documenting noteworthy levels of heterosexual anal intercourse come from yet another source. Gynecologists and venereologists are aware of significant rates of rectal gonorrhea and other STDs in women presenting at women's clinics as well as in private practice. Menda, Chulani, Yawalker, and Kulkarni (1971), in India, noted that of 30 patients with anorectal STDs, 25 were men, 1 a boy, and 4 women. Inasmuch as anal intercourse was acknowledged by these patients, the investigators assumed that gonorrhea was contracted through sexual contact.

In two separate studies, researchers at the CDC in Atlanta also described female anal gonorrhea. Schmale, Martin, and Domescik (1969) examined and interviewed 369 women sequentially presenting at an STD clinic. Of 206 women with gonorrhea, 109 (53.1%) had rectal gonorrhea. In a subset of 112 women, 55 (49%) had rectal gonorrhea. Seven of these women (9%) had rectal gonorrhea *alone* (i.e., were not cervically, urethrally, or vaginally infected), decreasing the likelihood that they became anally infected, for example, by vaginal seepage and cross-contamination rather than through anal intercourse. McLone, Scotti, Mackey, and Hackney (1968) studied 85 infected women patients at an STD clinic; 43 (50.6%) of the women had rectal gonorrhea and 4 had rectal gonorrhea alone.

The federal government data from the CDC show rather higher levels of rectal gonorrhea than observed by most other investigators. Nevertheless, those who look characteristically detect rectal gonorrhea. Berger and colleagues tested cervical, rectal, and oral smears, from nearly 2,000 women (Berger, Keith, & Moss, 1975; Keith, Berger, & Moss, 1976). Of 191 women with gonorrhea, 49 (26%) had rectal gonorrhea; of those 49 with rectal gonorrhea, 14 (28.6%) had rectal gonorrhea only. Blount and Holmes (1976) reviewed the location of primary syphilis chancres occurring in women throughout the United States during federal Fiscal Year 1974. Of 1,537 women, 40 (2.6%) had anal syphilis.

In a study of males with granuloma inguinale, Goldberg (1964) interviewed 273 infected male blacks in Jamaica and 250 in Chicago about heterosexual rectal coitus:

[The data presented] show that heterosexual rectal coitus is not a rare event. In the Kingston sample, 24.4 percent of the patients had had heterosexual rectal coitus at least once and in the Chicago sample, 25 percent. In both groups only about 20 percent of these indulged "occasionally" or "regularly." It is thus apparent that, while heterosexual rectal intercourse represented only a relatively small percentage of the total sexual activity of the persons concerned, it occurs frequently enough to account for the transmission of the disease. (p. 142)

Goldberg concluded that "Rectal coitus, both hetero- and homosexual in nature, is probably much more common than most societies are willing to admit" (p. 144).

Similarly, in a series of studies in Norfolk, Virginia, Pariser and Marino (1970) compared rectal gonorrhea among women sampled from family clinics, private practices, and obstetric and gynecology clinics. They found a wide range of incidence of rectal gonorrhea, commonly between 5 and 10% and, in one population:

Routine swab of the rectal mucosa revealed that 62 of 307 females had positive cultures—an incidence of 20.1% . . . the cervical culture was negative in 20% of the female patients having positive rectal cultures . . . when questioned, most of the females with positive cultures admitted rectal intercourse with a male partner. (Pariser & Marino, 1970, p. 200)

Physicians often discount Pariser and Marino's 20% figure, Dans (1975), for example, stated that: "the 20% figure is considered to be excessive" (p. 106). Dans and others have, however, made invalid comparisons. Pariser and Marino's 20% figure is the ratio of women with rectal gonorrhea only to those with rectal plus genital gonorrhea. Dans and other critics, on the other hand, improperly compared this ratio with others' published data in which the ratio of *women with rectal gonorrhea only* to *women with any form of gonorrhea* had been made (e.g., cervical plus rectal plus urethral plus vaginal, in all possible combinations).

Voeller (1988) reviewed 30 published studies of rectal gonorrhea in populations of women. Twenty-seven studies reported women with gonorrhea restricted to the rectum. Voeller found that if the comparisons are put on the correct, comparable footings to those of Pariser and Marino, 7 of the 27 studies revealed gonorrhea in only the rectum at levels of 19% or higher, several being even higher than Pariser and Marino's results, and as high as 29%.

Jones et al. (1985), in Indiana, described rectal chlamydial infection in 64 of 1,227 (5.2%) women tested, of whom 2.7% had negative genital cultures; 4.1% of the women gave a history of anal intercourse. Higher prevalences of rectal chlamydia in women (11.8% of 110 women) were reported by Darougar, Kinnison, and Jones (1971), as well as by Stamm et al. (1982), who found rectal infection in 32 of 150 (21%) women at an STD clinic; over half of these women acknowledged anal intercourse. Oriel described anal viral disease in women as well. He reported eight female patients with *anal* warts who lacked *genital* warts. Five of the eight women acknowledged anal intercourse (Oriel, 1971).

Anorectal STDs in women have usually been attributed to vaginal discharges spreading to the anus and infecting it (Bergman, Bruns, & Mikulicz, 1905; Blau, 1961; Oriel, 1971) or to "the careless use of toilet paper" (Brunet and Salberg, 1936). Judson (1984) has more recently argued this viewpoint, indicating that,

> Approximately 40 per cent of heterosexual women with endocervical gonorrhea will have infection of the anorectum, which I believe is most often attributable to the closeness of anal and vaginal orifices and contiguous spread of infectious secretions, and not to anal intercourse. (p. 182)

Judson offers no support for his view, nor do Coghill and Young (1989) for their similar viewpoint.

Inconsistent with such an interpretation are the data presented above from the work of Schmale et al. (1969), McLone et al. (1968), Pariser and Marino (1970), and Oriel (1971), and those reviewed by Voeller (1983,

1988). In each case, a significant incidence of rectal STD was found despite the absence of cervical pathogens. Nor are routine failures of the testing methodology sufficient to account for these cases (see, e.g., Schmale et al., 1969, p. 314). That laboratory error in *vaginal* culturing of *Neisseria* could occur in 20% of the women in Pariser and Marino's (1970) study seems improbable, the more so inasmuch as most of these women acknowledged rectal intercourse.

Recently, Voeller analyzed 38 published reports in which 32 found the anorectal site to be the *sole region which was infected with gonorrhea*; the sole-site percentages ranged from as low as 2% to as high as 28%. Fifteen of the 32 reports documenting sole-site occurrence recorded percentages of 10% or greater for the fraction these represented of all cases where any rectal infection was involved.

Although in a minority of his professional colleagues, Jensen, too, has suggested that rectal gonorrhea in females is caused by penile-anal coitus (T. Jensen, 1953). Perine and Osoba (1984), in discussing lymphogranuloma venereum, stated, "In women, the rectal mucosa can also be directly inoculated with chlamydia during anal intercourse, or it can be contaminated by migration of infectious vaginal secretions or by lymphatic spread from the cervix and posterior vaginal wall" (p. 284).

In the present writer's view, the data reviewed in this chapter, indicating the presence of rectal infections *without concurrent cervical or vaginal infection*, argue more strongly that female rectal gonorrhea, and probably other STDs, is linked with rectal intercourse than linked with seepage-contamination from the vagina, although the existence of each route may sometimes occur. The fact that several research teams obtained acknowledgment of anal intercourse from the women they interviewed further buttresses this perspective. It should, of course, be acknowledged that various STD agents, or even strains of one agent, may have different infectivities. This may be true in comparing agents with one another and in comparing target tissues' and organs' susceptibilities.

Summation of Heterosexual Anal Coitus Data

Individually, many (not all) of the pieces of evidence presented here are too small or too specialized in their sampling range to permit very solid conclusions to be drawn about the extent of heterosexual anal intercourse. However, collectively, they sum to an impressive data base. The U.S. geographic range of the studies cuts from the East to the West coasts, and from the northern Midwest to San Antonio, Texas. The surveyed populations are socioeconomically diverse both from study to study and within many, such as the large Texas investigations conducted by Bolling and those of Hunt and of Bell et al. They are racially diverse from study to study, and often within studies. The Hunt survey was a "scientifically balanced" survey with respect to most standard population survey attributes, although nearly all such surveys lack important resolving power so far as ethnic, racial, and many other dimensions are concerned. Future surveys will need to include readily understood anatomical and behavioral illustrations and language, and use the ordinary (street) language of the subjects being interviewed.

In summary, until large, carefully thought out, unified national surveys are conducted (and none adequate to the task is even in the planning stage), it is difficult to know just how extensive heterosexual anal intercourse is. Meanwhile, the AIDS and STD crises rage, and some estimate of the extent of anal intercourse seems crucial. Despite the sampling and interviewing flaws built into most existing surveys, research to date—gathered under difficult societal strictures—nevertheless permits us to see some broad brush strokes that reveal common sexual practices in our culture that are not adequately recognized or addressed. Kinsey Institute staff members Reinisch, Sanders, and Ziemba-Davis estimated that 39% of American women have engaged in at least one experience of heterosexual anal intercourse (Reinisch et al., 1988). Based on the review of the research presented in this chapter, a conservative figure of *at least 10%* seems an appropriate *lower* limit for the number of sexually active American women who engage in anal intercourse with some regularity.

U.S. Population Size for Homosexual and Heterosexual Anal Intercourse

According to the 1987 U.S. census figures (U.S. Bureau of the Census, 1988), the American female population between ages 15 and 64 years is about 81.6 million. Based on the estimate that 10% of women engage with some frequency in anal intercourse, roughly 8 million women are involved. Such figures are necessarily rough approximations, not only because of limits on confidence in the survey data, but because no refinement of the calculation has been made to adjust for celibacy, age-related variation in annual total sexual outlet, and so forth. Nor do these numbers adequately acknowledge sexuality among females on either side of the age range selected. Nevertheless, it seems reasonable to state that quite a few million U.S. women are potentially at risk of contracting the AIDS virus from male spouses through anal intercourse. These male spouses may be intravenous drug users, bisexuals, hemophiliacs, and the like.

This startling number of women at risk through anal intercourse probably exceeds the number of homosexual men at similar risk, judging from estimates of the size of the male population participating in same-gender sex. The former Director of the Kinsey Institute, Prof. Paul Gebhard, has estimated that 9% of American males have substantial homosexual experience (Voeller, 1990). This translates to about 7 million homosexually or bisexually active males among the 79.6 million U.S. men between the ages of 15 and 64. Although subject to some methodological limitations, a more recent Kinsey-linked survey suggested that the percentage may be lower, between 3.3 and 6.2% (Fay, Turner, Klassen, & Gagnon, 1989; Turner, 1989), still further increasing the heterosexual female–to–homosexual/bisexual male ratio for anal intercourse.

In addition, only a portion of gay men engage regularly in anal intercourse, many preferring oral sex or mutual masturbation, for example. The actual percentages and frequencies of practicing anal intercourse, and the extent to which the practice is *receptive* anal intercourse, vary considerably from one gay subculture to another, and differ among racial and ethnic groups. Highest levels are reported among self-identified gays, and among younger gays given to a wider range of experimentation and to a lower level of "gender role playing" than characterized an earlier generation of homosexual males. Thus, the selection of research populations even from among self-identified gays, let alone those from different, sometimes more diverse, subcultures, may lead to the observation of differences in reported levels of anal intercourse and of differences in the extent of participation by individual men in *both* receptive and insertive anal intercourse.

As Tripp (1975) pointed out: "Certainly oral and anal techniques are common in homosexuality, but in large segments of its population there is a distinct, often exclusive, preference for various forms of mutual masturbation and femoral (between thighs) intercourse" (p. 94). Ford and Beach (1951) stated that, "Only 16 percent of [American homosexuals and bisexuals] use anal and interfemoral copulation" (p. 126). McCary (1973) claimed anal sex between males is the preferred sexual practice in under 50% of them, an observation in keeping with the results of A. P. Bell and Weinberg's (1978) sizeable sampling and that of Henry (1941). In Hunt's (1974) careful study, he reported that anal intercourse, "commonly thought to be all but universal among homosexual males, has been experienced as inserter by only 20 percent of all males with any homosexual experience, and by 18 percent as insertee" (p. 318).

All these findings are in contrast with other data. These differences are probably a consequence of the differing subcultures, ages, and ethnic backgrounds of the informants and the limited access researchers have to populations of homosexual and bisexual men outside gay social and political institutions and venues.

Among 4,437 gay males recruited for the nation's largest prospective AIDS epidemiological study, located at four centers across the United States, 87% of the men reported engaging in receptive anal intercourse during the preceding 2 years, as mentioned by Detels (see Chapter 1). Again, however, the recruiting methods used in this study necessarily elicited urban, gay-identified populations of subjects particularly deeply concerned about the risk of AIDS, and therefore almost certainly non-representative of homosexual males nationally. In a similar study, Jeffries et al. (1985) compared histories of anal intercourse in HIV seronegative and seropositive homosexual men. Of 141 seropositive men, 56% acknowledged insertive anal intercourse and 65% receptive, whereas 51% of seronegative men engaged in insertive intercourse and 40% in receptive. In Spada's (1979) survey of 1,038 gay men associated with the gay movement, 76% reported enjoying anal intercourse; 51% engaged both in receptive and insertive acts, 23% only in receptive, and 19% exclusively in insertive sex.

Burton, Burn, Harvey, Mason, and McKerrow (1986) reported that 53% of 310 gay men interviewed in London engage in anal intercourse. McWhirter and Mattison's (1984) study of 156 male couples included 71% who engaged in anal intercourse, the majority alternating between inserter and insertee roles. In a Pittsburgh AIDS-linked study,

> Nine hundred fifty-five (69 per cent) of the 1,384 men [interviewed] reported engaging in anal intercourse (either receptive and/or insertive) within the past six months. The vast majority of men reported engaging in both insertive and receptive anal intercourse with only 11 per cent reporting that they were exclusively insertive and 8 per cent that they were exclusively receptive. (Valdiserri et al., 1988, p. 802)

Table 19-3 presents data from Bell and Weinberg's interviews of nearly 700 San Francisco gay men, showing their attitudes toward anal sex and their history of participation in it during the preceding 12 months. Black males and white males showed distinct differences in each respect (Bell & Weinberg, 1978). A more recent AIDS-based research comparison of San Francisco homosexual men no longer documented a significant racial difference in anal intercourse (Samuel & Winkelstein, 1987).

In Australia, Connell and Kippax (1990), interviewing 535 gay males, found that 95% had tried anal intercourse and about half had practiced it in the past 6 months. About half of the men also ranked this sexual act as the most physically satisfying sex act of various practices. Of those who reported engaging in the practice, 72.9% did so often or occasionally as inserter with condoms, and 44.3% without condom. As often or occasional receptive partner, 72.8% did so with condoms, 42.3% without condoms (Connell & Kippax, 1990, Table 6, p. 193).

Fear of AIDS, coupled with safer sex educational campaigns heavily targeted at U.S. gay males, has led to substantial changes in sexual

Table 19-3
Insertive Versus Receptive Anal Coitus among San Francisco
Homosexual Men[a]

	White Males	Black Males
Performing anal intercourse	N = 575	N = 111
Never	22%	10%
Once or a few times	16	11
More than a few times	40	37
Once a week or more	22	42
Receiving anal intercourse	N = 574	N = 111
Never	33%	22%
Once or a few times	18	20
More than a few times	30	32
Once a week or more	19	26
Favorite sexual activity	N = 552	N = 108
Performing anal intercourse	26%	44%
Receiving anal intercourse	5	11

[a] Adapted from Bell and Weinberg (1978).

behavior. AIDS education campaigns have also affected the numbers of men engaging in receptive anal intercourse in Costa Rica (Mata & Herrera, 1988), Mexico (Carrier, 1989b), and Australia (Connell & Kippax, 1990).

Selected other cultures also differ in the extent and sort of homosexual activity. Among the Siwan of Africa, "all men and boys engage in anal intercourse . . . and males are singled out as peculiar if they do not indulge in these homosexual activities" (Ford & Beach, 1951, pp. 131–132) as is true in parts of New Guinea, among many Australian aborigines, and among native Amazonians (see reviews by Morin [1986], Ford & Beach [1951, pp. 131–132, 177–178], and Dynes [1990]).

In Costa Rica, 201 homosexual men of different classes acknowledged differing levels of receptive anal intercourse (71% "gay," 91% prostitutes, 72% prisoners). At least two thirds of each category engaged in both insertive and receptive anal intercourse (Mata, Ramirez, & Rosero, 1988). Similar information is found in Carrier's (1989b) studies in Mexico. Carrier noted:

Because of the lack of stigmatization in Mexico of the anal inserter participant in homosexual encounters, most Mexican males are not fearful of bisexuality . . . it appears that for any given age group, more sexually active single males in Mexico have had sexual intercourse with both genders than have Anglo-American males. (Carrier, 1989a, pp. 227–228)

Thus, to wrap this all together, it seems quite probable that considerably more heterosexual women engage in *some* form of penile-anal intercourse than do homosexual and bisexual men with one another—quite probably well over twice as many. If only *receptive* penile-anal intercourse is considered, the figures for the numbers of heterosexual women weigh in even higher yet than those for homosexual men. If one takes these somewhat fragile estimates to their numerical conclusions, the number of women engaging in receptive anal intercourse may be more than four times that of men, or even greater. To be sure, most of these women are likely to have (a) fewer partners, (b) a lower frequency of anal sex, and (c) a lower transmission rate to their partners than characterizes many of their homosexual or bisexual male counterparts.

Women as *"Cul-de-Sacs"* for HIV Transmission

The research of Padian and her associates (Padian, Carlson, et al., 1987; Padian, Marquis, et al., 1987), Gill et al. (1986), Steigbigel et al. (1987, 1988), Day et al. (1989), Sion et al. (1988), De Vincenzi (De Vincenzi, 1988; De Vincenzi & Ancelle-Park, 1989), and D'Arminio-Monforte et al. (1989), discussed above, instructs us that heterosexual anal intercourse is indeed a route of HIV infection in the United States and elsewhere. The numbers of women engaging in this sex act suggests that AIDS researchers should be much more aware of it than most are. Heterosexual anal intercourse could constitute an important "hot spot" for those predicting the course of the spread of AIDS. In looking for forthcoming explosions of HIV infection, however, it is important to understand patterns of behavior as well as patterns of previous HIV dispersal.

Patterns of the heterosexual spread of AIDS strikingly differ between Africa (and Belgium) and the Western industrialized nations. In most of Europe and North America, women appear to be more readily infected with HIV than are heterosexual males. While too few data exist in Western nations to say with certainty that heterosexual men are less susceptible than women to infection through penile-vaginal contact (see review by Padian, 1990), most researchers believe this to be the case, paralleling the lesser risk *insertive* partners face, compared to receptive partners, in homosexual anal coitus. Where presumed monogamy exists, the husbands of sero-positive (through blood transfusion) Saudi Arabian women have remained sero-negative for several years (Al-Nozha et al., 1990). By contrast, in Africa, both heterosexual males and females have become infected, apparently because of persistent genital STD ulcers—common in Africa but rarer in the West, where diagnosis and treatment are far more accessible. Genital ulcerative disease in Africa provides a route for a male's infection by an infected woman. The male then, in turn, can carry HIV to other females. Thus males and females each act as strong vectors for spreading the virus ("AIDS in," 1987; Cameron et al., 1989; Clumeck, Van de Perre, Carael, Rouvroy, & Nzaramba, 1985; Greenblatt et al., 1988; Johnson & Laga, 1990; Quinn, Mann, Curran, & Piot, 1986).

Similarly, spread of HIV among homosexual males has almost certainly been exacerbated as a consequence of many partners in anal intercourse serving both as recipients and donors of HIV, through participating sequentially as *both* anal receptor and inserter during anal coitus, as described above. Large percentages of self-identified gay males engage both in receptive and in insertive anal intercourse. Thus, the male infected via his rectum may in turn infect multiple partners through his semen when he participates in insertive anal intercourse (Voeller, 1988).

The pattern of HIV spread among heterosexuals in Western industrialized nations will be significantly different from either the heterosexual African experience or the "Western" homosexual experience. Neither is an adequate model for heterosexual spread of AIDS in industrialized nations; here males, through their semen, are the principal sexual propagators of AIDS. But precisely because of the lower frequency of sexual spread of HIV *to* heterosexual Euro-American males (i.e., their seeming lower sexual infectability), female partners infected anally (or any other way) are much less likely to communicate HIV sexually to their partners than are anal-receptive homosexual or bisexual males or heterosexual African males. Thus, Western women *tend* to be terminators in infectivity chains; homosexual men engaging in both receptive and insertive anal intercourse are chain initiators and propagators, as are anal-receptive bisexual men (Voeller, 1988). In short, Euro-American women engaging in anal sex will *tend* to be less effective in sexually spreading HIV than their homosexual counterparts, even though many more such women exist.

While the differential capacities of homosexual males and of females to continue an infectivity chain make the heterosexual spread of HIV through anal intercourse less likely on a per-sexual-contact basis, the sheer numbers of women engaging in anal intercourse, and the extent of anal intercourse with males who have at least some coitus with other males or with intravenous drug users, suggests that heterosexual anal intercourse is a substantial potential reservoir for AIDS.

To the extent that homosexual or bisexual men engage reciprocally in serving as insertive *and* receptive partner in anal coitus, alternating the roles, as described above for some American gay subcultures, they will be more likely to continue the HIV chain of infection than will women. To the extent that these sex roles are not mixed by homosexual and bisexual males, the spread will be reduced (Voeller, 1988; Wiley & Herschkorn, 1989). Slower spread of HIV in southeastern Europe, the

eastern Mediterranean, and the Middle East, for example, has been attributed to a greater degree of role separation by males there into those
males more exclusively engaging only in insertive anal coitus and those
engaging only in receptive coitus (Trichopoulos, Sparos, & Petridou,
1988).

Conclusions

The foregoing data and discussion indicate that anal sex among heterosexual Americans is a fairly well-kept secret, possibly involving about
16 million people or more. The 8 million women in these numbers are
at some risk for contracting the AIDS virus, especially in those parts of
the nation with high prevalence of the virus and where intravenous
drug addiction is high.

Other risks also exist from anal intercourse. Voeller (Voeller, 1983a,
1983b), has pointed out that use of oil-based lubricants to facilitate anal
intercourse may be linked with the higher incidence of anal cancer in
homosexual males than is found in heterosexual males, although other
factors such as repeated infection with STDs, including herpes and pap-
illoma viruses, may well also be of significance . . . facts that need to
be communicated to both homosexual and heterosexual individuals and
couples.

In terms of other cultures and societies around the world, where heterosexual anal intercourse is practiced (e.g., Brazil, with the third highest number of reported AIDS cases in the West, and many Hispanic and
Arab countries) for pleasure, for contraception, to avoid menstrual
blood, or to preserve vaginal "virginity" for marriageable women (Parker, 1987; Voeller, 1983, 1988; Wilems, 1953), health care providers
should be alert to the diverse medical and psychological needs of both
heterosexual women and men, and to recognize that these cultures are
potential AIDS "hot spots." This task will be the more difficult for Westerners inasmuch as some governments do not wish their countries to
seem sexually decadent by Western standards, and inasmuch as surveys
of sexual behavior are even more difficult to conduct wisely in other
cultures than in our own.

In any case, a nation that in direct mail campaigns can survey and
determine which *color* envelope is the most likely to be opened by an
addressee, or which of two television shows has more viewers (and
why), seems to find surveying such urgent matters as the extent of anal
intercourse beyond its capacity. In the face of death and of misguided
prevention planning linked to lack of adequate sexual information in
the midst of a sexual AIDS crisis, this irony is particularly dismaying.

In discerning the nature and extent of the sexual practices in America,
we need to set aside our Judeo-Christian discomfort with sexuality and

have ongoing, properly conducted sexual surveys—sensitive to the immense racial, ethnic, and other diversities of our people, to their inhibitions, to their innumerable English and other language diversities, and to their conceptualizing dissimilarities. Until that occurs our laws will fail in equity and wisdom, our national understanding and tolerance will be attenuated, and our health and health policies will remain naive and dangerous—as is currently the status quo.

References

Abramson, P. R. (1988). Sexual assessment and the epidemiology of AIDS. *Journal of Sex Research, 25,* 323–346.

Abramson, P. R. (1990). Sexual science: Emerging discipine or oxymoron? *Journal of Sex Research, 27,* 147–165.

Abramson, P. R., & Herdt, G. (1990). The assessment of sexual practices relevant to the transmission of AIDS: A global perspective. *Journal of Sex Research, 27,* 215–232.

Angew, J. (1982). Klismaphilia—A physiological perspective. *American Journal of Psychotherapy, 35,* 554–566.

AIDS in Africa. (1987). *AIDS-Forschung (AIFO) 1,* 5–25.

Al-Nozha, M. Ramia, S., Al-Frayh, A., & Arif, M. (1990). Female to male: an inefficient mode of transmission of human immunodeficiency virus (HIV). *Journal of Acquired Immune Deficiency Syndromes, 3,* 193.

Athanasiou, R., Shaver, P., & Tavris, C. (1970, July). Sex. *Psychology Today,* pp. 39–42.

Barton, S. E., Underhill, G. S., Gilchrist, C., Jeffries, D. J., & Harris, J. R. W. (1985). HTLV-III antibody in prostitutes. *Lancet, 2,* 1424.

Bateson, M. C., & Goldsby, R. (1988). *Thinking AIDS.* New York: Addison-Wesley Publishing Company, Inc.

Bell, A. P., & Weinberg, M. S. (1978). *Homosexualities: A study of diversity among men and women* (pp. 107–108). New York: Simon and Schuster.

Bell, R. R., Turner, S., & Rosen, L. (1975). A multivariate analysis of female extramarital coitus. *Journal of Marriage and the Family, 37,* 375–384.

Benjamin, H. B. (1969). Removal of exotic foreign objects from the abdominal orifices. *American Journal of Proctology, 20,* 413–417.

Berger, G. S., Keith, L., & Moss, W. (1975). Prevalence of gonorrhoea among women using various methods of contraception. *British Journal of Venereal Diseases, 51,* 307–309.

Bergman, E. V., von Bruns, P., & von Mikulicz, J. (1905). *A system of practical surgery* (W. T. Bull & E. M. Foote, Eds. and Trans.) (Vol. 5, p. 144). London: Williams and Norgate.

Blau, S. (1961). The venereal diseases. In A. Ellis & A. Abarbanel (Eds.), *The encyclopedia of sexual behavior* (Vol. II). New York: Hawthorn Books.

Blount, J. H., & Holmes, K. K. (1976). Epidemiology of syphilis and the non-venereal treponematoses. In: *The biology of parasitic spirochetes* (pp. 157–176). Washington, DC: U.S. Department of Health, Education and Welfare.

Bolling, D. R. (1976). Heterosexual anal intercourse: An illustrative case history. *Journal of Family Practice, 3,* 557–558.

Bolling, D. R. (1977). Prevalence, goals and complications of heterosexual anal intercourse in a gynecologic population. *Journal of Reproductive Medicine, 19*, 120–124.

Bolling, D. (1987, December). *Heterosexual anal intercourse: A common entity, perceived rarity, neglected patients and ostrich syndrome.* Paper presented at the 1987 Kinsey Institute Conference on AIDS and Sex: An Integrated Biomedical and Biobehavioral Approach, Bloomington, IN.

Bolling, D. R., & Voeller, B. (1987). AIDS and heterosexual anal intercourse. *Journal of the American Medical Association, 258*, 474.

Boyko, W. J., Schechter, M. T., Craib, K. J. P., Constance, P., Nitz, R., Fay, S., McLeod, A., & O'Shughnessy, M. (1986). The Vancouver lymphadenopathy-AIDS study: 5. Antecedent behavioral, clinical and laboratory findings in patients with AIDS and HIV-seropositive controls. *Canadian Medical Association Journal, 135*, 881–887.

Brecher, E. M. (1984). *Love, sex and aging. A Consumers Union report.* Boston: Little, Brown and Company.

Brunet, W. M., & Salberg, J. B. (1936). Gonococcus infection of the anus and rectum in women: its importance, frequency and treatment. *American Journal of Syphilis Gonorrhea and Venereal Disease, 20*, 37–44.

Burton, S. W., Burn, S. B., Harvey, D., Mason, M., & McKerrow, G. (1986). AIDS informatioan (letter). *Lancet, 2*, 1040–1041.

Cameron, D. W., Simonsen, J. N., D'Costa, L. J., Ronald, A. R., Maitha, G. M., Gakinya, M. N., Cheang, M., Ndinya-Achola, A. O., Piot, P., Brunham, R. C., & Plummer, F. A. (1989). Female to male transmission of human immunodeficiency virus type 1: Risk factors for seroconversion in men. *Lancet, 2*, 403–407.

Carrier, J. M. (1989a). Gay liberation and coming out in Mexico. In G. Herdt (Ed.), *Gay and lesbian youth.* New York: The Haworth Press, Inc.

Carrier, J. M. (1989b). Sexual behaviour and spread of AIDS in Mexico. *Medical Anthropology, 10*, 129–142.

Christenson, C. V. (1971). *Kinsey: A biography.* Bloomington: Indiana University Press.

Clumeck, N., Van de Perre, P., Carael, M., Rouvroy, D., & Nzaramba, D. (1985). Heterosexual promiscuity among African patients with AIDS. *New England Journal of Medicine, 313*, 182.

Coghill, D. V., & Young, H. (1989). Genital gonorrhoea in women: A serovar correlation with concomitant rectal infection. *Journal of Infection, 18*, 131–141.

Cohen, J. B., Hauer, L. B., Poole, L. E., Cracchiolo, B. M., Levy, A., & Wofsy, C. B. (1986, June). *Prevalence of AIDS antibody and associated risk factors in a prospective study of 400 San Francisco prostitutes and other sexually active women* (Abstr). The Second International Conference on AIDS, Paris.

Cohen, J. B., Hauer, L. B., Poole, L. E., & Wofsy, C. B. (1987, June). *Sexual and other practices and risk of HIV infection in a cohort of 450 sexually active women in San Francisco* (Abstr.). The Third International Conference on AIDS, Washington, DC.

Collier, A. C., Meyers, J. D., Corey, L., Murphy, V. L., Roberts, P. L., & Hansfield, H. H. (1987). Cytomegalovirus infection in homosexual men. *American Journal of Medicine, 82*, 593–601.

Comfort, A. (1972). *The joy of sex: A cordon bleu guide to lovemaking.* New York: Crown Books.

Connell, R. W., & Kippax, S. (1990). Sexuality in the AIDS crisis: Patterns of sexual practice and pleasure in a sample of Australian gay and bisexual men. *Journal of Sex Research, 27*, 167–198.

Connolly, D., Brennan, A., Winters, D., & Gocke, D. (1989, June). *Prevalence of anal intercourse in women at a counseling testing site.* Paper presented at the Fifth International Conference on AIDS, Montreal.

Cornthwaite, S. A., Savage, W. D., & Willcox, R. R. (1974). Oral and rectal coitus amongst female gonorrhoea contacts in London. *British Journal of Clinical Practice, 28,* 305–306.

Daling, J. R., Weiss, N. S., Hislop, T. G., Maden, C., Coates, R. J., Sherman, K. J., Ashley, R. L., Beagrie, M., Ryan, J. A., & Corey, L. (1987). Sexual practices, sexually transmitted diseases, and the incidence of anal cancer. *New England Journal of Medicine, 317,* 973–977.

Dans, P. E. (1975). Gonococcal anogenital infection. *Clinical Obstetrics and Gynecology, 18,* 103–119.

D'Arminio-Monforte, A., Galli, M., Vigevani, G. M., Vigano, P., Saracco, A., Rizzardini, G., & Valsecchi, L. (1989, June). *Clinical aspects in the first 81 AIDS cases among women in Milan* (Abstr). The Fifth International Conference on AIDS, Montreal.

Darougar, S., Kinnison, J. R., & Jones, B. R. (1971). Chlamydial isolates from the rectum in association with chlamydial infection of the eye or genital tract. I. Laboratory aspects. In R. L. Nichols (Ed.), *Trachoma and related disorders caused by chlamydial agents,* (pp. 501–506). Amsterdam: Exerpta Medica.

Darrow, W. W., Echenberg, D. F., Jaffe, H. W., O'Malley, P. M., Byers, R. H., Getchell, J. P., & Curran J. W. (1987). Risk factors for human immunodeficiency virus (HIV) infections in homosexual men. *American Journal of Public Health, 77,* 479–483.

Day, N. A., Des Londe, J., Houston-Hamilton, A., & Steiger, G. (1989). *A baseline survey of AIDS risk behaviors and attitudes in San Francisco's black communicites.* San Francisco: Center for AIDS Prevention and Education.

De Vincenzi, I. (1988, June). *Heterosexual transmission of HIV: A European community multicentre study* (Abstr). The Fourth International Conference on AIDS, Stockholm.

De Vincenzi, I., & Ancelle-Park, R. (1989). *Heterosexual transmission of HIV: A European study. II: Female-to-male transmission.* Poster presented at the Fifth International Conference on AIDS, Montreal.

Drew, W. L. (1982). Epidemiology of cytomegalovirus. In L. M. de la Maza & E. M. Peterson (Eds.), *Medical virology,* (pp. 183–195). New York: Elsevier Science Publishing.

Dynes, W. R. (Ed.). (1990). *Encyclopedia of homosexuality.* New York: Garland Publishing, Inc.

Fauconney, J. (1902). *Le masturbation et sodomie feminine par le docteur Caufevnon.* Paris: Administration de la Vie en Cullotte Rouge.

Fay, R. E., Turner, C. F., Klassen, A. D., & Gagnon, J. H. (1989). Prevalence and patterns of same-gender sexual contact among men. *Science, 243,* 338–348.

Fischl, M. A., Dickinson, G. M., Flannagan, S., & Fletcher, M. A. (1986, June). *Human T-lymphotrophic virus type III infection among female prostitutes* (Abstr.). The Second International Conference on AIDS, Paris.

Ford, C. S., & Beach, F. A. (1951). *Patterns of sexual behavior.* New York: Harper & Brothers, Publishers.

Friedman-Kien, A., Laubenstein, L., Marmor, M., Hymen, K., Green, J., Ragaz, A., Gottlieb, J., Muggia, F., Demopoulos, R., Weintraub, M., Williams, D., et al. (1981). Kaposi's sarcoma and *Pneumocystis* pneumonia among homosexual men—New York City and California. *Morbidity and Mortality Weekly Reports, 30,* 305.

Gebhard, P. H., Gagnon, J. H., Pomeroy, W. B., & Christenson, C. V. (1965). *Sex offenders: An analysis of types.* New York: Harper & Row, Publishers.

Gebhard, P. H., & Johnson, A. B. (1979). *The Kinsey data: Marginal tabulations of the 1938–1963 interviews conducted by the Institute for Sex Research.* Philadelphia: W. B. Saunders Company.

Gill, P. S., Levine, A. M., Ross, R., Aguilar, S., Karilo, M., & Rasheed, S. (1986, June). *Prevalence of antibody to HTLV-III in female prostitutes from Los Angeles* (Abstr). The Second International Conference on AIDS, Paris.

Goldberg, J. (1964). Studies on granuloma inguinale: vii. Some epidemiological considerations of the disease. *British Journal of Venereal Diseases, 40,* 140–145.

Gottlieb, M. S., Schanker, H. M., Fan, P. T., Saxon, A., & Weissman, J. D. (1981). *Pneumocystis* pneumonia—Los Angeles. *Morbidity and Mortality Weekly Reports, 30,* 250.

Greenblatt, R. M., Lukehart, S. A., Plummer, F. A., Quinn, T. C., Critchlow, C. W., Ashley, R. L., D'Acosta, L. J., Ndinya-Achola, J. O., Corey, L., Ronald, A. R., & Holmes, K. K. (1988). Genital ulceration as a risk factor for human immunodeficiency virus infection. *AIDS, 2,* 47–50.

Guerra, F. (1971). *The pre-Columbian mind: a study into the aberrant nature of sexual drives, drugs affecting behaviour, and the attitude towards life and death, with a survey of psychotherapy, in pre-Columbian America.* London: Seminar Press.

Guinan, M. E., & Hardy, A. (1987). Epidemiology of AIDS in women in the United States. *Journal of the American Medical Association, 257,* 2039–2042.

Haft, J. S. (1973). Foreign bodies in the rectum: Some psychosexual aspects. *Medical Aspects of Human Sexuality, 7,* 74–95.

Haft, J. S. (1976). Vaginal vibrator lodges in rectum. *British Journal of Medicine, 1,* 1625–1626.

Henry, G. W. (1941). *Sex variants: A study of homosexual patterns.* New York: Paul B. Hoeber, Inc.

Hite, S. (1976). *The Hite report: A national study of female sexuality.* New York: McMillan Publishing Company, Inc. [Anal sex: Questionnaire Version I, question #26, p. 403. Questionnaire II, question #24, p. 405. Questionnaire IV, question #24, p. xv. Oral sex: pp. 232–247. Hite's data appear on different pages in different printings of her book. Anal sex data appears on p. 427 in one edition and on pp. 618–619 in another.]

Hoyle, R. L. (1965). *Checan: Essay on Erotic Elements in Peruvian Art.* Geneva: Nagel Publishers.

Hunt, M. (1974). *Sexual behavior in the 1970s.* Chicago: Playboy Press.

Jaffe, L. R., Seehaus, M., Wagner, C., & Leadbeater, B. J. (1988). Anal intercourse and knowledge of acquired immunodeficiency syndrome among minority group female adolescents. *Journal of Pediatrics, 112,* 1005–1007.

Jeffries, E., Willoughby, B., Boyko, W. J., Schechter, M. T., Wiggs, B., Fay, S., O'Shaughnessy, M. O. (1985). The Vancouver lymphadenopathy-AIDS study: 2. Seroepidemiology of HTLV-III antibody. *Canadian Medical Association Journal, 132,* 1373–1377.

Jensen, G. D. (1975). Anal coitus by married couples. *Medical Aspects of Human Sexuality, 9,* 115.

Jensen, T. (1953). Rectal gonorrhoea in women. *British Journal of Venereal Diseases, 29,* 222.

Johnson, A. M., & Laga, M. (1990). Heterosexual transmission of HIV. In N. J.

Alexander, H. L. Gabelnick, & J. Spieler (Eds.), *The heterosexual transmission of AIDS*. New York: Alan R. Liss Publishing Company.

Jones, R. B., Babinovitch, R. A., Katz, B. P., Batteiger, B. E., Quinn, T. S., Terho, P., & Lapworth, M. A. (1985). *Chlamydia trachomatis* in the pharynx and rectum of heterosexual patients at risk for genital infection. *Annals of Internal Medicine, 102,* 757–762.

Judson, F. N. (1984). Sexually transmitted viral hepatitis and enteric pathogens. *Urologic Clinics of North America, 11,* 177–185.

Katchadourian, H., & Lone, D. T. (1975). *Fundamentals of human sexuality* (2nd ed.). New York: Holt, Rhinehart and Winston.

Katzneslson, S., & Drew, L. (1984). Efficacy of the condom as a barrier to the transmission of cytomegalovirus. *Journal of Infectious Diseases, 150,* 155–157.

Keith, L., Berger, G. S., & Moss, W. (1976). Cervical gonorrhea in women using different methods of contraception. *Journal of the American Venereal Disease Association, 3,* 17–19.

Kingsley, L. A., Detels, R., Kaslow, R., Polk, B. F., Rinaldo, D. R., Chmiel, J., Detre, K., Kelsey, S. F., Odaka, N., Ostrow, D., Van Raden, M., & Vischer, B. (1987). Risk factors for seroconversion to human immunodeficiency virus among male homosexuals. *The Lancet, 1,* 345–348.

Kinsey, A. C., Pomeroy, W. B., & Martin, C. E. (1948). *Sexual Behavior in the human male.* Philadelphia: W. B. Saunders Company.

Kinsey, A. C., Pomeroy, W. B., Martin, C. E., & Gebhard, P. H. (1953). *Sexual behavior in the human female.* Philadelphia: W. B. Saunders Company.

Kronhausen, P., & Kronhausen, E. (1960). *Sex histories of American college men.* New York: Ballantine Books.

Lang D. J., Kummer, J. F., & Hartley, D. P. (1974). Cytomegalovirus in semen: Persistence and demonstration in extracellular fluids. *New England Journal of Medicine, 291,* 121–123.

Leal de Santa Inez, A. (1983). *Habitos e atitudes sexuais dos Brasileiros.* São Päulo: Editora Cultrix.

Levin, R. J., & Levin, A. (1975, September). Sexual pleasure: The surprising preferences of 100,000 women. *Redbook,* pp. 51–58.

Lorian, V. (1988). AIDS, anal sex, and heterosexuals. *The Lancet, 2,* 1111.

Marino, A. W. M., & Mancini, H. W. N. (1978). Anal eroticism. *Surgical Clinics of North America, 58,* 513–518.

Masters, W. H., & Johnson, V. E. (1979). *Homosexuality in perspective* (pp. 84–86). Boston: Little, Brown and Company.

Masters, W. H., Johnson, V. E., & Kolodny, R. C. (1986). *Masters and Johnson on sex and human loving.* Boston: Little, Brown, and Company.

Mata, L., & Herrera, G. (1988). AIDS and HIV infection in Costa Rica—a country in transition. *Immunology and Cell Biology, 66,* 175–183.

Mata, L., Ramirez, G., & Rosero, L. (1989). Tipologia y conducta de riesgo de infeccion con el virus de la immunodeficiencia humana (HIV), de hombres homosexuales de Costa Rica, 1985–1987. *Revista Costarricense de Ciencias Medicas, 8.*

McCary, J. L. (1973). *Human sexuality: Physiological, psychological, and sociological factors* (p. 368). New York: D. Van Nostrand Company.

McCary, J. L. (1978). *McCary's human sexuality* (3rd ed.). New York: Van Nostrand Reinhold Company. [Oral sex: pp. 158–159, 245, 266–267. Anal sex: pp. 149, 337–338]

McLaws, M-L., Oldenburg, B., Ross, M. W., & Cooper, D. A. (1990). Sexual behaviour in AIDS-related research: Reliability and validity of recall and diary measures. *Journal of Sex Research, 27*, 265–281.

McLone, D. G., Scotti, A. T., Mackey, D. M., & Hackney, J. F. (1968). Cephaloridine treatment of gonorrhoea in the female. *British Journal of Venereal Diseases, 44*, 220–222.

McWhirter, D. P., & Mattison, A. M. (1984). *The male couple: How relationships develop.* Englewood Cliffs, NJ: Prentice-Hall, Inc.

McWhirter, D. P., Sanders, S. A., & Reinisch, J. M. (1990). *Homosexuality/heterosexuality: concepts of sexual orientation.* New York: Oxford University Press.

Meares, E. M. (1975). Urethral infection after anal intercourse. *Journal of the American Medical Association, 232*, 549–550.

Menda, R. K., Chulani, H. L., Yawalkar, S. J., & Kulkarni, B. S. (1971). Venereal disease of the anal region. *Diseases of the Colon and Rectum, 14*, 454–459.

Mintz, L., Drew, W. L., Miner, R. C., & Braff, E. H. (1983). Cytomegalovirus infections in homosexual men: An epidemiological study. *Annals of Internal Medicine, 99*, 326–329.

Morin, J. (1986). *Anal pleasure & health: A guide for men and women* (2nd ed., pp. 10–14). Burlingame, CA: Yes Press.

Neubardt, S. (1974). Anal coitus. *Medical Aspects of Human Sexuality, 8*, 74–76.

Nichols, M., Paroski, P., Sampson, G., Leibel, J., & Kennedy, M. J. (1989, June). *Preliminary analysis of factors associated with compliance with guidelines for prevention of sexual transmission of HIV in heterosexual couples.* Poster presented at the Fifth International Conference on AIDS, Montreal.

Oriel, J. D. (1971). Anal warts and anal coitus. *British Journal of Venereal Diseases, 47*, 373–376.

Osmond, D., Bacchetti, P., Chaisson, R. E., Kelly, T., Stemper, R., Carlson, J., & Moss, A. R. (1988). Time of exposure and risk of HIV infections in homosexual partners of men with AIDS. *American Journal of Public Health, 78*, 944–948.

Padian, N. (1990). Heterosexual transmission: Infectivity and risks. In N. J. Alexander, H. L. Gabelnick, & J. Spieler (Eds.), *The heterosexual transmission of AIDS.* New York: Alan R. Liss Publishing Company.

Padian, N., Carlson, J., Browning, R., Nelson, L., Grimes, J., & Marquis, L. (1987, June). *Human immunodeficiency virus (HIV) infection among prostitutes in Nevada* (Abstr). The Third International Conference on AIDS, Washington, DC.

Padian, N., Marquis, L., Francis, D. P., Anderson, R. E., Rutherford, G. W., O'Malley, P. M., & Winkelstein, W. (1987). Male-to-female transmission of human immunodeficiency virus. *Journal of the American Medical Association, 258*, 788–790.

Pariser, H., & Marino, A. F. (1970). Gonorrhea—frequently unrecognized reservoirs. *Southern Medical Journal, 63*, 198–201.

Parker, R. (1987). Acquired immunodeficiency syndrome in urban Brazil. *Medical Anthropology Quarterly,* (NS) *1*(2), 155–175.

Perine, P. L., & Osoba, A. O. (1984). Lymphogranuloma venereum. In P. A. Mardh, K. K. Holmes, J. D. Oriel, P. Piot, & J. Schachter (Eds.), *Sexually transmitted diseases* (pp. 281–291). Amsterdam: Elsevier Biomedical Press.

Peterman, T. A., & Curran, J. W. (1986). Sexual transmission of human immunodeficiency virus. *Journal of the American Medical Association, 256*, 2222–2226.

Peterson, J. R. (1983). The Playboy reader's sex survey. *Playboy, 30*, 108.

Plaut, M. E. (1982). *The doctor's guide to you and your colon: A candid guide to our #1 hidden health complaint.* New York: Harper & Row.

Polk, B. F., Fox, R., Brookmeyer, R., Kanchanaraka, S., Kaslow, R., Vischer, B., Rinaldo, C., & Phair, J. (1987). Predictors of the acquired immunodeficiency syndrome developing in a cohort of seropositive homosexual men. *New England Journal of Medicine, 316*, 61–66.

Quinn, T. C., Mann, J. M., Curran, J. W., & Piot, P. (1986). AIDS in Africa: An epidemiologic paradigm. *Science, 234*, 955–963.

Reinisch, J. M., Hill, C. A., Sanders, S. A., & Ziemba-Davis, M. (1990). Sexual behavior among heterosexual college students. *Focus, 5*, 3.

Reinisch, J. M., Sanders, S. A., & Ziemba-Davis, M. (1988). The study of sexual behavior in relation to the transmission of human immunodeficiency virus. *American Psychologist, 43*, 921–927.

Reinisch, J. M., Sanders, S. A., & Ziemba-Davis, M. (1990). Self-labeled sexual orientation, sexual behavior, and knowledge about AIDS: Implication for biomedical research and education programs. In S. J. Blumenthal, A. Eichler, & G. Weissman (Eds.), *Proceedings of NIMH-NIDA Workshop on Women and AIDS: Promoting healthy behaviors.* Washington, DC: American Psychiatric Press (in press).

Roman, M., Charles, E., & Karasu, T. B. (1978). The value system of psychotherapists and changing mores. *Psychotherapy: Theory, Research and Practice, 15*(4), 409–415.

Rosen, R., & Rosen, L. (1981). *Human sexuality.* New York: Random House.

Rubenstein, C., & Tavris, C. (1987, September). Special survey results: 26,000 women reveal the secrets of intimacy. *Redbook*, pp. 147–215.

Samuel, M., & Winkelstein, W. (1987). Prevalence of human immunodeficiency virus in ethnic minority homosexual/bisexual men. *Journal of the American Medical Association, 257*, 1901–1902.

Savitz, L., & Rosen, L. (1988). The sexuality of prostitutes: Sexual enjoyment reported by "street walkers." *Journal for Sex Research, 24*, 200–208.

Schmale, J. D., Martin, J. E., & Domescik, G. (1969). Observations on the culture diagnosis of gonorrhea in women. *Journal of the American Medical Association, 210*, 312–314.

Sinistrari de Ameno, L. V. (1879). *De Sodomia: Tractatus in quo exponitur doctrina nova de sodomia foeminarum a tribadismo distincta.* Paris: Apud Isidorum Liseux.

Sion, F. S., Morais De Sa, C. A., Rachid De Lacerda, M. C., Quinhoes, E. P., Pereira, M. S., Galvao Castro, B., & Castilho, E. A. (1988, June). *The importance of anal intercourse in transmission of HIV to women* [Abstr]. The Fourth International Conference on AIDS, Stockholm.

Sodomy and the married man. (1969). *University of Richmond Law Review, 3*, 344–347.

Spada, J. (1979). *The Spada report: The newest survey of gay male sexuality.* New York: Signet Book, New American Library.

Stamm, W. E., Quinn, T. C., Mkrtichian, E. E., Wang, S. P., Schuffler, M. D., & Holmes, K. K. (1982). *Chlamydia trachomatis* proctitis. In P. A. Mardh, K. K. Holmes, J. D. Oriel, P. Piot, & J. Schachter (Eds.), *Sexually transmitted diseases* (pp. 111–114). Amsterdam: Elsevier Biomedical Press.

Steigbigel, N. H., Maude, D. W., Feiner, C. J., Harris, C. A., Saltzman, B. V., Klein, R. S., et al. (1987, June). *Heterosexual transmission of infection and disease by the human immunodeficiency virus (HIV)* [Abstr]. The Third International Conference on AIDS, Washington, DC.

Steigbigel, N. H., Maude, D. W., Feiver, C. J., Harris, C. A., Saltzman, B. V., Klein, R. S., et al. (1988, June). *Heterosexual transmission of HIV infection* (Abstr). The Fourth International Conference on AIDS, Stockholm.

Tavris, C., & Sadd, S. (1977). *The Redbook report on female sexuality.* New York: Delacorte Press.

Trichopoulos, D., Sparos, L., & Petridou, E. (1988). Homosexual role separation and spread of AIDS. *The Lancet, 2,* 965–966.

Tripp, C. A. (1975). *The homosexual matrix.* New York: McGraw-Hill Book Company.

Turner, C. F. (1989). Research on sexual behaviors that transmit HIV: Progress and problems. *AIDS, 3*(suppl. 2), S63–S69.

U.S. Bureau of the Census. (1988). United States population estimates, by age, sex, and race: 1980–1987. In *Current population reports: Population estimates and projections* (Series P-25, No. 1022, Table 2, p. 22). Washington, DC: Author.

Valdiserri, R. O., Lyter, D., Leviton, L. C., Callahan, C. M., Kingsley, L. A., & Rinaldo, C. R. (1988). Variables influencing condom use in a cohort of gay and bisexual men. *American Journal of Public Health, 78,* 801–805.

Voeller, B. (1983a). *Anal sex, rectal cancer and oil based lubricants.* Mariposa Occasional Paper #1A. New York: The Mariposa Education and Research Foundation.

Voeller, B. (1983b). Anorectal cancer and homosexuality. *Journal of the American Medical Association, 249,* 2459.

Voeller, B. (1983c). *Heterosexual anal intercourse.* Mariposa Occasional Paper #1B. New York: The Mariposa Education and Research Foundation.

Voeller, B. (1988). *Heterosexual anal intercourse: An AIDS risk factor.* Mariposa Occasional Paper #10. Los Angeles: The Mariposa Education and Research Foundation.

Voeller, B. (1990). Some uses and abuses of the Kinsey Scale. In D. P. McWhirter, S. A. Sanders, & J. M. Reinisch (Eds.), *Homosexuality/heterosexuality: Concepts of sexual orientation.* New York: Oxford University Press.

Voeller, B. (1991). AIDS and heterosexual anal intercourse. *Archives of Sexual Behavior* (in press).

Wilems, E. (1953). The structure of the Brazilian family. *Social Forces, 31,* 339–345.

Wiley, J. A., & Herschkorn, S. J. (1989). Homosexual role separation and AIDS epidemics: Insights from elementary models. *Journal of Sex Research, 26,* 434–449.

Winkelstein, W., Lyman, D. M., Padian, N., Grant, R., Samuel, M., Wiley, J. A., Anderson, R. W., Lang, W., Riggs, J., & Levy, J. A. (1987). Sexual practices and risk of infection by the human immunodeficiency virus. *Journal of the American Medical Association, 16,* 321–325.

Witz, M., Shpitz, B., Zager, M., Eliashiv, A., & Dinbar, A. (1984). Anal erotic instrumentation: A surgical problem. *Diseases of the Colon and Rectum, 27,* 331–332.

Wofsy, C. (1987). Human immunodeficiency virus infection in women. *Journal of the American Medical Association, 257,* 2074–2076.

Wolfe, L. (1980, September). The sexual profile of that Cosmopolitan girl. *Cosmopolitan,* pp. 254–265.

Wolfe, L. (1981). *The Cosmo report.* New York: Arbor House.

Wyatt, G. E., Peters, S. D., & Guthrie, D. (1988). Kinsey revisited, part I: Comparisons of the sexual socialization and sexual behavior of white women over 33 years. *Archives Sexual Behavior, 17,* 201–239.

20

Immunology of the Male Reproductive Tract: Implications for the Sexual Transmission of Human Immunodeficiency Virus

Deborah J. Anderson, Hans Wolff, Wenhao Zhang, and Jeffrey Pudney

Sexually transmitted viruses have emerged as a major health problem (Alexander & Anderson, 1987). Currently the most pathogenic venereal virus is the human immunodeficiency virus (HIV), the apparent cause of AIDS. This virus was initially confined to limited numbers of individuals in high-risk groups in North America: homosexual men, intravenous drug users, Haitians, and hemophiliacs (*"Pneumocystis carinii,"* 1982; Smath et al., 1983; Vieira, Frank, Spira, & Landesman, 1983). However, HIV has been epidemic with heterosexual transmission patterns in Central Africa since the late 1970s (Brun-Vezinet et al., 1984; Piot, Quinn, Taelman, et al., 1984; Van de Perre et al., 1984), and the virus has recently penetrated into the heterosexual population in North America (Centers for Disease Control, 1986). It has been projected that as many as 150,000 women and 160,000 to 800,000 heterosexual men in the United States are infected with the virus (Centers for Disease Control, 1986). Although HIV can be isolated from many bodily fluids of AIDS patients (Gallo, Salahuddin, Popovic, et al., 1984; Groopman et al., 1984; Ho et al., 1984; Vogt et al., 1986), epidemiologic studies indicate that the principal mode of transmission of the virus is through sexual contact (Peterman & Curran, 1986). Bidirectional heterosexual transmission of HIV in the United States has been established (Centers for Disease Control, 1985; Groopman, Sarngadharan, Salahuddin et al., 1985; Kreiss, Kitchen, Prince, Kasper, & Essex, 1985; Redfield et al., 1985), although transmission from men to women occurs at a higher frequency than transmission from women to men (Padian, 1987). Transmission of HIV to sexual partners occurs with varying degrees of efficiency (Padian,

1987), indicating that as-yet undefined cofactors may be involved in the sexual transmission of HIV.

This chapter reviews information from the literature and recent data from our laboratory concerning the mechanisms of HIV transmission through semen. Limited information exists on the presence of CD4$^+$ lymphocytes and macrophages, the principal HIV host cells, in the male reproductive tract and semen. The possibility that other cell types in the male reproductive tract host HIV remains virtually unexplored. The testis is an immunologically privileged site, and it remains to be determined whether cytotoxic T cells, natural killer cells, or systemic or locally produced anti-HIV antibodies can influence HIV infection in the testis and elsewhere in the reproductive tract. It is also not known whether HIV infects sexual partners as free virus, as virus bound to or incorporated into spermatozoa, through infected white blood cells, or by other unknown mechanisms. It is not known why some HIV-infected men are more infectious than others.

In this chapter the following hypotheses will be developed and explored:

1. Male reproductive tract immunosuppressive factors and immunological barriers create an environment that harbors viruses and other infectious organisms, including HIV.
2. Substantial numbers of known HIV host cells (CD4 lymphocytes and macrophages) are present in the human epididymis and semen. These cells may be principal host cells of HIV in the male reproductive tract, and may also play a role in the sexual transmission of HIV.
3. Reproductive tract inflammation is associated with increased numbers of HIV host cells in semen and may be a cofactor in the sexual transmission of HIV.
4. Male reproductive tract secretions influence biosynthetic functions of lymphocytes and may affect HIV production in certain reproductive tissues and semen.
5. Activation of macrophages and CD4 lymphocytes in the reproductive tract by sperm or microbial antigens may be a factor in HIV infection and transmission.

The Immunological Status of the Male Reproductive Tract and Semen

Immunological Barriers and Immunological Mediators in Male Reproductive Tissues

The male reproductive tract is an immunologically privileged region. Allografts survive for prolonged periods in the testicular interstitium, possibly as a result of the immunosuppressive effects of testosterone

(Aaron, Marescaux, & Petrovic, 1957; Ferguson & Scothorne, 1977; Head, Neaves, & Billingham, 1983; Maddocks, Oliver, & Setchell, 1984). Late-stage testicular germ cells and spermatozoa, which develop at puberty after immunological self-tolerance is established, express unique differentiation antigens that can stimulate autoimmune responses, and are also protected from the immune system by the blood-testis barrier, a physical permeability barrier formed by tight junctions between Sertoli cells (Fig. 20-1). Sertoli cells completely surround the testicular seminiferous tubules and normally prevent cellular and high-molecular-weight soluble immunological mediators from infiltrating into the germ cell compartment (Gilula, Fawcett, & Aoki, 1976).

Immunohistological studies of human testes reveal that numerous macrophages and scattered lymphocytes are present within the interstitial spaces between the seminiferous tubules, but are not found within intact seminiferous tubules of normal testes (El-Demiry et al., 1985; Pollanen & Niemi, 1987). Lymphocytes and macrophages were described within some seminiferous tubules in a recent immunohistological study of testicular biopsies from infertility patients (El-Demiry et al., 1985), indicating that the blood-testis barrier is permeable in some men. Autoimmunity, aging, and viral infections have been associated with breakdown of the blood-testis barrier (Johnson, Zane, Petty, & Neaves, 1984; Kocen & Critchley, 1961; Tung, Yule, Mahi-Brown, & Listrom, 1987). It is worth noting that many HIV-positive men demonstrate evidence of antisperm immunity (Wolff, Schill, Meurer, Helm, & Guertler, 1986), and because of their immunosuppressed state may develop opportunistic reproductive tract infections, both of which are factors that could compromise the blood-testis barrier.

High-molecular-weight soluble immunological mediators such as immunoglobulins and complement components are normally excluded from the germ cell compartment of the testis by the tight junctions between Sertoli cells (Dym, 1973). Similarly, viruses are too large to passively infiltrate through the intact blood-testis barrier. Active transport of viruses through Sertoli cells or by other mechanisms such as macrophage traffic into the germ cell compartment has not been described. Sertoli cells are active phagocytes and could potentially ingest HIV particles and transfer viral DNA or RNA to developing germ cells, which are attached via gap junctions (Russell, 1977). Such possible mechanisms of HIV infection in the testis have yet to be investigated. It can be predicted, however, that if established in the testis, HIV would benefit from immunological protection afforded by local immunosuppressive factors and the blood-testis barrier.

Elsewhere in the male genital tract sperm are separated from the immune system by an epithelial cell layer (Hamilton, 1975; see Figure 20-2 for a review of the anatomy of male reproductive tract organs). Epi-

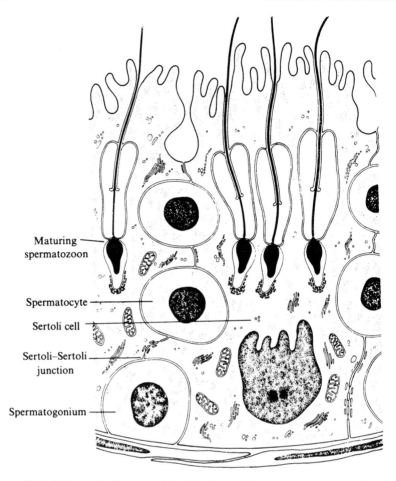

Maturing spermatozoon

Spermatocyte

Sertoli cell

Sertoli–Sertoli junction

Spermatogonium

Figure 20-1. Schematic diagram of the blood-testis barrier surrounding the seminiferous tubules of the testis, formed by Sertoli cells that are interconnected with multiple parallel ridges of tight junctions. (Reprinted with permission from Alexander N. J. & Anderson D. J. [1984]. Immunologic factors in reproductive fitness. In C. R. Austin & R. V. Short [Eds.], *Reproduction in mammals* [Vol. 4]. Cambridge, England: Cambridge University Press.)

thelial cells are interconnected by tight junctions at their apical surfaces; these tight junctions must be dissociated for cellular traffic to occur across this cell layer. Numerous leukocytes have been observed in electron microscopic studies of the rete testis and epididymis (Dym & Romrell, 1975; Hoffer, Hamilton, & Fawcett, 1973; Holstein, 1978; Wang & Holstein, 1983). Recent immunohistological studies indicate that a significant number of CD4$^+$ lymphocytes and macrophages are present in this region of the reproductive tract (Richie, Hargreave, James, & Chish-

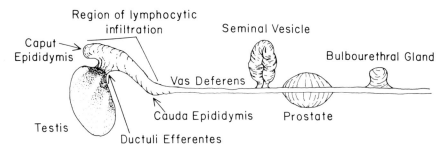

Figure 20-2. Schematic diagram of the anatomy of the human male genital tract.

olm, 1984; Taylor-Emery, Battaile, & Anderson, 1986). Evidence of infiltration through the epididymal epithelial layer into the sperm compartment by lymphocytes and macrophages in normal men is provided by studies of these cell types in semen. Appreciable numbers of macrophages and lymphocytes are present in normal human semen samples (Fig. 20-3); however, semen from vasectomized men contains less than one tenth the number of white blood cells seen in normal semen (Olsen & Shields, 1984; described in more detail below). These data indicate that a majority of white blood cells in normal semen originate in the testis or epididymis, regions upstream of the vasectomy site. Relatively few lymphocytes and macrophages have been visualized by immunohistological technique in the human prostate (El-Demiry et al., 1985); however, prostatic secretions can contain high levels of white blood cells in cases of prostatitis (Schaeffer, Wendel, Dunn, & Grayhack, 1981).

Lymphocytes of the CD8 (suppressor/cytotoxic) phenotype have also been detected in the human epididymis, and it has been proposed that these cells serve an immunoregulatory role in this region (Richie et al., 1984). Animal neonatal thymectomy and cell transfer studies have demonstrated that active T cell–mediated immunosuppression plays a role in preventing orchitis and epididymitis in mice and rats of certain genetic backgrounds (Sakaguchi, Fukuma, Kuribayashi, & Masuda, 1985; Taguchi & Nishizuka, 1981). CD8 suppressor T lymphocytes can regulate the production of HIV by infected CD4 lymphocytes (Walker, Moody, Stites, & Levy, 1986), and may play a role in regulating HIV infection in the male reproductive tract.

Antibody levels are reported to be undetectable or low in testicular and epididymal fluids (Koskimies, Kormano, & Lahti, 1971; Weininger, Fisher, Rifkin, & Bedford, 1982; Wong, Tsange, Fu, et al., 1983), indicating that circulating anti-HIV antibodies may not have free access to infected cells or free virus that may be present in these regions of the reproductive tract. Immunoglobulins (Igs) of the IgG class can, however, enter the ejaculate through transudation from serum into the prostate

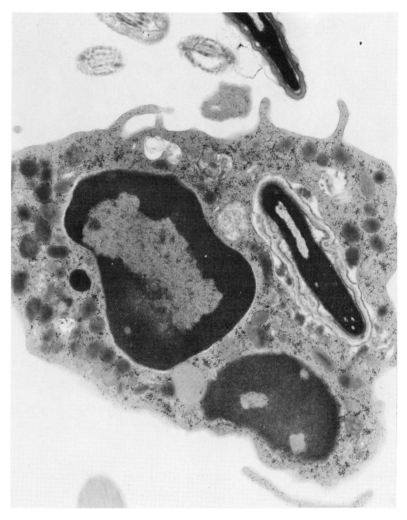

A

Figure 20-3. A: electron micrograph of a macrophage from human semen containing a sperm head in an autophagic vacuole. (×27,500.) (*Figure continues.*)

(Rumke, 1974). Furthermore, evidence is increasing that antibodies are produced locally in the male reproductive tract (Witkin, Zelikovsky, Good, & Day, 1981; Wolff & Schill, 1985, 1988). It is possible that antibodies to HIV and other pathogens are locally produced within the reproductive tissues in some infected men, and that these antibodies play a significant role in virus management and transmission. Depending on antibody class and specificity, antibodies directed against HIV could neutralize virus, bind complement, and kill infected cells, or serve

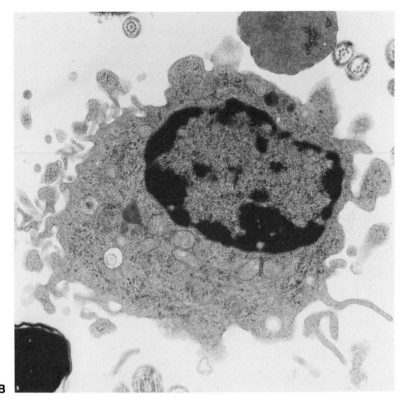

B

Figure 20-3. (*Continued*) B: electron micrograph of a cell of lymphoid morphology detected in human semen. (×21,000.)

as opsonins promoting the phagocytosis of free or cell-associated virus by macrophages and polymorphonuclear leukocytes. Such mechanisms could inhibit or promote HIV infection.

Lymphocytes and Macrophages in Semen

We have recently applied a sensitive immunohistological technique to quantitate and characterize lymphocyte subclasses and macrophages in human semen (Wolff & Anderson, 1988). The results of this study are summarized in Table 20-1. Lymphocytes, macrophages, and granulocytes were detected in most human semen samples. Granulocytes usually predominated, followed by macrophages, then by T lymphocytes. However, in some individuals macrophages or lymphocytes predominated (Wolff & Anderson, in press b). Individuals with evidence of local inflammation as detected by a seminal plasma granulocyte elastase assay had markedly increased numbers of all white blood cell types, including

Table 20-1
Prevalence of White Blood Cell Subpopulations in the Semen of Fertile Donors and Infertility Patients

		Fertile Donors ($n = 17$)	Infertility Patients ($n = 51$)
All white blood cells (HLe-1)	Detectable in	17 (100%)	51 (100%)
	Median	170,000	1,035,000
	Mean ± SD	1,690,894 ± 4,803,946	7,199,090 ± 19,419,682
	range	8,970–20,520,000	43,120–104,580,000
Granulocytes (Dako-MI)	Detectable in	17 (100%)	51 (100%)
	Median	100,000	537,600
	Mean ± SD	1,348,160 ± 4,655,858	5,407,075 ± 16,226,341
	range	6,250–19,950,000	31,787–91,507,500
Monocytes/macrophages (Dako-macrophage)	Detectable in	17 (100%)	51 (100%)
	Median	51,900	228,823
	Mean ± SD	175,897 ± 279,863	978,557 ± 771,101
	range	2,800–997,500	10,395–8,123,750
CD4$^+$ lymphocytes (Leu-3a&3b)	Detectable in	12 (71%)	40 (78%)
	Median	4,140	14,167
	Mean ± SD	9,205 ± 13,392	155,504 ± 550,184
	range	ND[a]–51,900	ND–3,869,250
CD8$^+$ lymphocytes (Leu-2a)	Detectable in	9 (53%)	36 (71%)
	Median	2,243	17,200
	Mean ± SD	6,250 ± 13,118	86,113 ± 208,967
	range	ND–56,666	ND–1,273,000
B lymphocytes (Dako-pan B)	Detectable in	4 (24%)	30 (59%)
	Median	ND	6,400
	Mean ± SD	2,737 ± 6,663	88,534 ± 411,109
	range	ND–25,200	ND–2,964,750

[a] ND = not detectable; leukocyte numbers were calculated per ejaculate.
From Wolff, H., and Anderson, D.A. (1988). Immunohistologic detection and quantitation of leukocyte subpopulations in human semen. *Fertility and Sterility 49*, 497–504.

macrophages and CD4$^+$ lymphocytes (Wolff & Anderson, in press a, in press c; Table 20-2). Factors such as venereal infections or sperm autoimmunity, which can cause genital tract inflammation and are associated with HIV seropositivity, may therefore be associated with elevated numbers of HIV host cells in reproductive tissues and semen, and may be cofactors influencing the infectivity of semen. Our immunohistological study has also provided quantitative evidence that semen from most vasectomized men contains significantly reduced levels of total white blood cells, CD4 lymphocytes, and macrophages compared to semen from intact men (Table 20-3).

It is possible that lymphocytes and/or macrophages in the reproductive tract or semen are activated by foreign sperm differentiation antigens or products of microorganisms. Seminal macrophages often appear as activated multinucleated giant cells containing large amounts of phagocytosed material, and we have detected interleukin-2 (Il-2) receptor, transferrin receptor, and HLA-Dr-positive cells in semen from some men, particularly men with evidence of local inflammation (Wolff & Anderson, in press c). This may be relevant to HIV transmission since activation enhances the production of HIV by lymphocytes and macrophages (Folks, Justement, Kinter, Dinarello, & Fauci, 1988; Folks et al., 1986).

Effects of Male Reproductive Tract Secretions and Seminal Plasma on Lymphocyte and Macrophage Activity

When added to tests of immune function in vitro, human seminal plasma affects the activities of most cell types that participate in immune responses and host defense, including T cells, B cells, natural killer (NK) cells, macrophages, and polymorphonuclear leukocytes (PMNs). The reported effects of seminal plasma on immune functions are varied and include:

1. Inhibition of the ability of T and B lymphocytes to proliferate in response to mitogen or antigen challenge (Lord, Sensabaugh, & Stites, 1977; Pitout & Jordan, 1976; Prakash, Coutinho, & Moller, 1976; Stites & Erickson, 1975).
2. Impairment of the ability of macrophages and PMNs to recognize target antigens by means of surface cytophilic antibodies (Brooks, Lammel, Petersen, & Stites, 1981).
3. Impairment of the ability of NK cells and cytotoxic T cells to recognize and destroy tumor target cells (James & Szymaniec, 1985; Rees, Vallely, Clegg, & Potter, 1986; Tarter, Cunningham-Rundles, & Koide, 1986).
4. Inhibition of phagocytosis by macrophages and PMNs (James et al., 1983; Stankova, Drach, Hicks, Zukoski, & Chvapil, 1976).

Table 20-2
Increase of White Blood Cells and Potential HIV Host Cells in Inflammatory Semen Samples

		Granulocyte Elastase in Semen of Infertility Patients (n = 105)		
		<250 ng/ml (no inflammation) n = 50 (48%)	250–1000 ng/ml (intermediate) n = 38 (36%)	>1000 ng/ml (inflammation) n = 17 (16%)
Total WBC per ejaculate	Median	520,625	964,236	19,800,000
	Mean ± SD	1,199,004 ± 1,660,989	4,167,688 ± 8,481,922	27,429,185 ± 29,563,911[a]
	range	27,520–9,487,059	91,000–41,773,875	1,876,875–104,580,000
Monocytes/macrophages per ejaculate	Median	134,565	225,568	2,594,000
	Mean ± SD	316,492 ± 465,213	721,255 ± 2,073,860	3,289,076 ± 2,271,332[a]
	range	5,050–2,070,000	34,000–12,790,000	469,333–8,124,000
CD4+ lymphoid cells per ejaculate	Median	14,100	17,723	82,900
	Mean ± SD	42,642 ± 66,384	175,603 ± 642,132	540,752 ± 1,210,172
	range	ND[b]–316,236	ND–3,906,000	14,235–3,869,000

[a] Total white blood cell, CD4+ lymphocyte and monocyte/macrophage numbers in the subgroup with > 1000 ng elastase/ml semen were significantly higher than in the non-inflammatory subgroup with < 250 ng elastase/ml semen (unpaired student's t-test; p < .005).
[b] Not detectable.

Table 20-3
Total Leukocyte and HIV Host Cell Numbers in Semen Samples
from Normal Donors and Vasectomized Men

		Normal Donors (n = 20)	Vasectomized Men (n = 6)
Total leukocytes[a] (anti-HLel)	Median	279,900	37,500
	Mean ± SD	5,284,732 ± 16,301,974	84,677 ± 114,300
Monocytes/macrophages (Dako-macrophage)	Median	89,881	26,550
	Mean ± SD	564,467 ± 1,607,546	55,645 ± 82,752
CD4+ lymphocytes (anti-leu 3a + b)	Median	6,145	0
	Mean ± SD	115,994 ± 441,911	2,223 ± 5,233

[a] Numbers are given per ejaculate.

In addition, human seminal plasma inhibits the action of complement (C1 and C3 components), thereby reducing the lytic effectiveness of antibodies in target cell destruction (Petersen, Lammel, Stites, & Brooks, 1980; Tarter & Alexander, 1984).

In vivo studies of seminal plasma immunosuppression, although limited, have demonstrated systemic immunosuppressive effects of seminal plasma or seminal plasma components (Alexander, Tarter, Fulgham, Ducsay, & Novy, in press; Anderson & Tarter, 1982; Hurtenbach & Shearer, 1982; Kuno, Ueno, & Hayaishi, 1986; Richards, Bedford, & Witkin, 1984). On the basis of extensive in vitro studies it can be predicted that seminal plasma immunosuppressive factors could suppress several levels of immunological mechanisms operative in defense against venereal infectious organisms, including HIV. Various immunomodulating factors in human reproductive tract fluids and seminal plasma have been identified (Alexander & Anderson, 1987). Studies in mice indicate that a variety of immunosuppressive factors with diverse activities are produced in the epididymis, prostate, and seminal vesicles (Anderson & Tarter, 1982).

Studies in our laboratory have recently focused on short-term effects of human seminal plasma (HSP) on lymphocyte and macrophage functions. Lymphocytes remain viable (membrane intact) for several days following short-term exposure to fresh undiluted HSP, or when cultured in the presence of 10% HSP. However, following exposure to HSP, activated peripheral blood lymphocytes lose their ability to synthesize DNA, RNA, and protein (Fig. 20-4A). This paralytic effect of HSP is irreversible: concanavalin A–activated lymphocytes that are incubated

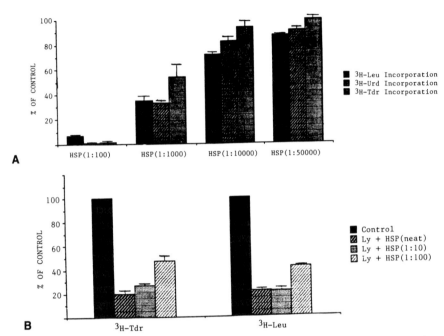

Figure 20-4. A: suppressive effect of human seminal plasma (HSP) on protein, RNA, and DNA synthesis by activated lymphocytes. Phytohemagglutinin-activated lymphocytes were cultured in various dilutions of HSP for 3 days; protein, RNA, and DNA synthesis were measured on day 3 of culture by incorporation of ^{3}H-radiolabeled leucine (^{3}H-Leu), uridine (^{3}H-Urd) or thymidine (^{3}H-Tdr). B: the suppressive effect of human seminal plasma on DNA and protein synthesis in activated lymphocytes is irreversible. Phytohemagglutinin-activated lymphocytes (Ly) were cultured in HSP for 1 hour, washed extensively, and cultured for 3 days before addition of ^{3}H-Tdr or ^{3}H-Leu.

for 1 hour in HSP, washed five times, and established in tissue culture are likewise unable to proliferate or synthesize protein, DNA, or RNA (Fig. 20-4B). These data indicate that HIV integration or production, events that require protein, DNA, and/or RNA synthesis, may be affected in lymphocytic HIV host cells by factors in human male reproductive tract secretions or seminal plasma (Zhang & Anderson, 1988b).

To investigate whether seminal plasma proteases or other factors affect the expression of surface molecules involved in immunological interactions (i.e., HLA-A,B,C, HLA-Dr, Il-2 receptor) or lymphocyte-HIV interactions (i.e., CD4 antigen), we used radioimmunoassay to measure the levels of these molecules on the surface of activated lymphocytes before and after exposure to HSP. No significant difference was seen between surface antigen expression on untreated cells and cells treated for 5 minutes to 1 hour with HSP (Zhang & Anderson, 1988b). However, HSP exposure did affect the levels of expression of Il-2 receptor on long-term cultured cells, and also degraded soluble Il-2 (Zhang & Anderson, 1988c).

Effects of HSP on macrophage phagocytosis have been conflicting both in the literature and in our laboratory. We found that diluted HSP markedly decreased macrophage phagocytic activity. However, exposure of macrophages to undiluted fresh HSP does not inhibit and may even increase macrophage phagocytosis (Zhang & Anderson, 1988a). Macrophages in human semen actively engulf latex beads hours after ejaculation (Fig. 20-5). Data from our laboratory indicate that human macrophages are also attracted to HSP in in vitro chemotaxis assay systems (Fig. 20-6). These studies indicate that, in contrast to lymphocytes, several macrophage functions may be unaffected or enhanced by seminal plasma. Lymphocytes may be unable to integrate or produce HIV in the presence of male reproductive tract secretions or semen; however, macrophages appear to be capable of phagocytosing. HIV and migrating through tissues following exposure to these factors. It was pointed out in the previous section that macrophages outnumber T lymphocytes in human semen. The data indicating differential effects of seminal plasma on lymphocyte and macrophage function further implicate macrophages as significant mediators of HIV transmission through semen.

Expression of the CD4 HIV Receptor on Spermatozoa and Other Cells in the Male Reproductive Tract

In rodents it has been demonstrated that endogenous epididymal retroviruses bind to spermatozoa and can be conveyed through intercourse to female sexual partners (Kiessling, Crowell, & Connell, 1987; Levy, Joyner, & Borenfreund, 1980). Recently evidence was presented by Ashida and Scofield (1987) that human spermatozoa express a CD4-like antigen. Therefore, the possibility exists that human spermatozoa bind HIV and play a key role in its transmission. To address this issue, we have recently performed a series of immunohistological, radioimmunoassay, and electron microscopic studies to detect expression of the CD4 HIV receptor on spermatozoa and other cells in the human male reproductive tract and semen (Wolff, Zhang, & Anderson, 1988).

Immunoperoxidase studies of sections of human testis and epididymis revealed no binding of anti-CD4 monoclonal antibodies to any cell types except occasional cells of lymphoid and macrophage morphology observed in interstitial tissues of the testis and epididymis. Testicular germ cells and testicular and epididymal sperm were negative for CD4 antigen expression, as was the epididymal epithelium. Immunoperoxidase studies of over 200 human semen samples have likewise failed to detect CD4

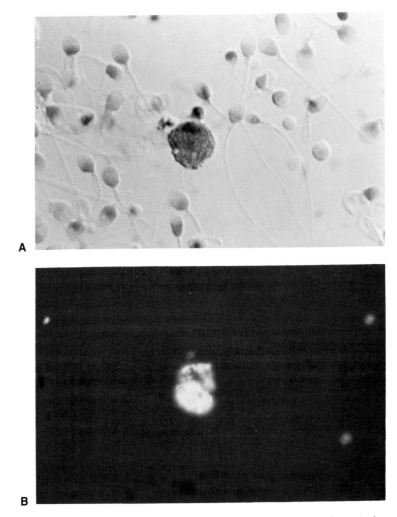

Figure 20-5. Ejaculated macrophages are phagocytic. A: macrophage in human semen identified by immunoperoxidase technqiue. (×4,000.) B: the same macrophage viewed under a fluorescence microscope revealing ingestion of fluoresceinated beads that had been added to the ejaculate 2 hours previously. (×4,000.)

antigen expression by sperm, even in individuals with leukocytospermia. CD4 antigen expression was consistently observed on cells of lymphoid and macrophage morphology in these samples. In addition to these studies, we used fluorescence and radioimmunoassay techniques to detect CD4 antigen expression on unfixed purified human ejaculated

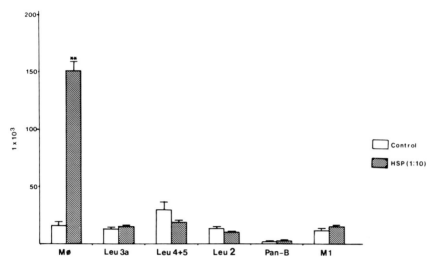

Figure 20-6. Human seminal plasma is a chemoattractant for human monocytes/macrophages. Human peripheral blood leukocytes were characterized and quantified by immunoperoxidase technique following migration through a filter in a Boyden chemotaxis chamber. Cells of the monocyte/macrophage phenotype ($M0^4$) demonstrated significant migration toward seminal plasma. Cells of T lymphocyte (Leu 4 and 5b[+]), helper T lymphocyte (Leu 3a and 3b[+]), suppressor cytotoxic T lymphocyte (Leu 2a[+]), B lymphocyte (Pan-B[+]) or granulocyte (M1[+]) phenotypes were not significantly attracted by seminal plasma.

spermatozoa. Neither of these techniques detected CD4 antigen expression on human spermatozoa.

We have performed preliminary experiments to directly address the question of whether HIV binds to human spermatozoa. Washed motile spermatozoa were incubated at 37°C in supernatants from HIV-infected Jurkat cell cultures that contained high levels of viable intact HIV (2×10^5 counts/million per milliliter in reverse transcriptase assay). Following incubation with HIV, sperm were washed three times in Hank's balanced salt solution and processed for immunohistology and electron microscopy. Analysis of spermatozoa using either technique failed to detect specific HIV binding to human spermatozoa (Fig. 20-7), whereas HIV particles were found to be associated with a subpopulation of human peripheral blood lymphocytes that were incubated in the HIV supernatant as a positive control. These data do not conclusively rule out HIV binding to spermatozoa, because in vitro conditions are different from in vivo conditions, and only one strain of HIV was used (another strain or variant of this strain may bind to sperm). However, they indicate that human ejaculated spermatozoa do not as a rule bind HIV via high-affinity receptors.

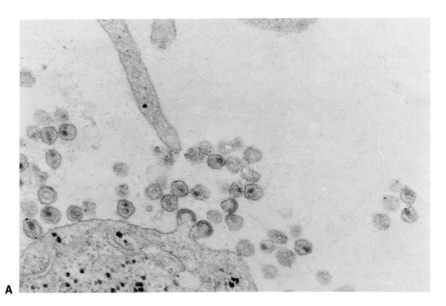

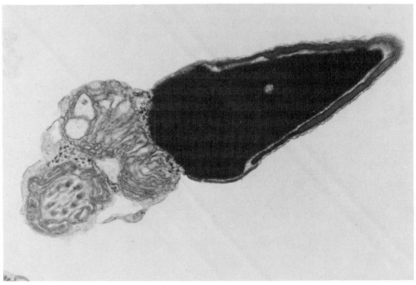

Figure 20-7. Motile human spermatozoa do not bind HIV in vitro. A: electron micrograph of HIV-transfected Jurkatt cell line, which was used as a source of HIV for this study. (×52,000.) B: electron micrograph of motile spermatozoa recovered from supernatant of HIV-infected cell culture. No attachment of HIV to the sperm surface was observed. (×29,250.)

Mechanisms can be proposed for nonspecific attachment of HIV to sperm; it is possible that HIV attaches to spermatozoa through mechanisms such as transglutaminase cross-linking of glutamine-rich proteins to the surface of sperm (Moore et al., 1987; Mukherjee, Agrawal, Manjunath, & Mukherjee, 1983). It is also possible that HIV is incorporated into germ cell DNA during spermatogenesis. Spermatozoa can migrate as far as the peritoneal cavity in women after intercourse, and are probably phagocytosed by macrophages (HIV host cells) that line the reproductive tract and peritoneal cavity (Hill & Anderson, 1988). Furthermore, evidence has been presented that human spermatozoa are also able to penetrate other cell types (Bendich, Borenfreund, & Sternberg, 1974), including activated T cells (Ashida & Scofield, 1987). These potential mechanisms of sperm-borne HIV transmission have not been adequately explored.

Summary and Conclusions

The mechanisms underlying the sexual transmission of HIV are largely unknown. The blood-testis barrier and immunosuppressive factors present in male reproductive tract secretions may prevent host immune defense mechanisms from functioning normally to eliminate HIV in certain regions of the male reproductive tract. Recent evidence indicates that CD4$^+$ lymphocytes and macrophages infiltrate the human rete testis and epididymis and are present in variable numbers in semen. They have also been detected within the blood-testis barrier of some infertility patients; likewise these cells may have access to the germ cell compartment in some HIV-positive men, because a significant percentage of these individuals have germ cell autoimmunity or opportunistic reproductive tract infections. The CD4$^+$ lymphocytes and macrophages residing in male reproductive tissues are likely the principal host cells of HIV in this region since the CD4 HIV receptor is not detectable on other cell types, including germ cells in the testis, epididymis, and prostate. Numbers of potential HIV host cells (CD4$^+$ lymphocytes and macrophages) are significantly increased in semen from men with evidence of genital tract inflammation, indicating that local infection or other causes of inflammation may be a factor in the efficiency of sexual transmission of HIV.

Potential HIV host cells in semen are markedly reduced after vasectomy in healthy men. These data suggest that vasectomy could reduce the risk of HIV transmission through semen, although it should be cautioned that prostatitis, local infection, or other types of genital tract inflammation could increase the numbers of HIV infected cells and/or free virus in vasectomy semen. Seminal plasma is chemotactic for macrophages and may attract macrophages to the insemination site in semen recipients, thereby providing a mechanism of transfer of virus from semen to recipient HIV host cells. Possible transmission of HIV by spermatozoa has not been substantiated by in vitro studies, but further studies should be performed in this area. Male reproductive tract antigens and secretions may modulate the infectious potential of HIV host cells in male reproductive tissues and semen.

Acknowledgments

This research has been principally supported by the Fearing Research Laboratory Endowment, grant AI23669 from the National Institutes of Health, and grant CSA88020 from the United States Agency for International Development, Contraceptive Research and Development Program. Drs. Wolff and Zhang were supported by fellowships from the Max Kade Foundation and World Health Organization, respectively.

References

Aaron, M., Marescaux, J., & Petrovic, A. (1957). Greffes homoplastiques chez mammiferes. *Colloques Internationales du CNRS, 78*, 25–33.

Alexander, N. J., & Anderson, D. J. (1987). Immunology of semen. *Fertility and Sterility, 47*, 192–205.

Alexander, N. J., Tarter, T. H., Fulgham, D. L., Ducsay, C. A., & Novy, M. J. (1987). Rectal infusion of semen results in transient elevation of blood prostaglandins. *American Journal of Reproductive Immunology and Microbiology, 15*(2), 47–51.

Anderson, D. J., & Tarter, T. H. (1982). Immunosuppressive effects of mouse seminal plasma components in vivo and in vitro. *Journal of Immunology, 128*, 535–539.

Ashida, E. R., & Scofield, V. L. (1987). Lymphocyte major histocompatibility complex-encoded class II structures may act as sperm receptors. *Proceedings of the National Academy of Sciences of the United States of America, 84*, 3395–3399.

Bendich, A., Borenfreund, E., & Sternberg, S. S. (1974). Penetration of somatic mammalian cells by sperm. *Science, 183*, 857–859.

Brooks, G. F., Lammel, C. J., Petersen, B. H., & Stites, D. P. (1981). Human seminal plasma inhibition of antibody complement-mediated killing and opsonization of *Neisseria gonorrhoeae* and other gram-negative organisms. *Journal of Clinical Investigation, 67*, 1523–1531.

Brun-Vezinet, F., Rouzioux, C., Montagnier, L., Chamaret, S., Gruest, J., Barre-Sinoussi, F., Geroldi, D., Chermann, J. C., McCormick, J., & Mitchell, S. (1984). Prevalence of antibodies to lymphadenopathy-associated retrovirus in African patients with AIDS. *Science, 226*, 453–456.

Centers for Disease Control. (1985). Heterosexual transmission of human T-lymphotropic virus type III/lymphadenopathy associated virus. *Morbidity and Mortality Weekly Report, 34*, 561–563.

Centers for Disease Control. (1986, August 11). *AIDS weekly surveillance report* (p. 1). Atlanta: Author.

Dym, M. (1973). The fine structure of the monkey (*Macaca*) Sertoli cell and its role in maintaining the blood-testis barrier. *Anatomical Records, 175,* 639–656.

Dym, M., & Romrell, L. J. (1975). Intraepithelial lymphocytes in the male reproductive tract of rats and rhesus monkeys. *Journal of Reproduction and Fertility, 42,* 1–7.

El-Demiry, M. I. M., Hargreave, T. B., Busuttil, A., James, K., Ritchie, A. W. S., & Chisholm, G. D. (1985). Lymphocyte subpopulations in the male genital tract. *British Journal of Urology, 57,* 769–774.

Ferguson, J., & Scothorne, R. S. (1977). Extended survival of pancreatic islet allografts in the testis of guinea pigs. *Journal of Anatomy, 124,* 1–8.

Folks, T. M., Kelley, J., Bann, S., Kinter, A., Justement, J., Gold, J., Redfield, R., Sell, K. W., & Fauci, A. S. (1986). Susceptibility of normal human lymphocytes to infection with HTLV-III/LAV. *Journal of Immunology, 136,* 4049–4053.

Folks, T. M., Justement, J., Kinter, A., Dinarello, C. A., & Fauci, A. S. (1988). Cytokine-induced expression of HIV-1 in a chronically infected promonocyte cell line. *Science, 238,* 800–802.

Gallo, R., Salahuddin, S. Z., Popovic, M., Shearer, G. M., Kaplan, M., Haynes, B. F., Palker, T. J., Redfield, R., Oleske, J., Safai, B., White, G., Foster P., & Markham, P. D. (1985). Frequent detection and isolation of cytopathic retroviruses (HTLV-III) from patients with AIDS and at risk of AIDS. *Science, 224,* 500–503.

Gilula, N. B., Fawcett, D. W., & Aoki, A. (1976). The Sertoli cell occluding junctions and gap junctions in mature and developing mammalian testis. *Developmental Biology, 50,* 142–168.

Groopman, J. E., Salauddin, S. Z., Sarngadharan, M. G., Markham, P. D., Gonda, M., Sliski, A., & Gallo, R. C. (1984). HTLV-III in saliva of people with AIDS-related complex and healthy homosexual men at risk for AIDS. *Science, 226,* 447–449.

Groopman, J. E., Sarngadharan, M. G., Salahuddin, S. Z., Buxbaum, R., Huberman, M. S., Kinniburgh, J., Sliski, A., McLane, M. F., Essex, M., & Gallo, R. C. (1985). Apparent transmission of human T-cell leukemia virus type III to a heterosexual woman with the acquired immunodeficiency syndrome. *Annals of Internal Medicine, 102,* 63–66.

Hamilton, D. W. (1975). Structure and function of the epithelium lining the ductuli efferentes, ductus epididymidis and ductus deferens. In R. O. Greep & E. B. Astwood (Eds.), *Handbook of physiology,* (Vol. 5, p. 259). Washington, DC: American Physiological Society.

Head, J. R., Neaves, W. B., & Billingham, R. E. (1983). Immune privilege in the testis. I. Basic parameters of allograft survival. *Transplantation, 36,* 423–431.

Hill, J. A., & Anderson, D. J. (1988). Immunological mechanisms of female infertility. In P. Johnson (Ed.), *Bailere's clinical immunology and allergy.*

Ho, D. D., Schooley, R. T., Rota, T. R., Kaplan, J. C., Flynn, T., Salahuddin, S. Z., Gonda, M. A., & Hirsch, M. S. (1984). HTLV-III in the semen and blood of a healthy homosexual man. *Science, 226,* 451–453.

Hoffer, A. P., Hamilton, D. W., & Fawcett, D. W. (1973). The ultrastructure of the principal cells and intraepithelial leucocytes in the initial segment of the rat epididymis. *Anatomical Records, 175,* 169–202.

Holstein, A. F. (1978). Spermatophagy in the seminiferous tubules and excurrent ducts of the testis in rhesus monkeys and man. *Andrologia, 10,* 331–352.

Hurtenbach, U., & Shearer, G. M. (1982). Germ cell-induced immune suppression in mice: Effect of innoculation of syngeneic spermatozoa on cell-mediated immune responses. *Journal of Experimental Medicine, 155,* 1719–1729.

James, K., Harvey, J., Bradbury, A. W., Hargreave, T. B., Cullen, R. T., & Donaldson, K. (1983). The effect of seminal plasma on macrophage function—a possible contributory factor in sexually transmitted disease. *AIDS Research, 1,* 45.

James, K., & Szymaniec, S. (1985). Human seminal plasma is a potent inhibitor of natural killer cell activity in vitro. *Journal of Reproductive Immunology, 8,* 61–70.

Johnson, L., Zane, R. S., Petty, C. S., & Neaves, W. B. (1984). Quantification of the human Sertoli cell population: Its distribution, relation to germ cell numbers, and age-related decline. *Biology of Reproduction, 31,* 785–795.

Kiessling, A. A., Crowell, R. C., & Connell, R. S. (1987). Sperm-associated retroviruses in the mouse epididymis. *Proceedings of the National Academy of Sciences of the United States of America, 84,* 8667–8671.

Kocen, R. S., & Critchley, E. (1961). Mumps epididymo-orchitis and its treatment with cortisone. *British Medical Journal, 2,* 20–24.

Koskimies, A. I., Kormano, M., & Lahti, A. (1971). A difference in the immunoglobulin content of seminiferous tubule fluid and rete testis fluid of the rat. *Journal of Reproduction and Fertility, 27,* 463–465.

Kreiss, J. K., Kitchen, L. W., Prince, H. E., Kasper, C. K., & Essex, M. (1985). Antibody to human T-lymphotropic virus type III in wives of hemophiliacs. Evidence for heterosexual transmission. *Annals of Internal Medicine, 102,* 623–626.

Kuno, S., Ueno, R., & Hayaishi, O. (1986). Prostaglandin E_2 administered via anus causes immunosuppression in male but not female rats: A possible pathogenesis of acquired immune deficiency syndrome in homosexual males. *Proceedings of the National Academy of Sciences of the United States of America, 83,* 2682–2683.

Levy, J. A., Joyner, J., & Borenfreund, E. (1980). Mouse sperm can horizontally transmit type C viruses. *Journal of General Virology, 51,* 439–443.

Lord, E. M., Sensabaugh, G. F., & Stites, D. P. (1977). Immunosuppressive activity of human seminal plasma. I. Inhibitor of *in vitro* lymphocyte activation. *Journal of Immunology, 118,* 1704–1711.

Maddocks, S., Oliver, J. R., & Setchell, B. P. (1984). The survival and function of isolated pancreatic islets of Langerhans transplanted into the testis of adult rats and their effect on the testis. *Inserum, 123,* 497–502.

Moore, J. T., Haagstrom, J., McCormick, D. J., Harvey, S., Madden, B., Holicky, E., Stanford, D. R., & Wieben, E. D. (1987). The major clotting protein from guinea pig seminal vesicle contains eight repeats of a 24-amino acid domain. *Proceedings of the National Academy of Sciences of the United States of America, 84,* 6712–6714.

Mukherjee, D. C., Agrawal, A. K., Manjunath, R., & Mukherjee, A. B. (1983). Suppression of epididymal sperm antigenicity in the rabbit by uteroglobin and transglutaminase in vitro. *Science, 219,* 989–991.

Olsen, G. P., & Shields, J. W. (1984). Seminal lymphocytes, plasma and AIDS. *Nature, 309,* 116–117.

Padian, N. (1987). Heterosexual transmission of acquired immunodeficiency syndrome: International perspectives and national projections. *Reviews of Infectious Diseases, 9,* 947–959.

Peterman, T. A., & Curran, J. W. (1986). Sexual transmission of human immunodeficiency virus. *JAMA, 256,* 2222.

Petersen, B. H., Lammel, C. J., Stites, D. P., & Brooks, G. F. (1980). Human seminal plasma inhibition of complement. *Journal of Laboratory Clinical Medicine, 96,* 582–591.

Piot, P., Quinn, T. C., Taelman, H., Feinsod, F. M., Kapita, B., Wobin, O., Mbendi, N., Mazebo, P., Ndangi, K., Stevens, W., Kalambayi, K., Mitchell, S., Bridts, C., & McCormick, J. B. (1984). Acquired immunodeficiency syndrome in a heterosexual population in Zaire. *Lancet, 2,* 65–69.

Pitout, M. J., & Jordan, J. H. (1976). Partial purification of an antimitogenic factor from human semen. *International Journal of Biochemistry, 7,* 149–155.

Pneumocystis carinii pneumonia among persons with hemophilia A. (1982). *Morbidity and Mortality Weekly Report, 31,* 365.

Pollanen, P., & Niemi, M. (1987). Immunohistochemical identification of macrophages, lymphoid cells, and HLA antigens in the human testis. *International Journal of Andrology, 10,* 37–42.

Prakash, C., Coutinho, A., & Moller, G. (1976). Inhibition of in vitro immune responses by a fraction from seminal plasma. *Scandanavian Journal of Immunology, 5,* 77–85.

Redfield, R. R., Markham, P. D., Salahuddin, S. Z., Sarngadharan, M. G., Bodmer, A. J., Folks, T. M., Ballou, W. R., Wright, D. C., & Gallo, R. C. (1985). Frequent transmission of HTLV-III among spouses of patients with AIDS-related complex and AIDS. *JAMA, 253,* 1571–1573.

Rees, R. C., Vallely, P., Clegg, A., & Potter, C. W. (1986). Suppression of natural and activated human antitumour cytotoxicity by human seminal plasma. *Clinical Experimental Immunology, 63,* 687–695.

Richards, J. M., Bedford, J. M., & Witkin, S. S. (1984). Rectal insemination modifies immune responses in rabbits. *Science, 224,* 390–392.

Richie, A. W. S., Hargreave, T. B., James, K., & Chisholm, G. D. (1984). Intraepithelial lymphocytes in the normal epididymis. A mechanism for tolerance to sperm autoantigens? *British Journal of Urology, 56,* 79–84.

Rumke, P. (1974). The origin of immunoglobulins in semen. *Clinical Experimental Immunology, 17,* 287–292.

Russell, L. (1977). Desmosome-like junctions between Sertoli and germ cells in the rat testis. *American Journal of Anatomy, 148,* 301–312.

Sakaguchi, S., Fukuma, K., Kuribayashi, K., & Masuda, T. (1985). Organ-specific autoimmune diseases induced in mice by elimination of T cell subset. I. Evidence for active participation of T cells in natural self-tolerance; deficit of a T cell subset as a possible cause of autoimmune disease. *Journal of Experimental Medicine, 161,* 72–87.

Schaeffer, A. J., Wendel, E. F., Dunn, J. K., & Grayhack, J. T. (1981). Prevalence and significance of prostatic inflammation. *Journal of Urology, 125,* 215–219.

Smath, C. B., Klem, R. S., Friedland, G. H., Moll, B., Emeson, E. E., & Spigland, E. (1983). Community-acquired opportunistic infections and defective cellular immunity in heterosexual drug abusers and homosexual men. *American Journal of Medicine, 74,* 433–441.

Stankova, L., Drach, G. W., Hicks, T., Zukoski, C. F., & Chvapil, M. (1976). Regulation of some functions of granulocytes by zinc of the prostate fluid and prostate tissue. *Journal of Laboratory Clinical Medicine, 88,* 640–648.

Stites, D. P., & Erickson, R. P. (1975). Suppressive effects of seminal plasma on lymphocyte activation. *Nature, 253,* 727–729.

Taguchi, O., & Nishizuka, Y. (1981). Experimental autoimmune orchitis after neonatal thymectomy in the mouse. *Clinical Experimental Immunology, 46,* 425–434.

Tarter, T. H., & Alexander, N. J. (1984). Complement-inhibiting activity of seminal plasma. *American Journal of Reproductive Immunology, 6*, 28–32.

Tarter, T. H., Cunningham-Rundles, S., & Koide, S. S. (1986). Suppression of natural killer cell activity by human seminal plasma in vitro: Identification of 19-OH-PGE as the suppressor factor. *Journal of Immunology, 136*, 2862–2867.

Taylor-Emery, S., Battaile, A., & Anderson, D. (1986). An immunohistological study of lymphoid cells in the human male reproductive tract. *Journal of Reproductive Immunology*, (Suppl), p. 71.

Tung, K. S. K., Yule, T. D., Mahi-Brown, C. A., & Listrom, M. B. (1987). Distribution of histopathology and Ia positive cells in actively induced and passively transferred experimental autoimmune orchitis. *Journal of Immunology, 138*, 752–759.

Van de Perre, P., Rouvroy, D., Lepage, P., Bogaerts, J., Kestelyn, P., Kayihigi, J., Hekker, A. C., Butzler, J. P., & Clumeck, N. (1984). Acquired immunodeficiency syndrome in Rwanda. *Lancet, 2*, 62–65.

Vieira, J., Frank, E., Spira, T. J., & Landesman, S. H. (1983). Acquired immune deficiency in Haitians, opportunistic infections in previously healthy Haitian immigrants. *New England Journal of Medicine, 308*, 125–129.

Vogt, M. W., Witt, D. J., Craven, D. E., Byington, R., Crawford, D. F., Schooley, R. T., & Hirsch, M. S. (1986). Isolation of HTLV-III/LAV from cervical secretions of women at risk for AIDS. *Lancet, 1*, 525–527.

Walker, C. M., Moody, D. J., Stites, P. P., & Levy, J. A. (1986). CD8[+] lymphocytes can control HIV infection in vitro by suppressing virus replication. *Science, 234*, 1563–1566.

Wang, Y. F., & Holstein, A. F. (1983). Intraepithelial lymphocytes and macrophages in the human epididymis. *Cell and Tissue Research, 223*, 517–521.

Weininger, R. B., Fisher, S., Rifkin, J., & Bedford, J. M. (1982). Experimental studies on the passage of specific IgG to the lumen of the rabbit epididymis. *Journal of Reproduction and Fertility, 66*, 251–258.

Witkin, S. S., Zelikovsky, G., Good, R. A., & Day, N. (1981). Demonstration of 11s IgA antibody to spermatozoa in human seminal fluid. *Clinical and Experimental Immunology, 44*, 368–374.

Wolff, H., & Anderson, D. J. (1988). Immunohistologic characterization and quantitation of leukocyte subpopulations in human semen. *Fertility and Sterility, 49*, 497–504.

Wolff, H., & Anderson, D. J. (In press a). Evaluation of granulocyte elastase as a seminal plasma marker for leukocytospermia. *Fertility and Sterility, 50*, 129–132.

Wolff, H., & Anderson, D. J. (In press b). Leukocytes in semen: their role in male infertility. Proceedings of the First Workshop on Biochemical Aspects on the Immunopathology of Reproduction. *International Journal of Andrology.*

Wolff, H., & Anderson, D. J. (In press c). Male genital tract inflammation is associated with increased numbers of potential human immunodeficiency virus host cells in semen. *Andrologia, 20*, 404–410.

Wolff, H., & Schill, W-B. (1985). Antisperm antibodies in infertile and homosexual men: Relationship to serologic and clinical findings. *Fertility and Sterility, 44*, 673–677.

Wolff, H., & Schill, W-B. (1988). Determination of sperm antibodies in human genital secretions by an ELISA technique. *Human Reproduction, 3*, 223–225.

Wolff, H., Schill, W-B., Meurer, M., Helm, K., & Guertler, L. (1986). Lymphadenopathy associated virus/human T-cell lymphotropic virus III antibodies in homosexual men with and without sperm antibodies. *Fertility and Sterility, 46*, 111–113.

Wolff, H., Zhang, W., & Anderson, D. J. (1988, June). *The CD4 antigen (HIV receptor) is not detectable on human testicular germ cells or spermatozoa.* Paper presented at the Fourth International Conference on AIDS, Stockholm.

Wong, P. Y. D., Tsang, A. Y. F., Fu, W., & Lau, H. K. (1983). Restricted entry of an anti-rat epididymal protein IgG into the rat epididymis. *International Journal of Andrology, 6*, 275–282.

Zhang, W. H., & Anderson, D. J. (1988a, June). *Effects of human seminal plasma on mechanisms of HIV infection.* Paper presented at the Fourth International Conference on AIDS, Stockholm.

Zhang, W. H., & Anderson, D. J. (1988b). Effects of short-term exposure to human seminal plasma on lymphocyte viability and function (Abstr.). *Journal of Andrology*, Suppl.

Zhang, W. H., & Anderson, D. J. (1988c). The immunosuppressive effect of human seminal plasma may be partially attributable to degradation and inactivation of the T cell growth factor Interleukin-2 (Il-2) (Abstr.). *Journal of Andrology*, Suppl.

VII
PREVENTION

21

Condoms—Manufacturing and Testing

Robert M. Nakamura

A recent cartoon caption sums up our present thinking about AIDS and condoms fairly concisely: "A tisket, a tasket, a condom or a casket . . . ." But how important are condoms? Why the sudden emphasis on this form of contraception? The condom industry is overwhelmed by tremendous increases in orders and, concurrently, the Food and Drug Administration has received dozens of new applications for approval from foreign condom manufacturers.

Condoms are now looked upon as one of the major means of preventing the spread of AIDS. For this reason, the current high visibility and popularity of condoms are not surprising. Because of this high visibility, they have come under careful scrutiny and have received poor marks on several counts. Specifically, there have been questions associated with breakage and, more recently, a recall of several condom lots has created confusion among the public.

What are the facts and how can we judge the claims made in the popular press? To have a greater understanding of the problems associated with condoms, the actual manufacture of condoms will be detailed as well as the steps taken by manufacturers to ensure the quality of their product. I will then discuss some of the problems that can arise after the condoms have been shipped to their distributors.

Although condoms have been available in various forms (e.g., animal caeca (intestinal skin), silk, and other materials) for several hundred years, production was minimal until the introduction of rubber by vulcanization in the mid-19th century (Dumm, Piotrow, & Dalsimer, 1974). Even though the rubber quality, and hence the condom, was question-

able, the reasonable price and ease in obtaining these latex rubber condoms led to their popularity. With stricter quality control during the
production of latex condoms, which began in the mid-1950s, the reliability and quality improved dramatically.

Manufacturing and Quality Control Steps

Forming and Shaping

Table 21-1 illustrates the manufacturing and quality control steps in condom production. First, raw latex is mixed with the necessary ingredients
to produce the strong and thin coat. Next, condom molds, called mandrels, are used to bind the latex and form the shape. This is a very
precise, rigorously controlled, continuous process that allows approximately 100 mandrels/minute to rotate through the latex solution and
obtain a uniformly thin coat. After this, the mandrels are raised from
the vertical position to a horizontal position, and as they proceed down
the assembly line, they go through an oven to cure (vulcanize) the latex.

At the next stage, the mandrels rotate while a rotating nylon brush,
almost at right angles at the top of the mandrel, rolls the top of the
condom down several centimeters to form the top rim of the condom.
After this, the condoms are pushed off the mandrel by another set of
nylon brushes.

After this point, the condoms are taken to another area where lubri-

Table 21-1
Flowchart for Manufacture and Quality Control of Condoms

Manufacturing Steps	Quality Checks and Testing
1. Raw materials	Raw materials tested
2. Forming/shaping	
3. Automated electronic testing	All condoms electronically tested
4. Performance and visual testing	Testing of random lot samples for: a. Water leakage b. Thickness and dimension c. Elongation and tensile strength d. Air inflation
5. Packaging	Finished product inspection: a. Marking b. Labeling c. Count and weight
6. Shipping	

cation or dry powder is added. The condoms are then rotated in a drum until the powder or lubricant is uniformly distributed on the condom.

Automatic Electronic Testing

At this stage, the condoms are brought to another location where each is tested electronically for leaks. First, workers visually inspect the condoms for obvious flaws incurred during the placement on the metal mandrels, and after those flawed condoms are discarded, the remaining condoms continue down the conveyor belt to be further tested. Most manufacturers then dip the condom in saltwater while applying an electrical charge to the mandrel. Conductance of current between the metal mandrel and the saltwater indicates a leakage of the condom (failure). Some manufacturers do not dip the condom into saltwater but use a charged multielement, brushlike device to perform this test.

As the condoms undergo the electronic testing, the ones that fail are diverted automatically into another channel and continue on the conveyor belt in a vertical position. The good condoms continue in the horizontal position. At the end of this stage, the good condoms are rolled off the mandrel in a form ready for packaging.

Performance and Visual Tests

From this point, a number of condoms are further tested by more rigorous criteria. These numbers are determined by the lot size and conform to the protocol established by the American Society for Testing Materials (ASTM). These test are, in sequence: the water leak test, the thickness and dimension test, and the elongation and tensile strength test.

The water leak test is standardized; 300 milliliters of water (about 10 fluid ounces) are poured into the condom and the condom tested visually for leaks. This test has as many variations as the number of people performing it. Some will close the open end by twisting it around their finger and physically squeeze the condom, or may roll it on a surface to check for obvious leaks both on the lower end and toward the open end. Others may just add the water and, after waiting 1 minute, check for water leakage by blotting the outer surface and examining the blotter.

The second test, for thickness and dimension, is necessary to ensure that the measurements are within the specifications of the product.

The third test, for elongation and tensile strength, uses a special instrument that stretches a segment cut from the condom until it breaks, and recording the force necessary to break it. The amount that the condom strip stretches (at least 600%) is referred to as elongation, and the force necessary to break it determines its tensile strength. The critical feature of this test is the ability to cut the condom segment in a uniform manner each time. The condom is laid flat and a strip is cut out from the area closest to the upper third of the condom from the open end.

Some manufacturers perform additional tests on their products. One of these tests is the air inflation test. This test, a measurement of strength, is the requirement of the International Organization for Standardization (ISO). It is used in many countries outside the United States in place of the elongation and tensile strength test. In this test, the condom is placed around a "tube," locked in place, and then inflated until it bursts. The testing instrument measures the total volume of air introduced into the condom and the pressure at the time of burst.

Packaging and Shipping

At this point in the manufacturing step, the condoms are prepared for packaging and final shipment. The condoms may undergo one more test for water leakage after packaging to ensure that the packaging machine was functioning properly. After this point, the more typical quality control steps—for example, checking for marking, labeling, and count and weight—are performed.

Summary

The description of the manufacturing process given here is typical of the manufacturers in the United States. For those abroad, similar requirements are met and, for those countries using the ISO standards, the air inflation test will be used in place of the test for elongation and tensile strength.

Therefore, key features of modern condom manufacture include electronic testing of each condom for pinhole leaks and additional statistical quality control tests for a set number of condoms from each lot.

During the evolution of these quality control schemes, the condom manufacturers kept samples from each lot and tested them over many years. They found that condoms stored under normal conditions are safe for at least 5 years and probably 10 years. Since most manufacturing sites for the major manufacturers are in or near major metropolitan areas in the developed nations, the consensus of the manufacturers was that the condoms were stable to the variations in climate.

After Manufacture

After the condom leaves the manufacturer's warehouse, the manufacturer has lost control of how its products will be used. Stories of condoms being left out under the direct sun for hours and days are not uncommon. Also, many of the warehouses where condoms are stored by the distributors are not air conditioned, and during the hot summer months the storage temperature may well exceed 120°F. Thus, even before we can concern ourselves with quality control factors associated with consumer usage, such as storage by the individual, types of lubricant used in conjunction with condoms, and actual misuse or abuse during the opening of the package, we must concern ourselves with the environmental factors related to condom failure.

Environmental Factors

TEMPERATURE

The poor tolerance of latex and its congeners to high temperatures is well known. In one case, the ISO air inflation test was used to test condoms shipped to Indonesia. Investigators from PIACT obtained samples from this lot, which they knew had been stored in their hot and humid warehouse and, upon testing, found that a large percentage of the condoms had lost their ability to expand, as shown in the lower volumes of air required to burst the condoms (Free, Hutchings, & Lubis, 1986).

AIR POLLUTION

An additional problem, which used to be regional but is now national, is air pollution. One member of the Air Quality Control Board in the Los Angeles office used to leave a few fresh rubberbands in a desk and check for their ability to stretch every few days. (W. Howell, Air Quality Control Board, personal communication, 1970). When the smog was bad, after a few days the rubberbands would easily break. Thus, warm or hot temperatures compounded with smoglike air pollution are well-known factors for weakening latex. In a study by Baker et al. (1988), condoms exposed to ozone levels corresponding to a stage 2 smog alert in Los Angeles for 48 to 72 hours significantly weakened (100% failure in air inflation test).

ULTRAVIOLET RADIATION

In another investigation by the group from PIACT, researchers were able to show that prolonged exposure to ultraviolet rays had a major effect on the breakage of condoms. After 6 to 8 hours of intense ultraviolet radiation, failure rates, determined by actual breakage during use, was nearly 30%.

Consumer Usage Factors

As in all over-the-counter products, the manufacturer must consider all possible routes for consumer-induced failures. With the greater appreciation and recognition of the value of the use of condoms in this "AIDS era," new problems with the use and abuse of condoms have arisen.

EDUCATION

Although the package is simple in appearance, none contains a good explanation of how to use the condom. Many manufacturers have begun to insert instructions on usage, but condom use still remains a mystery to most users. The first-time user, upon removing the condom from the package, is faced with the task of trying to figure out how to effectively use this rolled up piece of latex he is now holding in his hand. Many problems, such as tearing or puncturing the rim with a fingernail can be eliminated with proper technique. However, what are the proper techniques? Consumers need good, illustrated manuals that show how to use the product—from the opening of the package without damaging the condom to the actual placement and subsequent removal. These manuals should be made with direct input from the manufacturers, since they know how to handle their product in the best manner. There are too many manuals that ignore some of the simple techniques that are routinely used by the manufacturers when they handle the condoms. An additional set of guidelines for proper storage could eliminate many of the problems related to storage methods currently employed by consumers—for example, the practice of storing condoms in the wallet, which will place a tremendous amount of frictional stress on the package.

LUBRICANTS

The use of various lubricants with condoms has created a new set of problems. Petroleum products, such as vaseline, have been known to weaken latex. Thus, latex condoms are at risk of weakening by use of products made from petroleum (Voeller et al., 1989). At present, only the water-soluble lubricants appear to be safe when used together with the condom.

Summary

During the manufacturing process, condoms are subject to a variety of tests to determine the quality of the product. Each condom is electronically tested and tested for leakage, and random samples by predetermined quality-control protocols are tested. Thus, the products coming out of the shipping docks of the manufacturers are good, reliable condoms by all criteria set forth by domestic and international standards committees.

Since latex in the form of the finished product is technically still curing in the package, the condom actually reaches maximum strength in 3 to 6 months after manufacture (T. Enomoto, Fuji Latex, personal communication, 1979). The shelf life is generally accepted to be at least 5 years by the industry. In fact, I have been told that under the proper conditions condoms, when properly packaged, should be good for 10 years or more (T. Enomoto, Fuji Latex, personal communication, 1979). Some of the "real-world" problems reported in the popular press can be attributed to the fact that some stores have condoms on their shelves for longer than 5 years. Thus, consumers should only purchase condoms that display dates on their packaging. In other instances, the condoms may have been stored in warehouses that were not air conditioned and thus were subjected to high temperature during the summer months. Another possibility is that the condoms were stored in cities with high air-pollution levels, which generally coincides with the hot summer months and can compound the problem. In addition to these storage problems, if condoms were stored in areas of the warehouse under sky-lights, the ultraviolet rays from the sun can have an effect.

In all probability, no one factor is the sole culprit in condom failure problems. It is probably more realistic to say that it is the combination of many factors—environmental issues, consumer-related problems, and poor warehouse conditions. If we can minimize the environmental threats from the manufacturer to the distributors, and from the distributor to the final retail outlet, we can eliminate this set of concerns. But once the product is in the possession of the consumer, we must rely on education—whether on the package, as a package insert, or as literature available in the store. Consumers must be informed on the proper use of condoms and on how to store them in order to take advantage of the tremendous protection and functional simplicity of this device.

Acknowledgment

I wish to thank Mr. "Mickey" Saito of Circle Rubber for his patience and understanding during the course of my education in the manufacturing process of the condom.

References

Baker, R. F., Sherwin, R. A., Bernstein, G. S., Nakamura, R. M., Voeller, B., & Coulson, A. H. (1988). Precautions when lightning strikes during the monsoon: The effect of ozone on condoms [letter]. *JAMA, 260,* 1404–1405.

Dumm, J. J., Piotrow, P. T., & Dalsimer, I. A. (1974). The modern condom—a quality product for effective contraception. *Population Reports, Series H, Barrier Methods, No. 2.*

Free, M. J., Hutchings, J., & Lubis, S. (1986). An assessment of burst strength distribution data from monitoring quality of condom stocks in developing countries. *Contraception, 33,* 285–299.

Voeller, B., Coulson, A. H., Bernstein, G. S., & Nakamura, R. M. (1989). Mineral oil lubricants cause rapid deterioration of latex condoms. *Contraception, 39,* 95–102.

22

Barriers: Contraceptive and Noncontraceptive Effects

Gerald S. Bernstein

Barrier contraceptives have a unique place in the field of conception control. They are mentioned in the earliest medical writings and have been in use from before written history on through the modern era (Hines, 1970). Like most methods used to prevent pregnancy, barriers also have noncontraceptive effects that may be beneficial or injurious to health. One of these effects, the ability to deter the transmission of some sexually transmitted disease (STD) agents, has made barriers important in the quest to reduce transmission of the human immunodeficiency virus (HIV).

Elsewhere in this volume, other investigators have described laboratory studies of the effect of spermicides on HIV and the inability of HIV to penetrate intact latex condoms. At present, there are limited data about how successful barriers are in preventing transmission of HIV under conditions of clinical use (Fishl et al., 1987; Mann et al., 1987). However, studies of the use of barriers to prevent pregnancy and to reduce infection with STD agents other than HIV have produced a body of knowledge relevant to the prevention of spread of HIV. The purpose of this report is to discuss pertinent aspects of this information.

Types of Barriers

Table 22-1 describes the various types of barrier contraceptives. Vehicle barriers consist of a spermicide and a vehicle that is used to place the spermicide into the vagina to form a vehicle/spermicidal barrier over the cervical os. The solid mechanical barriers are used in conjunction with

Table 22-1
Types of Barrier Contraceptives

Condoms
Vaginal barriers
    Vehicle barriers
        1. Contraceptive jelly
        2. Contraceptive cream
        3. Aerosol foam
        4. Melting suppositories
        5. Foaming suppositories
    Solid mechanical barriers
        1. Diaphragms
        2. Cervical caps
    Medicated mechanical barriers
        Contraceptive vaginal sponge[a]

[a] The sponge is described as a spermicidal
vehicle but it can act as a mechanical barrier
if it covers the cervical os.

a vehicle barrier to shield the cervix behind a latex device containing a vehicle barrier. The medicated sponge is described as a vehicle for a spermicide but may also act as a mechanical barrier if it remains in front of the cervix (Bernstein & Nakamura, 1982).

Nonoxynol-9 (N-9) is the spermicide used in most contraceptive products available in the United States, although two preparations contain octoxynol, a nonionic surfactant similar to N-9, as the active agent.

Contraceptive Effects

Mechanisms of Action of Barriers

At first glance the mechanism of action of a barrier contraceptive is uncomplicated. Condoms retain the ejaculate within the device, thereby preventing semen from contacting the vagina or cervix. Vaginal contraceptives interact with semen within the vagina to prevent sperm from entering the cervix.

Ejaculation is a sequential event in which the first few drops of semen are forcibly ejected from the urethra while the bulk of the semen follows as a thick coagulum. Usually the first part of the ejaculate is thin, consists primarily of prostatic fluid, and contains most of the sperm in the specimen, while the second fraction is primarily seminal-vesicular fluid and is less rich in spermatozoa.

Near the time of ovulation increasing estrogen levels transform the cervical mucus from a viscous plug into a thin, watery, alkaline form that is highly receptive to spermatozoa. The cervical os dilates and the mucus spreads over the exocervix and may form a tongue that extrudes from the cervix into the vagina.

The mucus provides an excellent mechanism for trapping sperm. Sperm from the first portion of the ejaculate enter the mucus on contact and quickly ascend into the endocervical canal. If the semen forms a vaginal pool, sperm continue to migrate into the cervix until the vaginal acids overcome the buffering capacity of the seminal plasma and the spermatozoa in the vagina become immobile. Sperm transport thus occurs both in a rapid and in a slow phase, so that the maximum number of sperm in the cervix is reached in about 2 hours (Tredway et al., 1975). Sperm can survive within the mucus in the endocervical canal for 2 or 3 days. The cervix thus serves as a sustained release device allowing spermatozoa to migrate into the upper female reproductive tract for several days. Given favorable mucus, this allows flexibility in the timing between coitus and ovulation in order to achieve a pregnancy—that is, intercourse does not have to occur exactly at the time of ovulation.

Vehicle barriers interfere with sperm transport by means of the action of the blocking effect of the vehicle and the immobilization of sperm by the spermicidal agent. The product must inhibit the early *and* late phases of sperm transport to be effective. Solid mechanical barriers deflect semen away from the cervix and the vehicle barriers placed within the devices block spermatozoa that may progress around the end of the device toward the cervix. The blocking effect of the solid barrier requires that the device remain in place before the cervix. If the device is dislodged, the main contraceptive effect must come from the vehicle barrier alone.

Efficacy of Barriers in Preventing Pregnancy

A major advantage of barrier contraceptives is that, unlike hormonal contraceptives and intrauterine devices (IUDs), there are essentially no medical contraindications to their use apart from sensitivity to spermicides or other substances present in vehicle barriers, condoms, or vaginal mechanical barriers.

Unfortunately, barriers also have some disadvantages (Table 22-2). A major one is that failure rates tend to be higher for barriers than for oral contraceptives or IUDs. Table 22-3 demonstrates pregnancy rates for various methods of contraception that have been summarized from the literature (Hatcher et al., 1986). This table shows two types of information for each method: the lowest pregnancy rate expected when the method is used consistently and correctly at each exposure, and the results that might be expected in an average population of users. Barriers tend to have higher pregnancy rates in both categories and a larger difference between the two classifications. This wide range exists for all

Table 22-2
Disadvantages of Barrier Contraceptives

Generally less effective than oral contraceptives or IUDs.
Require strong motivation for successful use.
Require manipulation of genitalia.
Are considered to be messy.
Are sometimes inconvenient to use.
Some types must be used at or near the time of intercourse.
Taste considered by some to be unpleasant.

types of barrier contraceptives and can be attributed to motivational factors that lead to less than perfect use of the method.

Causes of Contraceptive Failure

Contraceptives may fail because they are sometimes biologically ineffective or because the user employs the method incorrectly or inconsistently. Pregnancies that occur even when a contraceptive is used consistently and correctly are called *method failures*, whereas pregnancies that occur because of poor compliance are called *patient failures*. In practice it is difficult to distinguish between the two types of failures, and the overall pregnancy rate, or use effectiveness, is the preferred measure. It is, however, important to consider both method and use effectiveness to provide information to consumers and to devise ways to improve the efficacy of barrier contraceptives.

Method failure may occur with vehicle barriers if the product does not completely block the entrance of sperm into the cervical mucus. This can be demonstrated in sperm transport studies in women who have postcoital tests done following intercourse at midcycle with a barrier contraceptive in place. The results are compared with those obtained when the same women have intercourse at midcycle without using a

Table 22-3
Pregnancy Rates for Various Types of Contraceptives[a]

Method	Lowest Failure Rate[b]	Typical Failure Rate[b]
Combination pills	0.5	2
IUD	1.5	5
Condoms	2	10
Diaphragms	2	19
Spermicides	3–5	18

[a] Reprinted with permission from Hatcher et al. (1986).
[b] Pregnancies per 100 users in first year of use (see text).

Table 22-4
Effect of Aerosol Foam on Sperm Transport
in Women[a]

Motile Sperm per HPF[b]	Number of Subjects in Each Group	
	Control	Foam
0	—	9
<1	—	2
>4	12	1

[a] Reprinted with permission from Bernstein, G. S. (1983).
[b] Mean number of sperm with progressive motility per high-powered field (HPF); Normal value >4.

barrier. (The participants are protected against pregnancy because they have had a tubal sterilization or are using an IUD.)

The postcoital test is done according to a standardized protocol (Davajan & Kunitake, 1969). Table 22-4 shows the effect of aerosol foam in preventing sperm from entering cervical mucus. There is a marked effect of the contraceptive except for one subject for whom the foam did not impede sperm transport. We may assume that this woman would not have been protected against pregnancy for this episode of intercourse, whereas the others would have a reduced chance of becoming pregnant.

Solid mechanical barriers may provide more effective cervical occlusion, but only if they are not dislodged during intercourse. If displacement does occur, only the vehicle barrier used with the mechanical device is left to protect the cervix.

Condoms are intended to keep the ejaculate out of the vagina entirely. They will fail to do this if they leak or rupture during intercourse, if the ejaculate regurgitates along the length of and over the top of the sheath, or if the condom slips off into the vagina during or afer coitus.

User failure may result from inconsistent use by poorly motivated individuals (*psychological failure*) or because the method is not used correctly (*methodological failure*) (Bernstein & Nakamura, 1979). While it may be difficult to separate these two causes, the distinction is important because incorrect use can be rectified by instruction, whereas poor motivation requires a more complex approach.

In situations in which a product is obtained from a health provider, the health worker must give explicit instructions and make certain the patient understands them. When contraceptives are purchased over the counter the instructions in the package insert must be very clear to assure proper use. Efficacy rates in clinical trials are usually higher than those attained when a product is used by the general public because of the process used to select study subjects and the intensive educational efforts of the study personnel. Within populations, factors such as age, marital status, level of education, and parity may also influence pregnancy rates.

Noncontraceptive Effects

There are two side effects of barriers that relate to AIDS.

Reduction of Transmission of STDs

Barrier contraceptives have the ability to reduce transmission of various STDs. The mechanism for this protection depend on the type of contraceptive used. Condoms keep organisms in infected semen from contacting the female genital tract. They also protect the male from contact with infected female genital fluids (Bernstein, 1990). Spermicides have the ability to inactivate most STD agents, including HIV, in vitro (Bolch & Warren, 1973; Hicks et al., 1985; North, 1990; Singh & Cutler, 1979, 1982; Voeller, 1985, 1986). In addition, the barrier effect of vehicle and mechanical barriers can reduce the transmission of organisms that can infect the female via the exocervix or endocervical canal.

Stone, Grimes, and Magder (1986) and North (1990) have summarized the data obtained from various clinical studies of the ability of different barrier contraceptives to prevent transmission of STDs. In keeping with the variable effectiveness of barriers in preventing pregnancy, the ability of barriers to prevent infection also varies widely. For example, the risk of condom users acquiring urethral gonorrhea ranged from 0.00 to 0.51 in various studies, whereas the risk of spermicide users acquiring a gonorrhea infection of the cervix ranged from 0.11 to 0.60.

More recently, Rosenberg, Rojanapith, Yakon, Feldblum, and Higgin (1987) reported that prostitutes in Bangkok who used a vaginal contraceptive sponge have a reduced risk of acquiring gonorrhea or chlamydia infections (0.31 and 0.67, respectively) compared to women not using the sponge. Also, Cramer et al. (1987) reported that barrier contraceptives substantially reduce the risk of tubal infertility.

The reasons for imperfect protection and variation between studies are similar to those that prevail where prevention of pregnancy is the measure of success—that is, inconsistent or imperfect use and biological failure of the method. The development of products that offer more efficient protection and are easier to use, as well as the provision of more effective instructions and educational programs, should improve the ability of barriers to prevent transmission of STD agents. Effectiveness will also depend on the motivation of individuals to use barriers as protection against STDs in contrast or in addition to using them to prevent pregnancy.

Trauma

Solid mechanical barriers have the capability of traumatizing the vagina. This has been shown for vaginal diaphragms (Widhalm, 1979) and a particular type of cervical cap (Bernstein et al., 1982). Any object placed in the vagina has the potential of traumatizing the vagina or cervix via either pressure exerted by the device or trauma caused by efforts to insert or remove it.

The mechanism of transmission of HIV from infected semen is still not certain. Current evidence indicates the virus is present in the white cell fraction of the ejaculate (Anderson et al., 1990; Ho et al., 1984; Zagury et al., 1984). Disruption of the genital epithelium would provide a portal of entry of infected cells into the circulation of the recipient. It is important, therefore, that any barrier used as part of an AIDS prevention program be atraumatic.

At present the condom seems best suited to prevent heterosexual transmission of HIV because it prevents contact with infected semen or cervical/vaginal fluids. Additional information is required about the effect of other barriers. Although spermicides inactivate HIV in vitro, these studies have been done primarily with free virus rather than with infected cells, which seem to be the main vector of transmission. Mechanical barriers that cover the cervix would be protective if HIV can gain access to the circulation via the surface of the cervix or if transmission can occur through the endocervical canal. Such protection may also reduce female-to-male transmission by shielding the male from infected cervical mucus. Until more information is available, however, condoms, perhaps supplemented by spermicides, should be used in situations in which there is a risk of sexual transmission of HIV.

References

Anderson, D. J., Wolff, H., Pudney, J., Wenhau, Z., Martinez, A., & Mayer K. (1990). Presence of HIV in semen. In N. J. Alexander, H. L. Gabelnick, & J. M. Spieler (Eds.), *Heterosexual transmission of AIDS* (pp. 273–290). New York: Wiley-Liss.

Bernstein, G. S., Kilzer, L. H., Coulson, A. H., Nakamura, R. M., Smith, G. Bernstein, R., Frezieres, R., Clark, V. A., & Coan, C. (1982). Studies of cervial caps I. Vaginal lesions associated with use of the Vimule cap. *Contraception, 26,* 443–456.

Bernstein, G. S. (1990). Presence of HIV in cervical and vaginal secretions. In N. J. Alexander, H. L. Gabelnick, J. M. Spieler (Eds.), *Heterosexual transmission of AIDS* (pp. 213–224). New York: Wiley-Liss.

Bernstein, G. S., & Nakamura, R. M. (1979). Clinical trials of vaginal contraception. In G. I. Zatuchni, A. J. Sobrero, J. J. Speidel, & J. J. Sciarra (Eds.),

New developments in vaginal contraception (pp. 264–270). New York: Harper and Row.

Bernstein, G. S., & Nakamura, R. M. (1982). *Development and testing of vaginal contraceptives: Studies of the polyurethane vaginal contraceptive sponge.* (Final report to NICHD: Contract No. NO-1-HD-8-2857.)

Bolch, O. H., Jr., & Warren, J. C. (1973) In vitro effects of Emko on *Neisseria gonorrhoeae* and *Trichomonas vaginalis. American Journal of Obstetrics and Gynecology, 115,* 1145–1148.

Cramer, D. W., Goldman, M. B., Schiff, I., Belisle, S., Albrecht, B., Stadel, B., Gibson, M., Wilson, E., Stillman, R., & Thompson, I. (1987). The relationship of tubal infertility to barrier method and oral contraceptive use. *JAMA, 257,* 2446–2450.

Davajan, V., & Kunitake, G. M. (1969). Fractional *in-vivo* and *in-vitro* examination of postcoital cervical mucus in the human. *Fertility and Sterility, 20,* 197–210.

Fischl, M. A., Dickinson, G. M., Scott, G. B., Kimas, N., Fletcher, M. A., Parks, W. (1987). Evaluation of heterosexual partners, children, and household contacts of adults with AIDS. *JAMA, 257,* 640–644.

Hatcher, R. A., Guest, F., Stewart, F., Stewart, G. K., Trussell, J., Cerel, S., & Cates, W. (1986). *Contraceptive technology 1986–1987* (p. 102). New York: Irvington Publishers.

Hicks, D. R., Martin, L. S., Getchell, J. P., Heath, J. L, Francis, D. P., McDougal, J. S., Curran, J. W., & Voeller, B. (1985). Inactivation of HTLV)III/LAV-infected cultures of normal human lymphocytes in vitro. *Lancet, 2,* 1422–1423.

Hines, N. E. (1970). *Medical history of contraception* (pp. 10, 59). New York: Shocken Books.

Ho, D. D., Schooley, R. T., Rota, T. R., Kaplan, J. C., Flynn, T., & Salahuddin, S. Z. (1984). HTLV-III in the semen and blood of a healthy homosexual man. *Science, 226,* 451–453.

Mann, J., Quinn, T. C., Piot, P., Bosenge, N., Nzilambi, N., Kalala, M., Francis, H., Colebunders, R. L., Byers, R., Azila, P. K., Kabeya, N., & Curren, J. W. (1987). Condom use and HIV infection among prostitutes in Zaire. *New England Journal of Medicine, 316,* 354.

North, B. B. (1990). Effectiveness of vaginal contraceptives in prevention of sexually transmitted diseases. In N. J. Alexander, H. L. Gabelnick, J. M. Spieler (Eds.), *Heterosexual transmission of AIDS* (pp. 273–290). New York: Wiley-Liss.

Rosenberg, M. J., Rojanapith, A., Yakon, W., Feldblum, P. J., & Higgin, J. E. (1987). Effect of the contraceptive sponge on chlamydial infection, gonorrhea, and candidiasis. *JAMA, 257,* 2308–2312.

Singh, B., & Cutler, J. C. (1979). Vaginal contraceptives for prophylaxis against sexually transmissible diseases. In G. I. Zatuchni, A. J. Sobrero, J. J. Speidel, & J. J. Sciarra (Eds.), *New developments in vaginal contraception* (pp. 175–185). New York: Harper and Row.

Singh, B., & Cutler, J. C. (1982). Demonstration of a spirocheticidal effect by chemical contraceptives on *Treponema pallidum. Bulletin of the Pan American Health Organization, 16,* 59–64.

Stone, K. M., Grimes, D. A., & Magder, L. S. (1986). Personal protection against sexually transmitted diseases. *American Journal of Obstetrics and Gynecology, 155,* 180–188.

Tredway, D. R., Settlege, D. S., Nakamura, R. M., Motoshima, M., Umezaki,

C. V., & Mishell, D. R., Jr. (1975). Significance of timing for the post coital evaluation of cervical mucus. *American Journal of Obstetrics and Gynecology, 121*, 387–393.

Voeller, B. (1985). *Spermicides: an additional potential barrier to the sexual spread of the AIDS virus, LAV/HTLV-III.* Mariposa Occasional Paper #4B. New York: The Mariposa Foundation.

Voeller, B. (1986). *Nonoxynol-9 and prevention of the sexual spread of LAV/HTLV-III and other STD agents.* Second International Conference on AIDS, Paris. Abstract #555, p. 66.

Widhalm, M. V. (1979). Vaginal lesion: Etiology—a malfitting diaphragm? *Journal of Nurse-Midwifery, 24*, 39–40.

Zagury, D., Bernard, J., Leibowitch, J., Safai, B., Groopman, J. E., Feldman, M., Sarngadharan, M. G., & Gallo, R. G. (1984). HTLV-III in cells cultured from semen of two patients with AIDS. *Science, 226*, 449–451.

23

The Role of Barrier Contraceptives in the Prevention of AIDS: An Epidemiological Approach

Anne Hersey Coulson

The first Kinsey report, published when I was a teenager, was a cause célèbre. The report on females was eagerly awaited. Kinsey and his work were known at most levels of society, at least the name and the subject matter. I attended a small women's college in New York, at 116th and Broadway. It was then the habit for the administration to use most summers for extensive construction work on the academic buildings, and one fall we returned to find the schedule had not been met and that we had to share the space with the construction workers into the fall semester. A mark of the times was the hand-lettered sign placed on the construction shed by some hopeful worker that read "Kinsey Office Field Studies."

It is particularly appropriate that the role of barrier contraceptives in the prevention of AIDS be discussed at this conference on AIDS and Sex. Barrier contraceptives have, in the aggregate, a long association with sexual activity, both for the prevention of unwanted pregnancy and the prevention of sexually transmitted diseases. The specific barrier contraceptives of concern in this paper are condoms and spermicides. Both of these barriers, at least under ideal in vitro conditions, can prevent the passage of human immunodeficiency virus type 1 (HIV-1) (the condom) or can inactivate the free HIV-1 or its host cell on contact (spermicide) (Hicks et al., Chapter 24) (Hicks et al., 1988; Voeller et al., 1985, 1986, 1988). These barriers are effective in the same way against other sexually transmitted disease agents, which may be important in the progression of HIV-1 infection to AIDS (Coulson et al., 1988). Both barriers can be used in sexual intercourse, whether vaginal or anal.

354

Drs. Stevens (Chapter 2) and Detels (Chapter 1) have presented us with considerable information from two major studies of the natural history of AIDS identifying particular sexual activities, especially receptive anal intercourse, as particularly high risk (Stevens, Chapter 2; Detels, Chapter 1). Their findings suggest that barrier contraceptives, by protecting the receptive partner from semen and the insertive partner from vaginal or anal secretions, might be especially effective in reducing the exposure to HIV-1 or infection with it.

Dr. Voeller has reminded us that anal intercourse is also practiced by a sizeable fraction of heterosexual couples for contraception, for pleasure, or for other reasons. Thus receptive anal intercourse may be a risk factor for both sexes (see Voeller, Chapter 19).

Neither spermicides nor condoms are perfect contraceptives. Each has a failure rate in preventing pregnancy. How effective can these imperfect methods be in the prevention of infection with HIV-1, and is this protection worthwhile in the prevention of a fatal disease?

In this paper, a model is described for determining the theoretical effectiveness of condoms and spermicides, making certain assumptions about the prevalence of HIV-1 infection, the risk of infection, and characteristics of condoms and spermicides and their use.

It is generally agreed that AIDS is a highly if not completely fatal disease. It is further agreed that there is a high probability that persons who have the HIV-1 infection will go on to develop AIDS. There is disagreement about the proportion that will; a recent report suggest that all those infected will eventually develop AIDS (Hessol et al., 1987).

Under these conditions, it appears at present that the best public health response to the AIDS epidemic is one of prevention, and that the most effective form of prevention will be that which protects persons who are not infected with HIV-1 from becoming infected.

This paper is concerned with prevention in the epidemiological sense, and particularly with the usefulness of condoms and spermicides in improving that prevention. Prevention is defined here as any activity that reduces the risk of progress from one identifiable level of health or disease to a worse level (e.g., from health to infection, or from disease to death). It should be noted that this definition does not require the total elimination of a risk in a population, but simply a reduction. This concept underlies later discussion here about prevention techniques and their likely usefulness.

There are three levels of prevention defined in epidemiology: primary, secondary, and tertiary (Mausner & Bahn, 1974). For purposes of this discussion, however, I would like to replace those three with four levels for the prevention of AIDS. Those four levels are:

Prevention of exposure
Prevention of infection

Prevention of disease
Prevention of death

Each of these four levels is discussed below, followed by a more detailed consideration of the preventive role of barrier contraceptives. Before presenting those discussions, however, I would like to indicate briefly the rationale for adding a fourth category level of prevention to the three traditionally differentiated.

Traditionally, prevention of exposure and prevention of infection would be grouped together as "primary" prevention. I think it is useful in this context and in other contexts such as health promotion, to separate them. Each type or level of prevention relates to one of two distinct and somewhat contradictory messages that sometimes appear in public statements about AIDS and its transmissibility. The first message is that there are clearly high-risk groups; the infection is difficult to acquire except under particular circumstances more commonly encountered in those groups. The second message is that AIDS is a very democratic disease, no respecter of persons, and therefore the whole population is at some risk. Both messages say something important, but the apparent conflict may confuse the intended audience. The first message (high-risk groups) relates to the prevention of exposure: high-risk groups have a fairly high prevalence of HIV-1 infection and exposure is difficult to control. The second message (AIDS threatens the whole population) relates to the prevention of infection: the prevalence of HIV-1 infection may be low, but it is finite and sexual contact with an infected partner may lead to infection.

Levels of Prevention for AIDS

Prevention of Exposure

This level of prevention could theoretically be attained by various measures preventing any contact that might transmit the virus to another individual. Each of the methods commonly discussed for preventing exposure has or may have serious problems consequent to it—in essence, undesirable side effects.

One means of preventing exposure is for individuals to abstain from sexual practice or to engage only in truly safe sexual practices with no possibility of exchange of body fluids. However, either of these is difficult if not impossible to achieve and may indeed have undesirable social and psychological side effects. Sex is an important aspect of our humanity and cannot lightly be denied. In the case of heterosexual couples wanting children, totally safe sex is, of course, impractical.

A second means of preventing exposure is to limit intravenous (IV) drug use to situations in which sterile equipment is available, or for users of IV drugs to have personal equipment, for their sole use. Of course, IV drug abusers may find it particularly difficult to obtain new, clean paraphernalia. Furthermore, cultural practices among ethnic groups using recreational drugs may mandate sharing of equipment. There are even problems with medical use of IV medications and vaccines: provision of new equipment for each patient may be beyond available financial resources, especially in developing countries.

Another means of preventing exposure is to avoid the use of blood or blood products. Given recent findings that show the presence of infection long before seroconversion, testing of the blood supply does not seem as adequate as it did earlier. The avoidance of exposure may not be possible in medical and surgical situations where a balance of risks must be considered: the probability of death if blood or blood products are not used must be balanced against the possibility of developing AIDS later as a consequence of the use of that blood.

Breast-feeding is another possible means of exposure of infants, because HIV-1 has been demonstrated in the milk of HIV-1–positive mothers. However, as with blood and blood products, avoidance of exposure—that is, reliance on artificial foods instead of breast-feeding—may have more serious consequences for the infant, including increased infant mortality (Munyakho, 1988).

Prevention of Infection Given Exposure

This level of prevention calls on strategies of protection of the susceptible host. Traditionally, this has involved immunization against the disease in question, destroying the agent before it can infect, or preventing the agent from reaching a "safe" home in the uninfected host.

An immunization program offers probably the best hope for ultimate prevention of infection. However, an effective, safe vaccine for AIDS is still in the future, although several are being tested (Wainberg, Kendall, & Gilmore, 1988; Schild & Minor, 1990). None is presently available.

The agent can be destroyed by heat, detergents, and various other processes. Some blood products can be treated to destroy HIV-1; IV needles can be treated with chlorine bleach (Resnick, Veren, Salahuddin, Tondreau & Markham, 1986). Drs. Masters and Johnson presented information at this conference on the susceptibility of HIV-1 to acidic environments (Chapter 25) as also noted by Voeller, 1985. It is not possible to treat body fluids in situ, and the usefulness of some fluids might be compromised or destroyed by a process used to inactivate HIV-1.

Preventing HIV-1 from entering the negative host may be accomplished by barrier contraceptives, notably the condom, which is intended to prevent sperm from entering the cervix. Such a barrier should prevent passage of HIV-1. Dr. Ndinya-Achola presented data on the evident role of genital lesions in the spread of HIV-1 infection in Africa (see Chapter 9). Whatever the mechanism of transport of susceptibility implied by these data, the condom must have an important role in preventing contact with the lesion during sexual intercourse.

Spermicides have been demonstrated to kill free virus in vitro, at noncytotoxic concentrations (Hicks et al., Chapter 24). A model for the usefulness of condoms and spermicides in the prevention of HIV-1 infection is discussed later in this paper.

Prevention of Disease Given Infection

There are reports of follow-up studies indicating that HIV-1 infection will inevitably lead to AIDS (Hessol et al., 1987). Dr. Anderson and his colleagues have shown us an elegant model, indicating that the average time from infection to disease may well be between 9 and 10 years (see Chapter 6). Our experience with the disease is not long enough to confirm this; if the interval is indeed that long, the inevitability of progression may be supported. If progression is inevitable, then prevention at this level may not be feasible. There appear to be three possible prevention approaches at present: treatment, support of the infected individual to improve resistance, and prevention of infection with other strains of HIV-1 or infection with possible cofactors, such as cytomegalovirus or Epstein-Barr virus.

Treatment and support are outside the scope of this paper. Prevention of infection with other strains of HIV-1 or other agents may be improved with barrier contraceptives (e.g., condoms and spermicides). In a sense it might be argued that the prevention of reinfection and other infections is at the second level of prevention discussed above, except that the intent in this case is to prevent disease, given the initial HIV-1 infection, through modality of preventing other infections.

Prevention of Death Given Disease

It is certainly agreed that AIDS is a highly, if not completely fatal disease. The explosive growth of the epidemic, coupled with a reasonably long course from diagnosis to death, has made it difficult to estimate the true case fatality because the denominator of cases consistently outruns the numerator of deaths. Although many people apparently believe that the increasing disparity between numerator and denominator in the Centers for Disease Control data means that AIDS is becoming progressively less fatal, this is not true. Treatment, such as azidodeoxythimidine, would appear to be the only means of preventing death. Support of the system may prolong life; among the very few long-term survivors, rigorous personal health promotion is credited for that survival. Prevention of other infections may also have at least short-term effects, because death itself is likely to be caused by infections (or conditions) other than HIV-1.

Barrier Contraceptives: An Analytical Model

The usefulness of barrier contraceptives in the prevention of AIDS is of course restricted to the sexual transmission of AIDS, whether anal, oral, penile, or vaginal. The receptive partner may be protected from semen by the condom; the insertive partner may be protected from the vaginal or anal secretions or saliva by the condom. The spermicide may protect both partners by killing free HIV-1 and host cells carrying the virus in semen or vaginal or anal secretions (Voeller, 1985, 1986). By destroying sperm mobility, spermicides prevent any virus contained in or attached to sperm from being transported.

In developing a model to examine the probable effect of use of barrier contraceptives in preventing infection with HIV-1, one must start with a basic concept of unmodified risk, that is, the risk of HIV-1 infection without the use of condoms or spermicides. Determining this actual risk is not simple. It is necessary to estimate two parameters: the probability that the individual with whom one has contact has the virus and can transmit it, and the probability of infection resulting from contact with an infected individual. The first of these probabilities is difficult to estimate because the actual prevalence of HIV-1 infection is rarely known; also, individuals who do not currently have HIV-1 antibody may have and be able to transmit the infection. Seropositivity now seems to be a lower bound to the prevalence of infection in the population rather than an estimate. There is a wide variation in seropositivity from group to group: a high-risk group may have a seroprevalence of HIV-1 of 50% or more (Curran, et al., 1988), whereas that of general populations such as blood donors may be less than 1% (Ward, Kleinman, Douglas, Grindon, & Holmberg, 1987). Both of these prevalence estimates differ between groups.

The probability of infection given contact with an infected individual is even more difficult to estimate. Some studies indicate that transmission from a seropositive to a seronegative partner, given that the seronegative partner is the receptive partner, may take an average of more than 100 sexual encounters (Padian, 1990; Padian et al., 1987). There is a possibility that the high average number of required encounters may reflect a resistance of the uninfected partner rather than a strict probability of infection or a need for multiple exposures. However, it must be noted that following artificial insemination using semen from a single donor, four of the eight women exposed became seropositive for HIV-1, for an infection rate of 50% (Stewart et al., 1985).

Clearly, the more sexual partners the higher the probability of one or more of them being infected, whether the prevalence of infection is high or low, so long as there is some prevalence. Also, the more high-risk sexual encounters with infected partners the greater the probability of becoming infected, whatever the per-encounter probability of trans-mission may be.

We can examine the working of the barrier contraceptive prevention model by assuming various levels of prevalence of infection and risk of infection. New values for either or both of these parameters can be substituted in the model very easily. Two sets of assumptions are used here for illustrative purposes: 50% prevalence of infection and 1% risk of infection per encounter and 20% prevalence and 10% risk of infection per encounter. Using these numbers will allow the calculation of un-modified risks on which to base determination of barrier contraceptive efficacy.

Whether these illustrative estimates are high or low is certainly ar-guable, but it is also important. A shift in the estimate of the unmodified risk will change the calculations of the absolute effect of barrier contra-ceptives but not their relative impact. Thus the model serves an im-portant analytical purpose despite the apparent arbitrariness of certain values that may be used in that model.

The unmodified risk is defined as the risk without any intervention and is calculated by multiplying the prevalence by the risk of infection and expressing the result in terms of a "standard" denominator (e.g., 1,000, 10,000, 100,000) representing total high-risk sexual encounters. Thus the unmodified risk for the 50% prevalence–1% risk group is 500/100,000 or 1/200 (0.5 x 0.01 = 0.005, or 1/200). In the 20% prevalence–10% risk group, the unmodified risk is 2,000/100,000 or 1 in 50 encoun-ters.

This unmodified risk will be modified by factors representing the use patterns and effectiveness of condoms and spermicides. As indicated above, neither condoms nor spermicides are perfect, even in their orig-inal roles as contraceptives. Condoms may fail to prevent transmission in three ways. First, the condom may not be used at all. Obviously it can provide no protection in that case. Even consistent condom users report less than 100% use. Second, the user may not use the condom properly: it may be put on wrong, it may be stressed by being pulled on rather than rolled, the condom may be reused, or the wearer may fail to withdraw carefully to avoid spillage after ejaculation. Third, the condom may split, break, or leak in use because of defects, weaknesses induced by age of condom, improper storage, mishandling, use of oil-based lubricants, exposure to environmental pollutants, or other causes.

Assumptions may be made, or, better, data on aspects of condom use and materials may be applied to determine consistency of condom use, frequency of correct use, and the proportion of time the barrier is ef-fectively maintained. An optimistic combination of levels of these three factors would be condom use 95% of the time, condom properly used 90% of the time, and an effective barrier maintained 99% of the time.

These assumptions lead to an 85% decrease in risk; where the unmodified risk is 500/100,000 or 1/200, the modified risk drops to 75/100,000 or 1 in nearly 1,300 encounters. By contrast, in a "worst case" assumption, with condom use 25% of the time, only half the time correctly, and only 70% of the time maintaining an effective barrier, the modified risk drops only 9%, to 456/100,000 or 1 in 219.

While low consistency of use, improper use, and impaired condom quality markedly reduce the overall effectiveness of condoms in preventing AIDS, the model indicates that *any* use of condoms is better than *no* use. Of course, given the high case fatality of AIDS, and the importance of prevention, individuals would be well advised to use high-quality, carefully protected condoms, correctly and all the time. However, some lesser protection is still likely at lower levels of performance.

Spermicides have the same three general types of failures as condoms. An individual may fail to use the spermicide at all; may use it improperly by applying it incorrectly, applying too little or in the wrong places, or failing to reapply at appropriate intervals; or, least likely, the spermicide may be inactive. (With respect to using spermicide at all, some individuals may believe they are using spermicide when, in fact, they are using a simple lubricant without spermicidal or biocidal properties.)

The same kinds of assumptions can be made for spermicide use as for condoms, with similar modifications in risk. Optimistic assumptions for spermicide might be 95% use, 95% correct use, and 99% effectiveness of the spermicide, yielding a calculated reduction from the unmodified risk of 89% or a drop in risk (again designated in terms of cases for a "standard" number of high-risk encounters) from 500/100,000 or 1/200 to 55/100,000 or 1/1,818 encounters.

Condoms and spermicides have very different modes of action: condoms provide an impermeable skin or latex barrier to prevent semen from reaching the receptive partner and to keep that partner's body fluids from reaching the penis of the insertive partner. The spermicide acts to kill at least free virus in the semen either within the condom or in the rectum or vagina. Thus each method should have an independent effect in the prevention of infection.

For the optimistic assumptions discussed above, and for the 50% prevalence–1% risk group, the risk for condoms alone was 75/100,000 and for spermicide alone 55/100,000. Combining the effects for condom and spermicide, the risk drops to 8/100,000 or 1/12,500. The "worst case" assumption risk reduces from 455/100,000 for condoms alone and 450/100,000 for spermicide alone to 410/100,000 or 1/244 for both together.

In summary, condoms and spermicides may be imperfect and imperfectly used. They may not reduce the infection risk to zero, but can reduce the risk substantially, particularly if they are used consistently and correctly.

It should be noted that the model described here is an extremely simple one intended to demonstrate the theoretical effects of condoms and spermicides in the prevention of HIV-1 infection. The model indicates that, whatever the precise risks and probabilities of behaviors are, use of condoms and spermicides should be better than nonuse, and increases in consistency and correctness of effective barriers should lead to lower risk of infection. The risk reductions described here must be considered relative rather than absolute. To determine the exact prevalence, risk, barrier contraceptive use, and barrier effectiveness in specific populations will require special studies of those populations. Data are being accumulated in a number of ongoing studies that may be useful in making these determinations.

This model, like all epidemiological models, is intended to assess the risk modification in *populations*. The individual's chances are not assessed except in a probabilistic way. For example, the probability that a high-risk sexual encounter with an infected partner will result in infection of the uninfected partner may be quite small—perhaps 1%. This does not preclude some unfortunate person's becoming infected on his or her first encounter.

A number of other factors could have been considered for the model, but would have complicated it unduly for the present chapter. Variability in infectiousness from person to person, or within one person from time to time, may have considerable bearing on the risk of infection. There may be factors of susceptibility and resistance in the uninfected partner that also alter the risk of infection, particularly in the very long periods during which the uninfected person does not seroconvert. (These long periods would appear to decrease the risk for high-risk encounter.)

Summary

In this chapter, four levels of disease prevention were defined and discussed. For one of these levels—the prevention of infection given exposure—a probabilistic model for the efficacy of barrier contraceptives, primarily condoms and spermicides, in the prevention of HIV-1 infection was examined. Used in the model were various assumptions about the prevalence of infection and risk of acquiring infection per high-risk sexual encounter, and probabilities of barrier contraceptive use and effectiveness. Examples based on these assumptions were generated to show how the model can be applied to make useful inferences about the possibile benefits of barrier use under varying conditions.

Acknowledgments

None of us works in a vacuum alone. I am grateful to Drs. Bernstein and Nakamura, and Drs. Voeller and Hicks. I cheerfully absolve them, however, from blame for anything in this chapter with which they disagree. I would also like to credit and offer the same absolution to Dr. Jeffrey Perlman, Chief of the Contraceptive Evaluation Branch, Center for Population Studies, National Institute for Child Health and Human Development and our project officer on the UCLA AIDS/Barrier Contraceptive Study (Primary Prevention Trial of Barrier Contraceptives Against HIV-III Infection, NO1-HD-62934; R. Detels, Principal Investigator; A. H. Coulson, Co-Principal Investigator). I have learned a great deal from all of you, and I have kept adding to and subtracting from my text based on concepts and data presented here.

References

Coulson, A. H., Voeller, B., Bernstein, G. S., Smith, J., Hicks, D., Nakamura, R. M., Detels, R., & Perlman, J. (1988, June). *Effect of anti-HIV spermicides on other sexually transmitted disease (STD) agents.* Paper presented at the Fourth International Conference on AIDS, Stockholm.

Curran, J. W., Jaffe, H. W., Hardy, A. M., Morgan, W. M., Selik, R. M., & Dondero, T. J. (1988). Epidemiology of HIV infection and AIDS in the United States. *Science, 239,* 610–616.

Hessol, N., Rutherford, G. W., O'Malley, P. M., Doll, L. S., Darrow, W. W., Jaffe, H. W., et al. (1987, June). *The natural history of human immunodeficiency virus infection in a cohort of homosexual and bisexual men: A 7-year prospective study* (Abstr.). The Third International Conference on AIDS, Washington, DC.

Hicks, D. R., Voeller, B., Resnick, L., Cullman, L. C., Coulson, A. H., Cassity, C. L., Bernstein, G. S., & Nakamura, R. M. (1988, June). *Chemical inactivation of HIV-1 (HTLV-III and HB2) by contraceptives/spermicidal agents.* Paper presented at the Fourth International Conference on AIDS, Stockholm.

Mausner, J., & Bahn, A. (1974). *Epidemiology: An introductory text.* Philadelphia: W. B. Saunders Company.

Munyakho, D. (1988). Breastfeeding and HIV infection. *Lancet, i,* 1394–1395.

Padian, N. (1990). Heterosexual transmission: Infectivity and risks. In N. J. Alexander, H. L. Gabelnick, J. Spieler, (Eds.), *Heterosexual transmission of AIDS.* New York: Wiley–Liss.

Padian, N., Marquis, L., Francis, D., Anderson, R., Rutherford, G., O'Malley, P., & Winkelstein, W. (1987). Male-to-female transmission of human immunodeficiency virus. *JAMA, 258,* 788–790.

Resnick, L., Veren, K., Salahuddin, Z., Tondreau, S., & Markham, P. (1986). Stability and inactivation of HTLV-III/LAV under clinical and laboratory environments. *JAMA, 255,* 1887–1891.

Schild, G. C., & Minor, P. D. (1990). Modern vaccines: Human immunodeficiency virus and AIDS: Challenges and progress. *Lancet, 335,* 1081–1084.

Stewart, G. L., Tyler, J. P. P., Cunningham, A. L., Ban, J. A., Driscoll, G. L., Gold, J., & Lamond, B. J. (1985). Transmission of human T-cell lymphotropic virus type III (HTLV-III) by artificial insemination by donor. *Lancet, ii,* 581–584.

Voeller, B. (1985). *Spermicides: an additional potential barrier to the sexual spread of* the *AIDS virus*, LAV/HTLV-III. Mariposa Occasional Paper No. 4B. New York: The Mariposa Foundation.

Voeller, B. (1986a). *Nonoxynol-9 and prevention of the sexual spread of LAV/HTLV-III and other STD agents* (Abstr.). The Second International Conference on AIDS, Paris.

Voeller, B. (1986b). Nonoxynol-9 and HTLV-III. *Lancet, i,* 1153.

Voeller, B., Hicks, D., Coulson, A. H., Bernstein, G. S., Nakamura, R. M., Detels, R., & Perlman, J. (1988, June). *Testing HIV leakage through condoms.* Paper presented at the Fourth International Conference on AIDS, Stockholm.

Wainberg, M. A., Kendall, O., & Gilmore, N. (1988). Vaccine and anti-viral strategies against infection caused by human immunodeficiency virus. *Journal of the Canadian Medical Association, 138,* 797–807.

Ward, J., Kleinman, S., Douglas, D., Grindon, A., & Holmberg, S. (1987, June). *Epidemiologic characteristics of blood donors who have antibody to the human immunodeficiency virus.* Paper presented at the Third International Conference on AIDS, Washington, DC.

24

Chemical Inactivation of Human Immunodeficiency Virus Type 1 (HIV-1 Isolates HTLV-III and HB2) by Spermicides and Other Common Chemical Compounds

Donald R. Hicks, Bruce Voeller, Lionel Resnick, Sara Silva, Christopher Weeks, and C. L. Cassity

Human immunodeficiency virus type 1 (HIV-1) is cytopathic for T4 helper/T8 inducer T cells and is the etiological agent for AIDS (Broder & Gallo, 1984; Gallo et al., 1985; Klatzman et al., 1984; Popovic, Sarngadharan, Read, & Gallo, 1984; Sarngadharan, Popovic, Bruch, Schupbach, & Gallo, 1984). HIV-1 is known to be transmitted sexually and through the introduction of infected blood and blood products; it is not a casually transmitted infectious agent (Fauci et al., 1985; Friedland et al., 1986). HIV-1 transmission by human semen, whether as free HIV-1 particles, as HIV-1–infected cells, or as HIV-1 attached to cells contained within semen (Anderson, Chapter 20; Anderson et al., 1990; Miller & Scofield, 1990; Voeller, 1985, 1986c), is currently under investigation. Consequently, inactivation of HIV-1 is currently of considerable interest.

Chemical disinfectants, heat, and low pH have been reported to inactivate HIV-1 (Martin, McDougal, & Loskaski, 1985; Resnick, Veren, Salahuddin, Tondreau, & Markham, 1986; Spire, Barre-Sinoussi, Dormont, Montagnier, & Chermann, 1985; Spire, Montagnier, Barre-Sinoussi, & Chermann, 1984). The use of spermicides and chemical contraceptives, because of their reported virucidal and cytotoxic properties, is of extreme interest to those investigators involved in sexually transmitted disease prevention and control efforts (Hicks et al., 1985; Postic, Singh, Squeglia, & Guevarra, 1978; Singh, Cutler, & Utidjian, 1972;

Stone, Grimes & Magder, 1986; Voeller, 1984a, 1984b, 1985, 1986a, 1986b). Of particular interest is the additional protective margin that may be provided by spermicidal preparations when used in conjunction with a properly used and reliable condom. Among those individuals participating in sexual behaviors that have been classified as "high risk"[1] an effective sexual lubricant used in conjunction with a properly used and reliable condom could provide an additional weapon to contain the spread of HIV-1 infection, especially in the case of condom spillage or breakage (Voeller et al., 1985). Commercially available federally approved spermicides (nonoxynol-9 and octoxynol-9) are marketed in various preparations (spermicidal jelly, spermicidal sexual lubricant, and spermicidal cream) at concentrations of 0.5 to 5%. Additional research is currently in progress utilizing other contraceptive/spermicidal compounds against HIV-1 and a second type of HIV designated as type 2 (Clavel et al., 1986; Hicks et al., 1988; Resnick et al., 1990).[2]

Virucidal and Cytotoxic Effects of Contraceptive/ Spermicidal Compounds

Seven chemical compounds were examined for their virucidal and cytotoxic effects directed against HIV-1: nonoxynol-9 (N-9; 100% v/v); two N-9–containing compounds [N-9 Mariposa Sexual Lubricant[3] (5% N-9 v/v) and the N-9 Vironox-9 hand soap (2% N-9 v/v)]; octoxynol-9 (O-9; 100% v/v); spermicidal-germicidal compound A (SGCA; 0.04% v/v quaternary ammonium chloride); Silvadene (100% v/v silver sulfadiazine); and Betadine (0.04% v/v povidone-iodine). Highly concentrated viral preparations of HIV-1 isolate H9/HTLV-IIIB[4] and Hillcrest Biological's

---

[1] "High risk" is utilized here to describe insertional anal sex, the sexual behavior most highly related to acquiring an HIV-1 infection. This is not the usual labeling of high risk by the sexual preference of the participants because anal sex between heterosexual couples is frequent.

[2] HIV-2 was reported as another retrovirus causing AIDS in 1985 by French and Portuguese clinicians in Guinea Bissau, West Africa (Clavel et al., 1986). The major proteins of HIV-2 are 140K, 36K, 26K, and 15K versus the HIV-1 proteins, which are 110K, 41K, 24K, and 18-19K. Some slight cross-reactivity between HIV-1 and HIV-2 has been reported in the 26/24K protein region. HIV-2 does not react with the HIV-1 tests in use in the United States. Three imported cases of HIV-2 have been reported on the U. S. East coast, with no apparent entry into the U.S. population at this time.

[3] The 5% N-9 Mariposa Sexual Lubricant was developed by the Mariposa Education and Research Foundation of Topanga, CA as an experimental sexual lubricant of exact known formulation to test against the "AIDS" virus. It is not a commercially available product.

[4] HTLV-III is an HIV-1 isolate developed by the U.S. National Cancer Institute. HTLV-III was the general designation for all HIV-1 isolates used by scientists at the National Cancer Institute before HIV-1 became the standard nomenclature. H9 is the continuously HTLV-III–infected lymphocyte cell line developed from Hut 78 as a subclone by the same group (Gallo et al., 1985). HTLV-III is now used as an HIV-1 isolate designation: HIV-1 (HTLV-III).

isolate HB2[5] were prepared by sucrose banding of cultured supernatants. Serial tenfold dilutions of the HIV-1 stock were performed and
1.0 milliliter was used (in triplicate) to infect human lymphocytes. End
points were calculated by the methods of Reed-Meunch from the highest
dilution with detectable reverse transcriptase (RT) activity and cell cytopathic effect (CPE). The HIV-1 virus stock contained an infectious titer
of 8.5 log 10 tissue culture 50% infectious dose ($TCID_{50}$)/milliliter for
HIV-1 (HB2), and a titer of 10 log 10 $TCID_{50}$/milliliter was obtained for
HIV-1 (HTLV-III). The HTLV-III virus preparation contained a total virus
particle count of 1010.5/milliliter by electron microscopy. Viral stocks
were divided into aliquots and maintained at $-80°C$. Each test compound was toxicity tested against polybrene-treated 3-day phytohem-
agglutinin (PHA)-stimulated fresh human donor peripheral blood lymphocytes and polybrene-treated "HC" (continuous cell line)
lymphocytes derived as a Hut 78 subclone. Lymphocyte toxicity was
operationally defined for this study as the concentration of the test substance that altered the cell count or cell viability (by trypan blue exclusion) by one standard deviation or more when compared to the average
of the untreated control cultures after 7 days at 37°C.[6]

To determine test compound effect on HIV-1 infectivity, the HIV-1
inoculum was thawed and exposed to different dilutions of the test
compound for 1 minute and 10 minutes at room temperture (23 to 27°C)
and 36°C. One milliliter of test compound–treated HIV-1 (HTLV-III) was
used in triplicate to infect HIV-1 antibody–negative (by Western blot)
3-day PHA stimulated fresh human donor peripheral blood lymphocytes
(PBLs) after the appropriate (nontoxic) dilution was performed. Concurrently, 1 milliliter of test compound–treated HIV-1 (HB2) HC-
adapted inoculum was used in triplicate to inoculate polybrene-treated
HC cells after the appropriate (nontoxic) dilution was performed. A stock
of untreated inoculum was used as a positive control. The PBLs were
maintained in medium BI[7] and the HCs were maintained in medium
AI, which is medium BI minus interleukin-2. Samples for RT testing

---

[5] HB2 is an HIV-1 isolate developed by Hillcrest Biologicals of Cypress, CA. Additionally,
Hillcrest Biologicals has developed an uninfected continuous lymphocyte cell line derived
as a Hut 78 subclone that has been adapted to HIV-1 (HB2) and is therefore easily infected
by HIV-1 (HB2). HIV-1 (HB2/HC) is the continuously infected designation.

[6] Lymphocytes are utilized in these experiments as amplifiers of infectivity. Should one
lymphocyte become infected with HIV-1, the amplification effect would result in detectable
RT during the culture period.

[7] Medium BI is: RPMI 1640 + 10% heat inactivated fetal bovine serum + 5% interleukin
2 + 25 millimolar hepes buffer + 50 micrograms/milliliter gentamicin + 2 millimolar L-
glutamine + 0.025 grams/milliliter polybrene. This medium is used to culture separated
human PBLs. Medium AI is medium BI minus interleukin-2, and is used to culture continuous lymphocyte cell lines.

were removed at 7-day intervals for 28 days and the media were replaced and fresh appropriately treated target cells were added at those times. A viable cell concentration of 0.5 to 1.0 $\times$ 10^6 cells/milliliter was maintained.

Infectivity was established by monitoring the cultures for characteristic cytopathic effect by light microscopy of Wright-Giemsa–stained cells. Supernatant fluids were assayed for particle associated RT activity using Mg^{2+} as a divalent cation, oligopoly-(rA)·p(dT)12–18 as the template primer representing viral polymerase activity, and oligo-dT-poly-(dA) as primer-template representing cellular polymerase activity as previously described (Gallo et al., 1985; Hicks et al., 1985).

HIV-1 was inactivated and an appropriate quantitation of the reduction of viral infectivity was determined after exposure to the test compounds. Virus infectivity decreased at least 6 log 10 $TCID_{50}$, when compared to the positive controls, at room temperature and 36°C after 1 minute of exposure to the test compounds at the following final concentrations (v/v): 0.005% N-9 and 0-9; 0.1% Mariposa Sexual Lubricant; 0.25% Vironox-9 hand soap; 0.1% SGCA; 0.01% Silvadene; and 0.5% Betadine. These results indicate that each of the test compounds exhibits a substantial virucidal effect.

Discussion

Currently, the U. S. Surgeon General is recommending the use of nonoxynol-9 in conjunction with a condom, a useful and practical additional layer of protection. Compounds that offer rapid in vitro inactivation of HIV-1 at low concentrations plus cytotoxicity for sperm and human lymphocytes and macrophages at commercial concentrations offer hope for decreasing HIV-1 transmission from human semen and vaginal fluids under more practical circumstances. In these experiments, we have shown that HIV-1 is rapidly inactivated as a free virus particle and is theoretically inactivating at slightly higher concentrations either within or attached to cellular constituents of infectious human semen and vaginal fluids.

Spermicides, when utilized in conjunction with a properly used and reliable condom, may provide a decreased risk of infection with HIV-1 when compared to a properly used and reliable condom alone. The protective role of such contraceptive compounds during condom usage (including leakage or breakage) as well as the reliability of currently available condoms should be assessed. Studies of spermicides in use or approved for use outside the United States may provide us with a more effective more easily tolerated spermicidal contraceptive than the Food and Drug Administration–approved spermicides (nonoxynol-9 and octoxynol-9) currently available over the counter in the United States.

References

Anderson, D. J., Wolff, H., Pudney, J., Zhang, W., Martinez, A., & Mayer, K. (1990). Presence of HIV in semen. In N. J. Alexander, H. L. Gabelnick, J. Spieler (Eds.), *Heterosexual transmission of AIDS*. New York: Wiley–Liss.

Broder, S., & Gallo, R. (1984). A pathogenic retrovirus (HTLV-III) linked to AIDS. *New England Journal of Medicine, 311*, 1292–1297.

Clavel, F., Gutard, D., Brun-Vzinet, F., Chamaret, S., Rey, M., Santos-Ferreira, M., Laurant, A., Dauguet, C., Katlama, C., Rouzioux, C., Klatzman, D., Champalimaud, J., & Montagnier, L. (1986). Isolation of a new human retrovirus from West-African patients with AIDS. *Science, 233*, 343–346.

Fauci, A., Masur, H., Gelmann, E., Markham, P., Hahn, B., & Lane, C. (1985). The acquired immunodefiency syndrome: An update. *Annals of Internal Medicine, 102*, 800–813.

Friedland, G., Saltzman, B., Rogers, M., Kahl, P., Lesser, M,., Mayers, M., & Klein, R. (1986). Lack of transmission of HTLV-III/LAV infection to household contacts of patients with AIDS or AIDS-related complex with oral candidiasis. *New England Journal of Medicine, 314*, 344–349.

Gallo, R., Salahuddin, S., Popovic, M., et al. (1985). Frequent detection and isolation of cytopathic retroviruses (HTLV-III) from patients with AIDS and at risk for AIDs. *Science, 224*, 500–503.

Hicks, D. R., Martin, L., Getchell, J., Heath, J., Francis, D. P., McDougal, J. S., Curran, J. W., & Voeller, B. (1985). Inactivation of HTLV-III/LAV-infected cultures of normal human lymphocytes by nonoxynol-9 in vitro. *Lancet, 2*, 1422–1423.

Hicks, D. R., Voeller, B., Resnick, L., Cullman, L. C., Coulson, A. H., Cassity, C. L., Bernstein, G. S., & Nakamura, R. M. (1988). Chemical inactivation of HIV-1 (HTLV-III and HB2) by Contraceptives/Spermicidal Agents (Abstr.). Fourth International Conference on AIDS, Stockholm.

Klatzmann, D., Barre-Sinoussi, F., Nugeyre, M., et al. (1984). Selective tropism of lymphadenopathy associated virus (LAV) for the helper/inducer T-lymphocytes. *Science, 225*, 59–63.

Martin, L. S., McDougal, J. S., & Loskaski, S. (1985). Disinfection and inactivation of the human T-lymphotrophic virus type III/lymphadenopathy-associated virus. *Journal of Infectious Disease, 152*, 400–403.

Miller, V. E., & Scofield, V. L. (1990). Transfer of HIV by semen: Role of sperm. In N. J. Alexander, H. L. Gabelnick, and J. Spieler (Eds.) *Heterosexual transmission of AIDS*. New York:Wiley–Liss.

Popovic, M., Sarngadharan, M., Read, E., & Gallo, R. (1984). Detection, isolation, and continuous production of cytopathic retroviruses (HTLV-III) from patients with AIDS and pre-AIDS. *Science, 224*, 497–500.

Postic, B., Singh, B., Squeglia, L., & Guevarra, L. (1978). Inactivation of clinical isolates of herpes virus hominis, types 1 and 2 by chemical contraceptives. *Sexually Transmitted Diseases, 5*, 22–24.

Resnick, L., Busso, M. E., & Duncan, R. C. (1990). Anti-HIV screening technology. In N. J. Alexander, H. L. Gabelnick, J. Spieler, (Eds.), *Heterosexual transmission of AIDS*. New York:Wiley–Liss.

Resnick, L., Veren, K., Salahuddin, S., Tondreau, S., & Markham, P. (1986). Stability of HTLV-III/LAV under clinical and laboratory environments and its inactivation by standard disinfectants. *JAMA, 255*, 1887–1891.

Sarngadharan, M., Popovic, M., Bruch, L., Schupbach, J., & Gallo, R. (1984). Antibodies reactive with human T-lymphotropic retroviruses (HTLV-III) in the serum of patients with AIDS. *Science, 224*, 506–508.

Singh, B., Cutler, J., & Utidjian, H. (1972). Studies on the development of a vaginal preparation providing both prophylaxis against venereal disease, other genital infections and contraception. II. In vitro effect of contraceptive and noncontraceptive preparations on *Treponema pallidum* and *Neisseria gonnorrhorae*. *British Journal of Venereal Diseases, 48,* 57–64.

Spire, B., Barre-Sinoussi, F., Dormont, D., Montagnier, L., & Chermann, J. (1985). Inactivation of lymphadenopathy-associated virus by heat, gamma rays, and ultraviolet light. *Lancet, 1,* 188–189.

Spire, B., Montagnier, L., Barre-Sinoussi, F., & Chermann, J. (1984). Inactivation of lymphadenopathy-associated virus by chemical disinfectants. *Lancet, 2,* 899–901.

Stone, K., Grimes, D., & Magder, L. (1986). Personal protection against sexually transmitted diseases. *American Journal of Obstetrics and Gynecology, 155,* 180–188.

Voeller, B., (1984a). *Nonoxynol-9 and the prevention of sexual spread of LAV/HTLV-III.* Mariposa Occasional Paper #2b. New York: Mariposa Education and Research Foundation.

Voeller, B. (1984b). *Nonoxynol-9 and the prevention of STDs.* Mariposa Occasional Paper #2a. New York: Mariposa Edcuation and Research Foundation.

Voeller, B. (1985). *Spermicides: an additional potential barrier to the sexual spread of the AIDS virus, LAV/HTLV-III.* Mariposa Occasional Paper #4B. New York: The Mariposa Foundation.

Voeller, B. (1986a). *Nonoxynol-9 and prevention of the sexual spread of LAV/HTLV-III and other STD agents.* [abstract] The Second International Conference on AIDS, Paris.

Voeller, B. (1986b, May). Nonoxynol-9 and HTLV-III. *Lancet,* p. 1153.

25

Sexual Physiology and AIDS Protection

William H. Masters and Virginia Johnson

Although the psychological, sociological, and epidemiological aspects of the effect of AIDS on this country's population are finally being researched in depth, the sexual physiology involved in the dissemination of this viral disease in men and women has received relatively little attention. This is particularly true if we consider two specific facets of sexual physiology that, once well defined, might provide an increased capacity for clinical control of the spread of HIV in both the heterosexual and homosexual populations.

In the human female, the development of specific measures for control of the pH of the vaginal environment during coitus might result in reduction in the rate of spread of HIV in the heterosexual population. In the human male, investigative attention might be productively focused on the potential for HIV contamination of the preejaculatory fluid emission. This mucoid body fluid, if infected with HIV, could be directly involved in the spread of AIDS in both the heterosexual and homosexual populations.

Vaginal pH Control

In the human female, the vagina provides an effective means of self-protection against most bacterial pathogens by maintaining an acid environment of such degree that these pathogens cannot survive if introduced to the vaginal environment. In other research ventures, Masters & Johnson Institute personnel have repeatedly recorded the normal pH range of the uncontaminated vaginal environment, from as low as a pH of 3.6 in the preovulatory phase of the menstrual cycle to as high as 4.5 in the postovulatory period.

In the laboratory, HIV has been reported destroyed at a pH range of 3.0 to 3.5 or below. However, it should be noted that in this laboratory experiment, a far greater viral concentration was employed than would be encountered in vivo in the human vagina or in seminal fluid (Martin, McDougal, & Luskoski, 1985). The significantly lower concentration of HIV usually present in the vagina might even be temporarily inhibited by a vaginal pH in the 3.5 range.

There are two well-established episodes in the menstrual cycle of the adult, sexually active female during which the pH of the vagina normally is elevated to 5.0 to 6.5 pH levels. First, menstrual flow routinely neutralizes vaginal acidity to approximately 5.5 to 6.0 pH levels, particularly during height of flow and if tampons are employed. Second, if a male partner ejaculates in the vagina, the pH is rapidly elevated to 6.0 to 6.5 levels and remains highly elevated for a minimum of 6 to 10 hours. Obviously, the seminal fluid bolus has significant natural buffering power. Since sperm will be immobilized in an environment with a pH of 5.0 or below, this physiological neutralization of the normal levels of vaginal acidity protects the sperm for conceptive purposes. If the ejaculate is contaminated by HIV, the virus will thrive and may be absorbed from the vagina at the pH levels (6.0 to 6.5) that are attained and passingly maintained by the buffering power of the seminal fluid bolus.

In addition, there are two more normally recurrent physiological responses during the menstrual cycle that may have at best a minimal influence on elevating vaginal pH: (a) the cervical mucus (pH 6.0 to 6.5) that may drain into the vagina, particularly during an ovulatory sequence; and (b) the lubricative fluid that develops as a transudate through the walls of the vagina in response to sexual stimulation. The extent and duration of the influence of these two physiological processes on neutralizing normal levels of vaginal acidity have not been adequately investigated to date.

If, during coitus, the pH of the vagina can be consistently maintained at a pH level of approximately 3.5 by the use of artificially buffered vaginal jelly, it is theoretically possible to reduce the infective potential of HIV-infected seminal fluid (Voeller, 1985). Alternatively, if the female partner is HIV positive, the potential for male partner infection during coitus could also be reduced by clinically constraining the vaginal pH to an approximate 3.5 level by routine use of the artificially buffered jelly during coitus.

Of course, there are investigative challenges in clinical sexual physiology that arise when attempting to resolve problems of vaginal pH control. Two clinical questions immediately come to mind. First, what would be the approximate duration of and degree of protection provided a female partner against the deposition of HIV-contaminated seminal fluid by a standardized intravaginal application of an artificially buffered vaginal jelly? Second, what would be the average time span required after insertion of the artificially buffered jelly to significantly neutralize an HIV-contaminated vaginal barrel for the male partner's protection during coitus?

There also are two immediate clinical concerns. First, how would women clinically tolerate repeated stabilizing of vaginal pH in the 3.5 range throughout the menstrual cycle? Will occasional symptoms of chemical irritation develop? If so, with what frequency? Second, how would the male partner tolerate such a pH-controlled vaginal environment? Would there be a significant degree of penile irritation? If so, with what frequency? These clinical concerns must constantly be kept in mind when developing a research program in vaginal pH control.

There are secondary clinical advantages that would immediately be returned from a standardized intravaginal application of an artificially buffered vaginal jelly with a controlled pH in the 3.5 range. If the inherent buffering power of the seminal fluid bolus can be rapidly neutralized, there obviously would be marked improvement in the contraceptive efficacy of such a pH-buffered jelly. The presence of nonoxynol-9 (the spermicidal agent present in contraceptive jellies) in the artificially buffered vaginal jelly would further elevate its contraceptive prowess. In addition, Hicks et al. (1985) and Voeller (1985) have clearly demonstrated that nonoxynol-9 has a direct lethal effect on HIV.

If any significant reduction in the rate of intravaginal HIV exchange between men and women can be established by routine use of a pH-controlled vaginal jelly, a small step will have been taken toward improvement in clinical protection against the spread of the virus.

Preejaculatory Emission

Frequently, a mucoid substance is excreted from the urethral meatus during late excitement and early plateau phases of the human male's sexual response cycle. This mucoid excretion reportedly has origin in Cowper's glands, which have egress to the penile urethra just distal to the external bladder sphincter.

There is marked variation between individual males in both regularity of occurrence and the fluid volume of the emission. Although in some men the preejaculatory emission may rival the actual ejaculate in total volume, usually the volume produced during any one sexual encounter is at the 1- to 2-milliliter level.

To date, there is no report in the literature of viral cultures taken from the preejaculatory emission of men with a positive HIV serum titer. Since the virus has been cultured from bodily fluids such as saliva and tears of men and women with positive HIV serum titers (Acheson, 1986), it is logical to presume that there might be a positive return from culturing the preejaculatory emission from HIV seropositive men. This presumption is given even more credence when it is recalled that there are a number of reports of the presence of a low concentration of sperm in the preejaculatory fluid.

The clinical application of this type of investigation in male sexual physiology is obvious. If positive HIV cultures are returned from the preejaculatory emission of men with HIV serum titers, the present concepts of "safer sex" techniques would have to be altered. It would no longer be considered adequate just to suggest that condoms be worn during sexual encounters. Suggestions for condom usage would have to be far more specific. Not only would the public have to be educated to realize that condoms should be worn during sexual encounters, but they would need to know specifically *when* during the sexual encounter condoms should be employed. Most sexually interacting couples wait until the immediacy of a mounting process before covering the penis with a condom. If there is HIV contamination of the preejaculatory emission, the sexual partner of the HIV-positive male would be at risk of infection well before there is routine placement of the condom.

Finally, women should not be placed at risk of male cooperation in the safer sex technique of condom control of the ejaculate. Occasionally condoms fail, leaving the dependent female unprotected. Of far more import, women are left totally unprotected against an HIV-contaminated ejaculate when the male partner refuses to use a condom for social, ethnic, or religious reasons. Whether or not the male partner uses a condom, the female partner is entitled to whatever protection is possible against HIV. The potential of vaginal pH control during coitus is one of these protective possibilities. The correct timing for instituting condom control is yet another.

References

Acheson, E. D. (1986). AIDS: A challenge for the public health. *Lancet, 1*, 662–665.

Hicks, D. R., Martin, L. S., Getchell, J. P., Heath, J. L., Francis, D. P., McDougal, J. S., Curran, G. W., & Voeller, B. (1985). Inactivation of HTLV-III/LAV infected cultures of normal human lymphocytes by nonoxynol-9 in vitro. *Lancet, 2*, 1422–1423.

Martin, L. S., McDougal, S., & Luskoski, S. (1985). Disinfection and inactivation of the human T-lymphotrophic virus type III/lymphadenopathy-associated virus. *Journal of Infectious Diseases, 152*, 500–503.

Voeller, B. (1985). *Spermicides: An additional potential barrier to the sexual spread of the AIDS virus, LAV-HTLV-III.* Mariposa Occasional Paper #4B. New York: The Mariposa Foundation and Research Foundation.

26

Women and Acquired Immunodeficiency Syndrome: Issues for Prevention

Margaret Nichols

Five Case Vignettes

Although as of 1987 women comprised only 7% of all the reported AIDS patients in the United States, they are an important potential source for heterosexual human immunodeficiency virus (HIV) transmission as well as the source of transmission to infants. In New York State in early 1988, 1 in 61 infants was born with the antibodies to the AIDS virus (Josephs, 1988). Moreover, recent trends indicate increases in sexually transmitted cases in U.S.-born women, suggesting an increase in the heterosexual transmission of the AIDS virus to women (Hardy & Guinan, 1987).

A few brief case vignettes culled from interviews and clinical records of clients from the Hyacinth Foundation AIDS Project in New Jersey will provide a picture of the diverse background of women with AIDS and a compelling portrait of the problems these women face.

Diane is a 32-year-old, strikingly beautiful black woman who grew up in Newark, NJ. By the age of 13 Diane was "shooting drugs," by 15 she was a street prostitute, and by the age of 22 she had two children for whom she was the sole source of financial support. Five years ago she stopped using drugs; last year she was diagnosed as having AIDS and *Pneumocystis carinii* pneumonia. She continues her job as a fork-lift operator but will not have the physical stamina to work much longer. Fortunately, Diane has a large extended family that is helping her care for her children. Diane has been celibate since her diagnosis. She assesses her chances of finding a new appropriate male partner at close to zero. She is probably correct; the stigma attached to AIDS for women is commonly far greater than that for men.

Sylvia was a black woman who divorced her first husband, a drug addict, 6 years ago and remarried a non–drug-using man shortly thereafter. Two years ago she began to exhibit vague symptoms that were incorrectly diagnosed as psychosomatic. It did not occur to either her or her physician that she might be HIV infected, and she was psychiatrically hospitalized with a diagnosis of bulimia because she was vomiting so frequently. Last spring Sylvia's ex-husband called her to inform her of his HIV-seropositive status. She died within a month of learning that her symptoms were HIV related, rather than psychiatric. Her grief-stricken second husband, himself now seropositive (apparently infected by Sylvia), is now caring for his two children from a previous marriage and the two children from her first marriage that Sylvia left behind.

Susan is a 29-year-old white working-class woman who also suffered symptoms that were misdiagnosed for a year. In her third month of pregnancy, Susan mentioned to her doctor that she had had several different sexual partners since adolescence. On the basis of this clue, her doctor finally referred her for HIV antibody testing, and Susan received a diagnosis of AIDS-related complex (ARC) when her test results were positive. Only at this point did her steady male sexual partner of 4 years, the father of her child, admit his past homosexual behavior. He told Susan that he had been afraid to tell her before this because he was afraid she would leave him. He also said that he never considered himself to be at risk for AIDS because "AIDS happens to homosexuals, not bisexuals." Bobby, the boyfriend, subsequently was diagnosed with ARC himself. Susan was told, in her fourth month of pregnancy, that if she carried her baby to term both she and the baby would probably die. Frantic, she was left on her own to seek an abortion and was unable to find a clinic that would agree to perform the more complicated second trimester procedure on an HIV-infected woman. In her fifth month, on the same day she finally found a clinic that would perform the abortion, Susan's baby moved inside her for the first time and she was unable to go through with the abortion. Susan's daughter was born in August and is healthy and thriving at this time. Susan, however, contracted toxoplasmosis in September. Now well, Susan, on her good days, thinks about having a second baby. She believes, incorrectly, that if her daughter remains healthy that that means a second child would necessarily also be healthy. On the other hand, if this child becomes ill, Susan will feel a desperate need to try to replace her sick baby with a well one. In addition, being a mother is life-affirming for Susan and helps her maintain a hopeful attitude toward her own illness, an attitude that ironically is psychologically adaptive in many ways although it could lead her to become pregnant for a second time.

Clarissa, 23, is a Puerto Rican woman who lives in Jersey City and whose first husband was an addict. Clarissa learned she was seropositive

because she was routinely tested during pregnancy. Her deep religios-
ity, among other things, prevented her from aborting this pregnancy
and her first child was born with AIDS. Soon after this birth, her first
husband died of AIDS. When Clarissa's little boy was a year old, she
met another man who miraculously, from Clarissa's point of view, fell
in love with her and moved in to live with her and support her despite
her seropositive status. Clarissa's new partner asked her to have a second child—he wanted her to have his baby—and Clarissa complied.
She never considered *not* complying, in fact, and left her fate, in her
view, in the hands of God. Clarissa is now seven months pregnant. Her
little boy died of AIDS 6 weeks ago.

Cindy is 20. She grew up in a poor white family where she was physically and sexually abused, and by the age of 12 she was a runaway
sleeping in abandoned cars. Her husband is an addict who was diagnosed as having AIDS when their baby was 5 months old. Cindy, Richard, and the baby have an income of $136 per month from local welfare
because Richard's Social Security Disability benefit was denied when
his clinic doctor filled out the paperwork incorrectly. Cindy and the baby
are healthy, but Cindy didn't want Richard to use condoms in their
sexual relationship even after his diagnosis. During group counseling,
Cindy verbalized her fantasy that, if she contracted AIDS, her abusive
and neglectful family might finally "come around" and care for her in
a way they never had loved her as a child. The group persuaded her
to give up this fantasy and she began to have safer sex with Richard.

Epidemiology of AIDS in Women

To some extent, the relatively low overall percentage of female AIDS
cases in the United States is misleading because it primarily reflects the
fact that AIDS in this country was first diagnosed among gay men. In
some parts of Africa, where heterosexual behavior appears to be the
dominant mode of transmission, 50% of cases are female. In this country,
areas in which AIDS is found among·intravenous (IV) drug users tend
to have higher percentages of women, reflecting not only AIDS among
female drug users but also heterosexual transmission from male addicts
to their female partners. New Jersey, with 17% of female AIDS cases,
has the highest percentage; New York State, another state with many
IV drug user cases, has 11% of female AIDS patients.

The 7% overall figure for women has remained stable since 1982. If
one analyzes risk factors for heterosexual men and women (Table 26-
1), it is clear that IV drug use is the primary transmission mode (52%)
for women, just as it is for men (68%). The most noteworthy differences
in risk factors between men and women are that a higher percentage of
women contracted AIDS through heterosexual sexual contact (21% ver-

Table 26-1
Heterosexual Men and Women with AIDS, by Sex and Transmission
Category, 1981–1986

Risk Group	Men	Women	M–F Ratio
IV drug user	68%	52%	4:1
Heterosexual contact	1%	21%	0.2:1
Endemic area	8%	6%	4:1
Hemophiliac	4%	<1%	33:1
Transfusions	6%	10%	2:1
Undetermined	12%	11%	3:1
Total cases	5,358	181	3:1

sus 1%). In absolute numbers, female sexual transmission cases outnumber male by a ratio of 5:1 (through 1986, 381 versus 75), although
we should note that the actual extent of sexual transmission as a mode
of transmission is somewhat obscured by the fact that individuals with
multiple modes of risk—say, IV drug use and heterosexual sexual risk—
are likely not to be categorized in the heterosexual sexual transmission
category.

It is difficult to accurately interpret these data. At least two theories
have been advanced to explain the disparate rates and numbers of U.S.
female versus male heterosexual sexual transmission. According to one
theory, the rates could be an artifact of the particular stage of the epidemic the United States was experiencing during the decade of the
1980s. That is, assume that AIDS first appeared in mass numbers in this
country in gay males—who primarily, though not exclusively, transmit
to other males—and IV drug users, 75 to 80% of whom are male. If this
is true, as it appears to be, then it could take a long time for the epidemic
to spread to large numbers of women, especially through sexual transmission. Furthermore, assume, as also appears to be true, that male
drug addicts have more female sexual partners and more female sexual
partners who are not themselves IV drug users (L. S. Brown, Addiction
Research and Treatment Corporation, Brooklyn, NY, personal communication, November 1987; Murphy, Brown, & Primm, 1987), as reported by Dr. Des Jarlais (Chapter 17). Thus, infected women, fewer in
number to begin with, would be slower to sexually transmit the disease
back to uninfected men than infected men would be to transmit it to
women.

A second theory on the disparate rates is related to the fact that female-
to-male vaginal intercourse may be a less efficient mode of transmission
than is male-to-female sexual transmission. This is plausible: semen may
contain higher quantities of HIV than do vaginal secretions, and the

Table 26-2
Risk Factors for Heterosexual Male Contacts[a]

Intravenous drug user	67%
Bisexual man	16
Hemophiliac	1
Other/unreported risks	16

[a] Data from Centers for Disease Control.

vagina/cervix/uterus appear to be better "portals of entry" than are the skin of the penis and the penile urethra. Some people argue that the apparent ease of female-to-male sexual transmission in Africa is affected by the common presence of open venereal lesions on the penises of African men, whose general health care is of much poorer quality than that of American men. Moreover, some women engage in receptive anal intercourse, a route known to have high efficiency of transmission, and one anatomically unavailable as a route for heterosexually active men. Obviously, these theories are neither exhaustive nor mutually exclusive.

Table 26-2 shows risk factors of the male partners who presumably infected women who contracted AIDS through heterosexual activity. As one would expect, the majority (67%) of these male partners appear to have been IV drug users, with a smaller percentage (16%) of "other/ unreported risks." Other data support the observation that many women appear not to know how they sexually contracted AIDS and/or do not consider themselves to be at risk, and later in this chapter we will discuss why this might be so.

If one analyzes the epidemiological data on women with AIDS in order to predict future trends, one sees that, over time, higher percentages of women have become sexually infected with AIDS in the United States, while relatively lower percentages have been infected by needle contact or are foreign-born heterosexual contact cases (Hardy & Guinan, 1987) (Table 26-3). There has been no corresponding increase in heterosexual transmission for men. Moreover, there is a 10-month doubling time for

Table 26-3
Significant Trends in AIDS in Women, 1982–1987 ($p < .05$)[a]

	1982	1986
Increase in U.S.-born heterosexual contacts	14.0%	26.0%
Increase in mean age	30 years	35.6 years
Decrease in female IV drug users	59.0%	48.0%
Decrease in foreign-born heterosexual women	18.0%	5.0%

[a] Data from Hardy & Guinan (1987).

the mostly female heterosexual transmission cases as opposed to a 14-month doubling time for gay male and intravenous drug user cases (Hardy & Guinan, 1987). In other words, we can predict for the future that heterosexual women who contract AIDS sexually will be one new rapidly multiplying risk group and could theoretically grow to comprise as many AIDs cases as we now see among gay men and drug users. Of course, if this happens we will see a corresponding rapidly occurring rise in the number of infants born infected—a phenomenon already occurring in New York.

When we look at other characteristics of women with AIDS, we see that AIDS is not spread evenly among racial or social classes, as it is, more or less, among gay men. Overall, both male and female heterosexuals with AIDS in the United States tend to be people of color—half are black and about one quarter Hispanic. As one would expect, children with AIDS, the children of these women, are also primarily nonwhite. Women with AIDS are young: AIDS affects women in their child-bearing years, with one third of female AIDS cases 29 years old or younger and another 45% between the ages of 30 and 39 years (Guinan & Hardy, 1987).

These data reflect the fact that there are at this point in time in the United States two major and only somewhat overlapping AIDS epidemics. One is among homosexually active men who are infecting mostly each other (although not exclusively) and whose behavior has changed so drastically (see Dr. McKusick's discussion in Chapter 7) that the *rate* of *newly* infected people in this group is decreasing in some areas. The second epidemic is among poor, minority people in urban centers where IV drug use and shared needle use is very common. This epidemic started with mostly male drug users, and some female drug users and the female drug users' children, but is spreading out to include women who are not drug users but who have been the sexual partners of drug users, and these women's children. If unchecked, and depending upon factors such as the ease of female-to-male transmission and rates of female partner change, this "second epidemic" could eventually outstrip and dwarf the first and become a disease of poor minority heterosexuals and their children who live in urban centers. This is the most likely and greatest possible spread to the so-called "heterosexual population." Whether this epidemiological spread continues unchecked into the white heterosexual middle class depends upon factors such as rates of sexual contacts across racial barriers (interracial pairings) and across social class.

Another source of spread to white middle-class heterosexuals is through the activity of functionally bisexual men, both men who label themselves bisexual (a very small percentage of all men), men who identify as heterosexual but who have clandestine homosexual activity, and men who identify themselves as gay but have some continuing heterosexual activity. The approximately 3% of women with AIDS who have contracted AIDS in this way are a harbinger of this trend, which through tertiary female-to-male sexual spread could introduce a new risk group—white middle-class heterosexuals whose only source of infection is sexual contact with other white non–drug-using heterosexuals. It is impossible to predict the rate or likelihood of substantial spread in this way at the current time, because there are too many unknown, unmeasured variables that could affect this transmission. Among these unknown variables is the true rate of bisexual activity in the United States, behavior discussed by Dr. Reinisch in Chapter 3 in this book.

Part of the difficulty faced in describing or predicting the course of the AIDS epidemic among women (and thus among children and non–IV drug-using heterosexuals) lies in the fact that there are so few cohort studies of women such as the ones in progress studying AIDS among gay men. In the absence of such studies, we must piece together a picture of women and AIDS from bits of data gathered in high-incidence, and some low-incidence, areas. Among the most relevant statistics are the following:

1. In New York City AIDs is the leading cause of death for all women ages 25 to 34 years (Josephs, 1987a, b).

2. In New York last year there was an 8% increase in new IV drug user AIDS cases over a base of 50% seroprevalence among addicts. If one compares that to 1% or less new seropositive conversions among gay men, one can see that AIDS is becoming more and more a heterosexual minority problem and less a gay male problem. Indeed, actual figures may even be higher than official Centers for Disease Control (CDC) statistics show. A recent analysis of IV drug user deaths in New York City led to the conclusion that there may be as much as 50% underreporting of cases of AIDS among IV drug users (Josephs, 1987a,b).

3. A sample of women with AIDS at Montefiore Hospital in New York showed that 35 women had a combined total of 52 dependent children. This underscores the fact that women with AIDS do not simply bear infected infants; they also have other children. In fact, the problem of "AIDS orphans" is a much bigger problem in numbers than that of pediatric AIDS cases (Friedland, 1985).

4. A sampling of seroprevalence statistics for women from various places in the United States (Table 26-4) shows rates of seropositivity varying from 0.5 to 5%, figures roughly comparable to some current rates in Africa.

Table 26-4
AIDS Seroprevalence Rates in Women

1. Jacksonville, FL prenatal clinic (Kaunitz et al., 1987): 2 of 302 women were seropositive, 1 had identified risk factors (6.7:1,000)
2. Alameda County, CA marriage licenses (Del Tempelis, Shell & Hoffman, 1987): 5.3:1,000
3. Alameda County, CA sexually transmitted diseases clinic (Del Tempelis et al., 1987): 6.7:1,000
4. Bronx, NY abortion clinic (Schoenbaum & Alderman, (1987): 20.6:1,000
5. San Francisco Women's Cohort Study (Cohen, Hauer, Poole, & Wofsy, 1987): 50:1,000

5. Women with AIDS die faster than men, even when the confounding factor of IV drug use is controlled (Kolata, 1987). Fischl et al. (1987), in Florida, reported that women with AIDS live an average of 7 months after diagnosis as opposed to a 12- to 14-month average for men. Harder reported that in San Francisco women live an average of 40 days after diagnosis as opposed to 1 year or more for men.

6. Women who are HIV-infected bear infected infants, but probably the percentage of infected infants is lower than the 50% rates first reported, perhaps as low as 30 to 35% (Peckham, Sentuvia, & Ades, 1987). Moreover, most pregnant seropositive women choose *not* to terminate their pregnancies, even when informed of the risk. A study in New York reported that of 80 pregnant seropositive women only 1 chose to abort (Josephs, 1987). These figures reflect the tremendous role that motherhood plays in the lives of women who do not have access to other social roles or to other sources of self-esteem, pleasure, and life satisfaction. Moreover, to many poor women whose lives are already full of risks, even the possibly 50:50 odds of having a well baby represent a comparatively acceptable risk. Coupled with these factors are variables such as religious beliefs about abortion and mistrust of doctors and health authorities who convey risk information.

Issues Influencing Sexual Transmission

Let us examine some of the research that has bearing specifically upon the *sexual* transmission of AIDS, through heterosexual sexual activity, to and from women. Four recent studies are of particular significance: the Downstate Study (Landesman, Minkoff, Holman, McCalia, & Sijin, 1987); the "partner studies" done by Fischl et al. (1987) and Padian et al. (1987); and Cohen et al.'s San Francisco Women's Cohort Study (1987).

In November 1987, Landesman et al. reported on a study of 602 newborn infants and their mothers delivered at Kings County Hospital in New York. The sample was almost entirely nonwhite, poor, and urban. The researchers conducted 15-minute interviews of the mothers to determine CDC-defined risk factors and compared that to the mothers' antibody status as ascertained from cord blood samples. As a result of the interviews, the researchers determined that 28% of the sample might be at risk, mostly because they were Haitian or sex partners of Haitians.

By contrast, blood samples showed 2% of these women to be infected. When the researchers tried to match the real seropositives to the interview-determined risk factors, 42% could not be classified. In other words, neither the woman nor the interviewer would have suspected her seropositive status. Moreover, of the seven women the interviewers picked as high-risk women, three of them did not recognize their own risk. Only the IV drug users recognized their own risk. In other words, the Downstate Study suggests that women still do not recognize their risk of exposure to AIDS from sexual contact: two thirds of these seropositives did not know they were at risk. This may be because the informational and educational messages we give women *mislead* them and in part it may be because women often do not know the sexual and drug use histories of the men with whom they have sexual relations.

The two most substantial "partner studies" of women done in the United States give us information about the relative risk of exposure to AIDS through various forms of heterosexual sexual contact. Both Fischl and Padian and their colleagues studied the steady sexual partners of men (and for Fischl, women as well) who had already been diagnosed as having AIDS or ARC at entry into their studies. (Both studies are somewhat flawed by the assumption that the partner first diagnosed with AIDS or ARC was the first to be infected and "gave" the virus to the other, an assumption that is not necessarily true.) Both studies followed partners over some time (2 years or more) and thus were able to ascertain, for nondiagnosed partners who were seronegative at entry, the approximate point at which they seroconverted. What do the results of these studies tell us?

1. Statistically, on the average it appears to take many exposures through vaginal sex with an infected man for a woman to become infected. It can sometimes take years of exposure for infection to occur. Twenty-three percent of Padian et al.'s female partners were infected at entry into the study, a rate similar to that obtained by Paul Paroski in a partner study at Woodhull Hospital in Brooklyn (personal communication, November

1987). Fourteen percent of Fischl et al.'s female partners were infected at entry into the study, and these women had been in relationships with their infected partners for nearly 6 years on average at the time the study began. In Fischl et al.'s study, most women who continued to engage in unprotected sex with their male partners eventually became infected but, again, it often took several years of repeated exposure.

2. Fischl et al. reported that seroconversion did not occur in 9 of the 10 couples who regularly used condoms after becoming enrolled in the study.

3. Vaginal sex is clearly sufficient for transmission. Anal intercourse probably is yet more efficient—Padian et al. estimate more than twice as efficient.

4. Padian et al. (as Dr. Detels reports for gay men, as well, in Chapter 1) found that oral sex did not appear to increase transmission risks significantly. Fischl et al. found a correlation between oral sex and seroconversion, but this variable was confounded because in her study couples who used condoms also did not engage in oral sex, whereas couples who did *not* use condoms did engage in oral sex. Thus oral sex may have been a mere marker for couples who engaged generally in "unsafe" sex.

5. One of the most striking findings is the lack of relationship between number of sexual partners and seroconversion. In fact, Padian et al. found that their seronegative women actually had more sexual partners since 1987 than did the seropositives.

The other study that gives us some in-depth information about sexual transmission to women is the San Francisco Women's Cohort study being conducted by Judith Cohen and her associates with 450 women (Cohen et al., 1987). This is a study of all seropositivity in women, not just infection through sexual transmission. In this study, women were interviewed extensively and also tested for HIV antibody. Moreover, their male partners were interviewed. Cohen et al. found an overall seroprevalence of 5% in their sample, half of which was linked to IV drug use on the part of the women. For the other half, sexual transmission cases, several findings are noteworthy.

1. Among women who acquired HIV through sexual transmission, their infection could invariably be traced to one partner—an ongoing steady male partner who had a history of IV drug use or who was bisexual. Again, the male partner's risk status was often unknown to the woman until the male partner was interviewed by the research team. Having multiple sexual partners was *not* correlated with seropositivity in this sample. In

fact, the same team tested a sample of prostitutes and, in every case, seropositivity was connected to IV drug use, not sexual transmission.

2. Anal sex (discussed in Chapters 1, 2, and 19) was not related to seropositivity—only vaginal sex was—but a high percentage (over 40%) of both seropositive and seronegative women had practiced anal sex. (Padian et al. also found that 31% of their sample practiced anal sex and, as in Dr. Reinisch's study reported in Chapter 3, they found that women who had had bisexual male partners were more likely to have had anal sex.)

3. The only other factor that correlated strongly with seropositivity was having sex during menstruation (Cohen et al., 1987; L. Poole, San Francisco General Hospital, personal communication, November 1987). This may have simply been a marker for high sexual frequency. On the other hand, there are other possible explanations for this finding that are related to the physiology of the vagina or the site of entry of the virus into a woman's system. In Chapter 25, Masters and Johnson discussed vaginal pH and its possible relationship to HIV infection; this may account for the menstruation finding. Alternately, since the cervical os is dilated at menstruation, it may be that entry at the cervix or uterus is more efficient than vaginal entry for the HIV virus.

This last hypothesis is worthy of more investigation. There is a bit of circumstantial evidence that hints at this theory—that is, data on donor insemination and AIDS. The well-known Australian report (Stewart et al., 1985) of four women who seroconverted after being inseminated by the same infected sperm donor contrasts with a study by Pies, Rskenazi, Newsletter, & Shepard, 1987) of lesbians inseminated by infected gay men. In the latter study, not one woman became infected, and it may be because these women tended to use a self-insemination method less likely to bring semen into contact with the cervical os or uterus than the medical methods used in the Australian report. There is a desperate need to follow up these data with biophysical research on site of entry for women, because it has such a direct bearing on prevention strategies. If the cervix or uterus are the prime "portals of entry," then barrier contraceptives such as the diaphragm, cervical cap, or contraceptive sponge might reduce risk of transmission (see Chapters 22 and 23 by Drs. Bernstein and Coulson). Women could then control their own risk reduction much more easily than at present, when the only method of risk reduction, the condom, is one that is used by the male partner.

All four of the studies discussed above suggest that women tend to become infected by steady male sexual partners rather than from multiple partners. This finding is worth examining in more detail, since it is in such contrast to the data on sexual transmission of AIDS among gay men. J. Wiley (University of California at Berkeley, personal communication, November 1987) has used Padian et al.'s data as well as data from two other partner studies in which number of sexual episodes was counted and has calculated the mean infectivity rate per exposure for vaginal intercourse (infected male partner to female partner) to be 0.001. This suggests two things: it may explain why it seems to take a long time in these partner studies for women to become infected, but it also may explain why multiple partners are less important in vaginal transmission to women. One implication of the infectivity rate per exposure is that until one gets above a number of sexual episodes equal to 1 divided by this coefficient, having multiple partners versus one partner does not make a great deal of difference in one's probability of becoming infected. If Wiley's figure is correct, until a woman has over 1,000 sexual episodes, it makes little difference whether she has those episodes with one or many partners, unless she can screen her one partner for risk factors better than she can screen her multiple partners. And as we have already observed, women do not seem to be able to screen their partners very well.

Just for comparison, contrast this figure with the rate for gonorrhea, where the comparable coefficient is about 0.5: in other words, for gonorrhea, anything over *two* sexual episodes makes multiple partners more risky. There are other factors that make multiple partners more risky; for example, multiple partners might put one at greater risk for other sexually transmitted diseases, and these in turn might be cofactors for HIV transmission.) Moreover, most heterosexual women are not nearly as sexually active as some gay men. In Fischl et al.'s study, for example, heterosexual women had an average of two to three sexual episodes per week, probably not nearly as high a frequency as that for some groups of gay men, especially early in the epidemic. Moreover, in practical terms women tend to practice serial monogamy and to have fewer partners than do men in general. For example, Padian et al.'s seropositive women had an average of 2.5 partners in the same time period during which the San Francisco Men's Cohort study found its sampling of gay men having 200 partners and its sampling of heterosexual men, 20.

If women tend to get AIDS from a steady partner whose risk factors are unknown to them, who then are these men? Some are IV drug users, as discussed earlier. Others are bisexual men. From the survey data presented by Reinisch et al. in Chapter 3, we know that perhaps as many as 25% of adult males are behaviorally bisexual. In other words, hidden bisexual behavior among apparently heterosexual men probably accounts for many of the "unknown" cases of sexual transmission to women. However, there is also quite a bit of hidden bisexual behavior among self-identified gay males. For example, consider the data from a prevention study conducted by Gay Men's Health Crisis that specifically recruited gay men (Quadland, Shattls, Schuman, Jacobs, & D'Eamo, 1987). Among a sample of 619 men, 92 individuals had collectively had 210 *different* female partners in the last 5 years.

Gaps in Knowledge and Need for Future Research

It is probably clear at this point that there are several glaring gaps in our knowledge of AIDS, and indeed of sexual behavior in general, that have direct bearing on the spread, especially the sexual spread, of HIV to and from women. Among the areas that need to be addressed are:

1. Research on the exact mechanism of transmission. If it were to be determined, for example, that the uterus rather than the vagina was the primary site of transmission, women could use barrier methods such as a diaphragm or contraceptive sponge for prevention. Currently, women must rely on a prevention technique—condoms—that is not under their own control.

2. Research on the sexual patterns of bisexual men to determine future risks of transmission to women through this population. In light of the high incidence of bisexual behavior in adult men found in all sexological research from the time of Kinsey to the present, and in the light of Padian et al.'s evidence regarding anal intercourse in female partners of these men, it is somewhat surprising that this transmission vector has not been more salient in the AIDS epidemic. Although there are theoretical explanations for this phenomenon, more research is needed to investigate the future potential of this mode as a source of transmission. Moreover, research is needed to investigate effective prevention techniques for bisexual men, most of whom do not self-identify as gay and may not be exposed to prevention messages promulgated within the gay community.

3. Research is needed to evaluate the efficacy of different prevention techniques targeted at women. Preliminary reports (J. Jackson, New Jersey Department of Health, personal communication, November 1987) suggest that information alone is not effective at behavior change, and field reports suggest strongly that techniques useful with gay men cannot be translated to women.

4. Closer analysis of data suggesting rather low efficiency of transmission, male to female, via vaginal intercourse. Some researchers have speculated that the research findings that point to this are actually an artifact of a phenomenon that has been

labeled the "super-spreader" effect. That is, it may be that the semen of some infected males is highly effective at transmitting AIDS whereas other infected men are virtually incapable of transmitting HIV to their sexual partners. This phenomenon might be responsible for data that suggest that vaginal inter-course is a fairly inefficient mode of transmission, when in fact the real picture may be closer to "all or nothing"—some males transmit very easily, others not at all.

5. Research on the sexual and drug use habits of the urban nongay populations most affected at present. For example, we need to know more about the sexual patterns of male and female drug users, the needle use habits of occasional drug users versus addicts, and the likelihood of sexual contact across class and racial lines. These data can help us carefully target prevention messages as well as to predict the likelihood of future spread of AIDS to other segments of the heterosexual population.

6. Research on female-to-male transmission. Reports from some sources (e.g., the military) are highly suspect (Potterat, Phillips, & Muth, 1987). While some female-to-male transmission seems to occur, we must know much more to ascertain how prevalent it actually is, and in this area in particular we need researchers highly skilled in sexual interview techniques. Reports from the field (J. French, New Jersey Department of Health, personal communication, November, 1987; Wallace, 1987) cast particular doubt upon men who claim to have contracted AIDS from pros-titutes but who may actually be concealing a bisexual back-ground.

7. Development and testing of prevention strategies aimed at men. As long as the major prevention tool for women (condoms) remains in the hands of men, it is critical to reach men with prevention messages.

Recommendations for Prevention: Interim Strategies

In the absence of comprehensive data, we must nevertheless increase and improve prevention efforts aimed toward women. We have no re-search investigating the efficacy of prevention strategies for women, but some survey reports, such as that done by Joyce Jackson in New Jersey, suggest that women in the minority urban populations who are most at risk have a good deal of knowledge about how AIDS is spread but are nevertheless not practicing safer sex techniques (i.e., not utilizing condoms). Interestingly, the only women showing significant behavior change are prostitutes: the CDC Multi-Center Prostitute Study (Centers for Disease Control, 1987) showed that 78% of prostitutes insisted that their male customers use condoms. However, even the prostitutes who made customers use condoms did not use condoms during sex with their boy friends or spouses.

Any prevention efforts targeted toward women must address the issue of why even women knowledgeable about AIDS do not use condoms for sex. The reasons are complex and probably not entirely clear, but they are bound up in sex role socialization and social class issues. To begin with, women in this culture are socialized to defer to men in the setting of limits for sex: men usually determine how, when, and how often sex will occur. For a woman to suggest condom use, she must first overcome her internal barrier to setting any sexual limits. Moreover, for women sex is less likely to fulfill a pleasure function than it is to fulfill more pragmatic functions. Among these is the role of sex as a barter exchange for the financial support of a male partner. A poor woman, in particular, often cannot afford to alienate the man who supports her and her children. Many urban poor minority women at risk report fears of abandonment or even physical assault if they suggest condom use, and these women cannot simply replace a recalcitrant partner with a more compliant one that easily. Related to this is the notion of relative risk and temporally close versus temporally distant risk: if a woman balances the risk of AIDS in the future against the immediate loss of food and shelter for her and her children, safer sex loses.

A major obstacle to prevention for women is our failure to convince men of the necessity to use condoms. Safer sex messages have been effective in the gay male community for a number of reasons, but among these is the fact that messages played more upon the desire to protect *oneself* than the altruistic desire to protect others. That is, gay men were told to use condoms because they risked AIDS from both insertive and receptive anal intercourse. Most heterosexual men have the (erroneous) impression that condoms are necessary only to protect their female partners, and thus we are relying solely upon an altruistic motive rather than addressing the motive of self-interest. This is doubly difficult because condom use often interferes with a rather rigid sexual repertoire among many heterosexuals, compared with the repertoire of substantial numbers of gay men. That is, many gay men have never been as focused upon anal intercourse as a sexual mode as heterosexuals are focused upon vaginal intercourse. Also, among drug users, we have obscured the real significance of sexual transmission by grouping those with multiple risk factors as "IV drug user" cases. In other words, a drug-abusing woman who contracts AIDS is presumed to have been infected through needle use, although she may indeed have been infected through sex. Because we view all cases of AIDS among drug users as needle related, we underemphasize the role of sexual transmission to the very population we would like to change their sexual habits.

It has been particularly difficult for (mostly) white professionals in the health care field to reach minority populations in an effective way. These very groups of people often have grave mistrust of a white bureaucracy, often based on many years of negative experiences with hospitals, social service agencies, and the like. Hyacinth Foundation staff members, for example, often are asked to respond to questions from black audiences about whether AIDS itself is a plot by the United States government to eradicate blacks. Herb Samuels, prevention consultant for the New York City Health Department, says that Health Department employees are sometimes physically assaulted in minority communities.

A not insignificant factor in the failure of women (and male heterosexuals) to use condoms is that most AIDS prevention messages emphasize multiple sexual partners as the chief risk factor. To most of the population, this phrase translates as "promiscuity," and is extremely perjorative. This message has two unfortunate effects. First, it implies that only "promiscuous" people get AIDS, and that is one reason why women do not see themselves to be at risk if their sexual pattern has been serial monogamy with a small number of partners. Second, the "promiscuity" message means that if a woman asks her male partner to use a condom, she is implying either that he has been "immoral" or that she has been. If she implies that she has been "sexually loose" she risks rejection from her male partner; if she suggests that he may have been she risks his anger and outrage. Moreover, women have been socialized to emotionally protect their male partners. Joyce Jackson describes women who say that they cannot ask their partners to use condoms for fear of hurting their feelings; these women truly are prepared to die for love.

In the absence of better research, and despite the complexity of the problem, what, then, can we suggest as interim prevention strategies for women? We can do the following:

1. Provide free or low-cost legal clean needles to drug users; half of women with AIDS are still drug users.
2. Focus prevention on the group that needs it the most: inner-city minority women of child-bearing age. Use people indigenous to the community to deliver prevention messages.
3. Emphasize the risk from ongoing steady sexual partners rather than multiple partners. Emphasize the length of time the HIV virus has been around, so that women who have been in monogamous relationships for several years do not believe themselves to be safe. Let women know that it *can* take many exposures to one infected partner before infection takes place, so that women in steady relationships where safer sex has not been practiced do not feel that it is "too late" to start condom use. "It's never too late" is a better message.

4. Target the two highest risk behaviors—anal and vaginal intercourse—rather than a laundry list of "possibly safe/possibly unsafe" behaviors. The less we ask people to change, the more likely they will be to change.
5. Target heterosexual men, not just women.
6. Consider race, class, language (e.g., Spanish), and age factors in designing prevention strategies; consider the roles and functions sex plays in the lives of women.
7. Suggest two barrier methods—for example, condoms plus spermicides with nonoxynol-9, diaphragm, etc. Suggest that if a woman absolutely cannot get her man to use a condom, she at least can use a method she herself can control.

We must learn more about AIDS and women, indeed about all aspects of AIDS. But in the meantime, we can prevent thousands of unnecessary deaths if we simply act upon the existing knowledge we already have.

References

Centers for Disease Control. (1987). Antibody to human immunodeficiency virus in female prostitutes. *JAMA, 257,* 2011–2013.

Cohen, J., Hauer, L. B., Poole, L. E., & Wofsy, C. B. (1987, June). *Sexual and other practices and risk of HIV infection in a cohort of 450 sexually active women in San Francisco.* Paper presented at the Third International Conference on AIDS, Washington, DC.

Del Tempelis, C., Shell, G., & Hoffman, M. (1987). Human immunodeficiency virus infection in women in the San Francisco Bay area. *JAMA, 258,* 474–475.

Fischl, M. A., Dickinson, G., Scott, G., Klimas, N., Fletcher, M. A., & Parks, W. (1987). Evaluation of heterosexual partners, children, and household contacts of adults with AIDS. *JAMA, 257,* 640–644.

Friedland, G. (1990). *Perinatal risks and incidence.* Unpublished manuscript.

Guinan, M. E., & Hardy, A. M. (1987). Epidemiology of AIDS in women in the United States, 1981 through 1986. *JAMA, 257,* 2039–2042.

Hardy, A. M., & Guinan, M. E. (1987, June). *AIDS in women in the United States.* Paper presented at the Third International Conference on AIDS, Washington, DC.

Josephs, S. (1987a, September 30). *Women with AIDS: The New York City experience.* Paper presented at Center for Population and Family Health Women and AIDS Symposium, New York.

Josephs, S. (1987b, November). Women with AIDS New York City. *City Health Information, 6*(no. 10).

Josephs, S. (1988, January 24). In *AIDS antibodies in New York infants.*

Kaunitz, A. M., Bower, J. L., Paryani, S. G., de Sausave, L., Sanchez-Ramos, L., & Harrington, P. (1987). Prenatal care and HIV screening. *JAMA, 258,* 2693.

Kolata, G. (1987, October 20). AIDS is killing women faster, researchers say. *New York Times.*

Landesman, S., Minkoff, H., Holman, S., McCalia, S., & Sijin, O. (1987). Serosurvey of human immunodeficiency virus infection in participants. *JAMA, 258,* 2701–2703.

Murphy, D. L., Brown, L. S., & Primm, B. J. (1987, June). *Patterns of heterosexual contacts among intravenous drug abusers: Implications for the heterosexual transmission of the human immunodeficiency virus.* Paper presented at the Third International Conference on AIDS, Washington, DC.

Padian, N., Marquis, L., Francis, D. P., Anderson, R. E., Rutherford, G., O'Malley, P., & Winkelstein, W. (1987). Male-to-female transmission of human immunodeficiency virus. *JAMA, 258,* 788–790.

Peckham, C. S., Sentuvia, Y. D., & Ades, A. E. (1987). Obstetric and perinatal consequences of human immunodeficiency virus (HIV) infection: A review. *British Journal of Obstetrics and Gynecology, 94,* 403–407.

Pies, C., Rskenazi, B., Newsletter, A., & Shepard, C. (1987, June). *Exposure to the AIDS virus through artificial insemination in a population of lesbians in California.* Paper presented at the Third International Conference on AIDS, Washington, DC.

Potterat, J., Phillips, L., & Muth, J. (1987). Lying to military physicians about risk factors for HIV infections. *JAMA, 257,* 1727.

Quadland, M., Shattls, W., Schuman, R., Jacobs, R., & D'Eamo, J. E. (1987, October). *The 800 men project* (unpublished paper). New York: Gay Men's Health Crisis.

Schoenbaum, E. E., & Alderman, M. H. (1987). Antibody to HIV in women seeking abortion in New York City. *Annals of Internal Medicine, 104,* 599.

Stewart, G., Cunningham, A., Driscoll, G., Tyler, J., Barr, J., & Gold, J. (1985). Transmission of human T-cell lymphotropic virus type III (HTLV-III) by artificial insemination by donor. *Lancet, 2,* 581–584.

Wallace, J. (1987, June). *HIV exposure in New York City streetwalkers (prostitutes).* Paper presented at the Third International Conference on AIDS, Washington, DC.

VIII
POLICY AND
EDUCATION

27

AIDS: National Policy and Education

The Rt. Rev. William E. Swing

During the presidential campaign, late in 1987, I had occasion to talk briefly with Vice President George Bush about AIDS. Having been his parish priest in Washington, D.C., for some years, I felt bold enough to volunteer to give him a briefing on the AIDS epidemic as I have experienced it in San Francisco. He quickly accepted my offer, and invited me to the White House.

Initially I planned to use our time together to do nothing but tell real stories of real people: the human dimension is the all-important dimension. However, the opportunity would be so golden that I wanted to do more. I'd wanted to offer a possible AIDS agenda to one who aspires to be the President of the United States. Once I got my parochial mind around that ambitious task, I recognized that I needed help. So I contacted a few of San Francisco's leading physicians in this field and invited them to my office—Marcus Conant and Donald Abrahms and Merv Silverman, among others. What I have to say in the rest of this chapter has greatly been influenced by these people.

After we got deeply into discussion, one of these physicians said: "What in the world am I doing offering advice to George Bush? I don't want him to be President." But, we all agreed that if one candidate could come out with an enlightened AIDS statement, the others would be challenged to do likewise and health in the midst of the epidemic would be well served.

My chapter is based on the assumption that there are three critical considerations for the President of the United States and his staff in regard to AIDS:

1. A sense of the scope of the President's role in the epidemic.
2. A sense of human suffering.
3. A positive plan of action to be presented to this nation.

Scope of the President's Role in the AIDS Epidemic

Whereas over 55,000 Americans died in the Vietnam War, the number of AIDS deaths since 1981 in our country, according to most projections, will exceed that number during 1988. Our national leaders will preside over human carnage from one disease, the likes of which has never been faced by our country. The President must come to grips with the epidemic in a sophisticated and caring way. This nation might well be facing enormous grief and subsequent mental health problems, might be hysterically consumed by fear, will be excruciatingly burdened by financial woes, and obviously will be absorbed with the political ramifications as the hundreds of legislative issues that face our state and federal lawmakers at present indicate. Our President does not have the luxury of expedient political posturing. The urgency of AIDS requires strong, clear leadership now.

If the shores of the United States were hit by a military invasion that would take the lives of tens of thousands, and threaten the lives of a million or more of us, it would be the solemn responsibility of the President to declare war. Unfortunately, the enemy has often been understood to be the person with AIDS instead of the virus itself. If we are at war, then it is the second war in a row for our nation in which some of the combatants have been spit upon on Main Street. The President of the United States has a responsibility to set the moral tone for waging war on AIDS, and this means having a doctrine of care for those who are suffering.

The Toll of Human Suffering

I asked a physician about the most important thing that the President could do in regard to this epidemic, and he said: "Have him go regularly to someone's room and watch that person die."

George and Barbara Bush had a 3-year-old daughter who died of leukemia years ago, and obviously the suffering still haunts them. As awful as that was, it is not the same as in an AIDS death. In deaths other than AIDS, it is not likely that your church will say that it was God's wrath that brought the disease. In other deaths, the parents rarely have to suffer in anonymity for fear that neighbors might hear about it. No other deaths carry the stigma and isolation and public scorn that an AIDS death does. Children with AIDS had their bedrooms fire-bombed in Arcadia, FL. A Sacramento nurse with AIDS lost her friends in the medical community. A 17-year-old girl in Fremont, CA, had one sexual encounter and now is a nonperson with AIDS . . . her high school even refuses to allow AIDS education. Neither the girl nor the disease are judged to be worthy of consideration by her community. An AIDS death is different.

"Have him go regularly to someone's room and watch that person die." Harsh words but seriously intentioned.

Interestingly enough, that is what the President's AIDS Commission is doing with Philip Strobel, aged 34, of Washington, DC. The Commission's Executive Director says: "We are very conscious of the incredible burden the disease places on the individual by watching it through Philip." The bottom line of the AIDS epidemic is not a projected body count or an insurance company's actuarial estimate; it is one person dying. If you've never met such a person, if you never enter into the person's plight, then the AIDS epidemic, for you, is something less than the human catastrophe that it is.

A Positive Plan to Be Presented to the Nation

According to medical polls, AIDS is now the number one medical fear in our nation—having passed cancer. In the face of this fear, it is crucial for the President of the United States to present to this nation a rational and competent plan for addressing the AIDS epidemic. Such a plan should cover at least these areas:

1. The President needs to promise management of the disease. It is not enough to point out that state and local governments share in the responsibility with the federal government. That's true, but no one can lead like a President. No one else has the Centers for Disease Control (CDC) and National Institutes of Health (NIH) as extensions of his or her office. The President needs to be aware of just how the CDC and NIH function and needs to allow CDC to do its job. He should surround himself with a few medical people who will see AIDS as primarily a medical, not a political, issue. Responsible central leadership is required in an epidemic.

2. The President needs to arrive at a wise budget figure for attacking AIDS. That budget figure would have to be realistic about money for care and treatment, money for research, and money for education. It will have to be in the billions.

3. The President needs to support nondiscriminatory legislation for people infected, or presumed to be infected, with the virus. Such people need to be assured that they will not lose their right to education, their insurance, their jobs, their homes, or

their health care if they are AIDS positive and have an opportunistic infection.

4. The President needs to back an education effort that will stop the disease. Chastity is to be encouraged and is the safest approach to sex in the present crisis. But chastity simply will not be a viable option for many. About 800,000 teenagers will become pregnant this year. Nothing could be more immoral for a government than to let young people die of ignorance—especially when the young people have been asking for AIDS education. Education must be explicit if it is going to face reality.

5. The President must come to grips with the quantity of treatment and care that are going to be needed. At present, there is only a brave minority in the public sector who are addressing this issue, and they have nearly exhausted their resources—before the epidemic explodes. Our nation has no long-term planning for providing care and funding care. This must change.

At this unique moment of history our nation has gone, in a few years, from having no policy and education regarding AIDS to our current position where we urgently are forced to make multiple decisions. The vast moral vacuum at the center of the these decisions will be filled quickly. And it looks, from Jesse Helms' legislative efforts, for example, that those who might well determine policy are those least involved in the epidemic. It is critical at this moment for the few people who work in the AIDS field to speak up—the public health officers, the researchers, the physicians, the clergy, the nurses, the caregivers, most of all people with AIDS and their loved ones. The democratic process only works when an issue is enjoined by responsible advocates.

IX
PERSONAL ASPECTS
OF AIDS

28

Emotional Impact of AIDS: Male Couples and Their Families

Andrew M. Mattison and David P. McWhirter

Background

In 1974, the authors began psychotherapeutic assessment and treatment of male couples who were presenting with a wide range of relationship issues that resembled those heard from married and cohabiting heterosexual couples. Since there was little information available at that time about male couples, the couples whom we were seeing in our practice were asking some basic questions. For example, they wondered how other male couples dealt with issues such as "coming out" to family and friends, developing their own traditions to symbolize their union, and combining money and possessions.

Because of the paucity of relevant information in the behavioral science literature, the authors conducted a study in which they interviewed 156 male couples who were together from 1 to 37 years. An important finding was the observation that, regardless of the differences from man to man, human relationships themselves form separate entities and pass through a series of developmental stages in much the same way that an individual grows and develops. As shown in Table 28-1 (McWhirter & Mattison, 1984), six stages were identified, with each possessing a unique set of characteristics comprised of both positive and negative factors. This theoretical model of a relationship's stages provides a helpful framework for both couples and the therapist in understanding how relationships and individuals change over time. Over one third of the sample had lived together longer than 10 years, and the day-to-day lives of these couples shared many similarities with married couples. It was

Table 28-1
Developmental Stages of the Male Couple[a]

Stage One: Blending (year 1)
Characteristics:
1. Merging
2. Limerence
3. Equalizing of partnership
4. High sexual activity

Stage Two: Nesting (years 2 and 3)
Characteristics:
1. Homemaking
2. Find compatibility
3. Decline of limerence
4. Ambivalence

Stage Three: Maintaining (years 4 and 5)
Characteristics:
1. Reappearance of the individual
2. Risk taking
3. Dealing with conflict
4. Establishing traditions

Stage Four: Building (years 6 through 10)
Characteristics:
1. Collaborating
2. Increasing productivity
3. Establishing independence
4. Dependability of partner

Stage Five: Releasing (years 11 through 20)
Characteristics:
1. Trusting
2. Merging of money and possessions
3. Constricting
4. Taking each other for granted

Stage Six: Renewing (beyond 20 years)
Characteristics:
1. Achieving security
2. Shifting perspectives
3. Restoring the partnership
4. Remembering

[a] Reprinted with permission from McWhirter and Mattison (1984).

established that male couples maintain long-term spousal relationships that are mutually rewarding and satisfying. These observations are further substantiated by an additional study from 1985 in which the authors interviewed 65 male couples living together from 20 to 50 years (Mattison & McWhirter, 1990).

Introduction

With the onset of the AIDS epidemic, the authors began seeing in their clinical practice male couples seeking assistance in coping with the diagnosis of human immunodeficiency virus (HIV) disease in one or both of the partners. The need for further information became apparent as couples asked how other male couples were dealing with the impact of an AIDS diagnosis on their relationships. They raised dozens of issues that every couple must deal with in beginning to understand and manage a life-threatening illness. In addition, there were issues that were fairly unique. For example, some couples were not "out of the closet" with their families, or more frequently, there was an unspoken agreement that the son's sexual orientation was not to be acknowledged or even discussed. Out of a need to elicit emotional support from family and gain acknowledgment of the spousal relationship, the couple had to find ways to tell the truth. Another problem was facing the various forms of discrimination many couples experienced as a result of the stigma connected with AIDS. At the AIDs and Sex Conference, Dr. June Osborn noted that she normally concludes her lectures about AIDS with the reminder to audiences that AIDS is a problem of the human family and that now certain members of that family are in trouble. In response to the needs we encountered, we decided to investigate the personal impact of AIDS on these members of the human family.

Over the course of 10 months we conducted videotaped interviews of 27 male couples in which one or both partners had AIDS. In addition, we interviewed 14 surviving partners—that is, men who had lost their lovers to AIDS. The participants were from New York, New Jersey, Texas, and California. We examined the emotional impact of AIDs on the family, beginning with the couple themselves or the surviving partner, their families of origin, their extended or gay families, and in some instances the health care workers who functioned as part of their families. We traveled to their home cities and utilized a recorded interview format, asking our participants to address a broad range of topics. Beginning with how they met, the individuals and couples reviewed the stories of the development of their relationships before the diagnosis of AIDS. We asked about their friends and their family life, their long-term plans and their dreams, as well as their fears and worries. We heard funny stories and poignant stories of their day-to-day lives before the diagnosis of AIDS. They shared with us the events that led to the diagnosis of AIDS, and finally they traced their lives living with the disease, relating countless issues raised and the impact that AIDS had on them as individuals and as couples, on their families, their friends, and sometimes their health care workers.

Methods

Using a descriptive research method for this phenomenological investigation, the authors developed interview-recording schedules that yielded objective and quantifiable information. However, this kind of descriptive information does not explain relationships, help make predictions, or allow investigators to seek meaning and implications. Therefore, we added questions that were open-ended and that allowed individuals the opportunity to explain and describe their experiences and feelings. These responses added something more to the descriptions.

Any researcher involved in a field study knows that more happens in the interview than is recorded. The complete picture of a couple that can be observed by the interviewers cannot be understood or presented by mere facts and figures. Therefore, at the outset we decided to videotape every interview.

Incorporating videotaping into the research method allowed the investigators to review the interviews on television monitors, to observe subtleties in interpersonal reactions, and to hear interactions that might not have been apparent during the formal interview. In addition, every interview was transcribed verbatim. Thus, the combination of a video review of each interview along with a reading review from the typed transcripts was employed in identifying themes as well as cataloguing the range of diverse responses. Another significant advantage of incorporating video technology into a research method is that brief video vignettes can be shown in the presentation of findings for illustration instead of employing the traditional written case examples (as was done at the AIDS and Sex conference). Most important, utilization of video presentation helps to humanize the data by bringing the participants of a study directly to the conference-room attendees.

The Interview

In reviewing the Kinsey interview strategy and rationale (Kinsey, Pomeroy, & Martin, 1948) and the author's previous work (McWhirter & Mattison, 1984), two distinct interview methods were identified: the research interview and the clinical interview. The research interview is intended to collect data that are reliable and reproducible and that are available for tabular presentation and statistical manipulation. The clinical interview is useful in collecting both objective information and less tangible data about the thoughts, feelings, and interactional components in the lives of the participants. We utilized both formats.

Participants

The authors contacted colleagues in five cities around the United States to assist in locating couples one or both of whom had AIDS as well as the surviving partners of persons who had died of AIDS. Because of the

Table 28-2
Participant Information

	Number	
Couples (N = 27)		
One partner with AIDS	19	
HIV antibody status of other partner:		
Positive		7
Negative		10
Not tested		2
Both partners with AIDS	5	
One partner with ARC and one with AIDS	2	
Both partners with ARC	1	
Surviving partners (N = 14)		
With AIDS	4	
With ARC	1	
With no diagnosis	9	
Antibody status:		
Postive		3
Negative		4
Not tested		2

sensitive nature of the study, the health status of participants, and the understanding that the interviews not only would be videotaped but also would be used for educational presentations, securing participants was a formidable task.

Before we made direct contact with a prospective participant, the referring colleague requested the couple's or individual's permission for us to contact them. Being contacted and reassured by a professional in their home city increased the probability of participation by those who otherwise would have been reluctant.

The sample represented a diversity of ages, educational backgrounds, employment, socioeconomic levels, and degrees of openness about homosexuality. The sample was predominantly white, with a small representation of blacks and Hispanics. Tables 28-2 and 28-3 provide information about the participants.

Findings

As we have stated, dozens of recurring themes were identified that dealt with the impact of AIDS on the couples, surviving partners, and families we interviewed. Six of the most commonly reported themes revealed in our study were: (a) consolidation of the relationship after the diagnosis, (b) interaction with the families of origin, (c) shifting perspectives following diagnosis, (d) developing and maintaining a support system, (e)

Table 28-3
Demographic Information

A. Length of couple's relationship at time of interview ($N = 27$)
   Range: 6 months to 23 years
   Mean: 6.8 years
   Median: 5.2 years

B. Length of relationship of surviving partners ($N = 14$)
   Range: 9 months to 20 years
   Mean: 4.3 years
   Median: 3.2 years

C. Geographical breakdown

Number of Participants	Locations
4	Dallas, TX
2	Newark, NJ
21	New York, NY
18	San Diego, CA
24	San Francisco, CA

relationship with the health care family, and (f) dealing with mourning and loss.

Consolidation of the Relationship After the Diagnosis

It is not uncommon that adversity draws a couple together. In the authors' original research with male couples (McWhirter & Mattison, 1984), the well-being of the relationship was found to be affected by threats from outside factors, such as problems with health, money, and occupation. These factors can interrupt the usual development of stages and cause the couple to pull together to meet the crisis. However, the opposite has also been shown to occur: the strains may cause a rift, an emotional distancing from one another and, at times, a relationship breakup.

Interestingly, a striking parallel was found between male couples with AIDS and Stage Six couples who had been together 20 years and longer: the consolidation observed in male couples with AIDS was similar to that of Stage Six couples. One partner observed, "When I got AIDS, it was like going from being 33 to 70 years old in a few months. My mother said, 'You and Dennis are going through in your lives now what your father and I are dealing with at our ages and after 35 years of married life together.'" With those couples who have AIDS, there is an emergence of the intense personal concerns and worries about the anticipated loss from death of the partner. This can either bring the partners much closer together and contribute to what we term "repartnering" with Stage Six couples, or it can increase the distancing.

Shortly after his lover David's diagnosis of AIDs, Sal vividly recalled details of long talks with David:

SAL: I remember telling him how much I loved him and always would no matter what happened. We had waded through a lot of storms in the past and we were in this one together and this is the worst one we would ever face. I knew he needed to hear this, but I needed to say it also. We needed each other badly and we stayed with each other from then on. Very seldom were we apart. It made our relationship all the more strong; it cemented it. And since that realization that I was losing him and was not going to be able to have him for the rest of our natural lives, it became very important that the time we did have together be very special.

Bruce and Scott were in their mid-20s and had been a couple for 2 years at the time of the interview. Interestingly, they began their relationship several months after Scott's diagnosis. They described matter-of-factly and in tandem how they pulled together to manage this illness in the context of their ongoing relationship:

BRUCE: We are talking about a basic human emotion that exists. I realize that's just love. When you love somebody, all that other stuff is incidental. Very often when we reach a crisis situation we talk about how much we love each other and that we'll get through it, and that's what we do. Love is always the answer, and it gets us through it.

SCOTT: Problems are temporary and love isn't; that's how you get through.

Danny and Louie had been a couple for 10 years when Louie was diagnosed as having AIDS. This had occurred 1 year before our interview. After two close calls with death over the last few months, Louie reported feeling much better. At the time of the interview, he was celebrating his 34th birthday that week. His lover Danny is healthy and an involved care partner. They talked about the pulling together experienced in managing their lives with this illness:

LOUIE: We learned to be there for each other because, despite the heartbreak and the tragedy of this illness, there is just a wonderful, beautiful intimate sharing that goes on when we both take care of each other. Because one of the things that happens is you learn that it's not that Danny is taking care of me because I'm sick. What really happens, and this is what couples really have to learn, is that you have to take care of each other.

DANNY: We were able to start building more quickly upon our relationship and what we were going to do for what we have left of it.

LOUIE: There's AIDS which is the physical illness that is affecting our relationship. It's making it more broken. So we really have to work at being closer and being more sensitive to each other's needs.

David and Thomas had been a couple for 5 years when David was diagnosed as having AIDS. Both were 30 years old. No longer able to work as a police officer because of neurological impairments, David was confined to home while Thomas continued his career in the clothing design industry. Thomas talked about his added appreciation of the specialness of the relationship shortly after David's diagnosis:

THOMAS: We always have had breakfast together every morning, but now it's really special every morning, you know. It's a ritual: It's an important part of our day that we come together to have breakfast every morning. I leave for work and he does chores around the apartment or contacts some of his buddies at the police station to follow up on some of his old projects. Every little thing matters, even a phone call during the day. We always did it. But now we appreciate it. If he calls me in the middle of the day, it's great. If I think of him I just pick up the phone and I call him where I might not have done that before he got sick. We have always bought each other little gifts for no special occasions. We always did this stuff, but we never thought about it. Now we do.

In addition to the positive and encouraging stories about couples finding a renewed closeness in their relationship, there were some negative reports as well. Convinced that his lover was the means of transmission of the AIDS virus, one of the participants who had developed AIDS angrily and dramatically terminated a relationship of 20 years. Another member of a couple commented that certain interpersonal problems that were present before the diagnosis persisted after the diagnosis was received.

ROB: Over the 11 years that we have been a couple, we admit that communication skills have not been at the top of our list. Neither of us are great talkers, and we generally pulled into ourselves when we became isolated from each other in times of stress or crisis. It was not surprising for us to do the same kind of thing when Phil was diagnosed with AIDS.

However, much more frequently we heard stories that resembled the Stage Six characteristic of relationship renewal. Frequently, we heard the expression of "pulling together" in an emotional way that bonded the relationship or strengthened it. There was generally a high degree of "taking stock of what we have with each other," and an exquisite appreciation of the relationship itself. Concomitantly, there was the sobering realization that what they had as a couple, what they prized the most—each other—they most would probably lose.

Interaction with the Families of Origin

After the diagnosis of AIDs, over 60% of the couples reached out to their mothers, fathers, brothers, sisters, and relatives for emotional support and, in some instances, day-to-day care. Before the diagnosis, most of the participants were "out of the closet" to family members and enjoyed their recognition of them as gay men and received varying degrees of support.

For some couples, the well wishes of family were appreciated, but it was the extended family or the couples' gay family that provided for their needs.

> LOUIE: So while my family has been very supportive, they've not been with me on a daily basis. It's not that I blame them, or find fault. However, I do recognize the value of my gay family who are those 10 or 15 people that are there, that worry that I'm getting a bronchoscopy on Monday and they are concerned about what I'm going to eat on Tuesday. That's what matters now. I'm less concerned with friends and family and people who send well wishes through the mail because the reality is when you're in bed, that doesn't get you fed. I love my family, but they're just not here.

> DANNY: I'm appreciative of Louie's family. Being able to talk to them about what's happening to our lives with this illness makes me feel more supported. I don't feel like I'm working against them. I feel like we're working together and it's just another bit of the cement that helps to hold our relationship together as opposed to pulling it apart. In that way, his family has helped keep us together because they are able to talk about it. It's not one more secret that you hide from them.

Even among those parents and other family members who have accepted the sexual orientation of their relative and have shown support of the spousal relationship with his partner, the diagnosis of AIDS itself can have negative repercussions for the couple.

Tony and Phil had been a couple for 3 years when Phil was diagnosed as having AIDS. They visited Phil's family at his parents' home to share the bad news with them.

> TONY: We were in our room in his parents' apartment and Phil asked me to tell his sister that he had AIDS. So I told his sister and added that he didn't want his mother to know just yet. She immediately told her mother, which was the right thing to do. We were going back to the room and I heard things being thrown around and carrying on and all kinds of hysteria. And I went outside and all our things were being thrown away. And his mother never touched him again after that. Once he was hospitalized, she never came into the room without putting on gloves and sterilizing herself and washing. She never kissed him again between that time and the day he died.

It wasn't until after Phil died that Tony realized that if Phil's parents had had accurate information about AIDS they probably would have offered the kind of support and caring for which their son was asking. Tony became very active in developing an information group among parents whose sons or daughters were stricken with AIDS.

Together for 11 years, Jerry and Ed enjoyed a good relationship with both of their families and alternated spending holidays with each family. Both Jerry and Ed took seriously their roles as uncles with their nieces and nephews. Nonetheless, they were worried that they might lose the mutually satisfying relationship with their families after Ed developed AIDS. That did not happen, and Jerry eagerly offered his suggestions to other couples with AIDS:

> JERRY: The hedging to tell your parents has to be the worst part of it. The anxiety and the build-up for us was awful. Once it's done, it's done. My advice to anybody who is in that situation is to get it out immediately because all it does is create a tension that is totally unnecessary.

> ED: They were incredible, absolutely incredible. When I was really sick the first week, they just stayed with me at the hospital. They sat in my room every day. They were just there for me. I'd be sleeping and wake up and they'd be there. We talked about everything and I knew that they'd always be there for me, but it was just so wonderful the way they did it.

Some couples who were not "out of the closet" to their families told us of the "double whammy" they had to deal with: first, sharing with their families that they were not "just roommates" but a committed gay couple, and second, that one of them had AIDS.

Together 8 years, Mark and Greg grew up in the same town in rural Iowa. As teenagers, their families had seen them as best friends and were nor surprised when they moved to California and set up a household together. Both allowed their parents to think they were "just roommates sharing a house." Two years before Greg was diagnosed as having AIDS, he "came out" to his family, and after a difficult period of adjustment, they accepted his "way of life." Mark had not "come out of the closet" to his family. He had shared with them neither the nature of his relationship with Greg, nor the fact that Greg had AIDS. In a straightforward manner, Mark commented, "If Greg does die from AIDS, I will tell my parents that he moved away."

More frequently we heard stories of an outpouring of emotional support from brothers, sisters, parents, and relatives—even from those couples who had not previously "come out of the closet" to family members. Families frequently have neither a social nor a psychological context in which to understand their son's homosexuality. Also, the gay son often fails to understand that he did not reach his current level of self-acceptance until he passed through all of the steps of "coming out," and the family too must go through the same process to reintegrate him into their daily lives. However, the diagnosis of a life-threatening disease can accelerate that process or make the issue secondary to fears of loss.

> GEORGE: Now that both Raymond and I have AIDS, the issues have shifted with our families. They want their sons to live. Somehow the gay political things became quite secondary. And I have in many ways found that AIDS in our life finally helped them to accept our gayness, in the sense that it's no longer an issue whether or not we're gay. It's an issue about whether or not we're going to live. And they want us to live. They want us to live.

Shifting Perspectives Following Diagnosis

Just as we found in couples together over 20 years, changing perspectives were high on the list. When asked what had changed in their lives since the diagnosis, it was not surprising to hear the most common response "everything." What appeared to be most striking was how quickly and dramatically the changes occurred. One person lamented that "We went from our honeymoon stage right to talking about wills and funeral plans." One surviving partner recalled how quickly and early in their relationship AIDS struck his lover:

> NICK: I only knew Steven for 6 months before he was diagnosed,

and for much of that 6 months he was ill. We met in April of '84. That Thanksgiving in November he chose to have Thanksgiving with just the two of us because he didn't feel well and he didn't want to invite his family or anybody that we would have to think about because he needed to rest and to have a quiet time with me. By Christmas, when we went to Arizona to be with my family, he was sick the whole time we were there, which I think was the onset of AIDS, although it looked like a terrible flu. We came back and he was very sick. He finished his degree that spring and was literally staggering through the last few days, and staggering through his graduation and was diagnosed within 3 weeks of getting his degree. After Steven's diagnosis, we spent weeks where every day we wept. We held one another and wept. It was a very intense grieving. And then there was a period when he was very angry and he slammed the phone at all of us. I was also angry at that same time about what this was doing to our lives, so we were both really angry and I can remember thinking how long can I live like this? How long can we live like this? If this lasts as long as the weeping lasted, it will be too awful. And it didn't last that long.

Some of the most poignant stories we heard were from couples in which both had AIDS.

JOE: After Dave had been sick for a year and a half, I began experiencing shortness of breath when I went for my daily swim. I knew immediately what I was dealing with. I waited a week to be sure that I wasn't just tired because I was working very hard on a project for a client and I thought maybe this was fatigue, but the second week I was having to work so hard to swim my mile that I said to myself, "You have a choice, kid, either get out of the pool now and admit that you have PCP [*Pneumocytis carinii* pneumonia] or swim until you have a heart attack. Because one or the other is going to happen today." So when I was diagnosed, I went into the hospital one day after Dave and we were in the same AIDS ward together. My diagnosis was a much more painful shock for Dave than it was for me. I had already faced that by facing Dave's. I never looked that demon anywhere except eye-to-eye. So when it was my demon, I wasn't nearly as frightened. But Dave was tremendously upset and hurt by my diagnosis, because he was by that time very sick, very dependent on me. I was essentially his security in the world and his stability and if I was now going to be subject to the whims of this illness, what was going to happen to him?

Another couple recalled changes happening several years before the diagnosis:

JIM: It wasn't until one of our close friends got sick with AIDS that it really sank in that this could happen to Rick and me. Over our last 9 years we had done lots of planning for our future which might very well not happen now. That was a sobering and worrisome realization. Changes started to occur from that moment onward. A few months later I started to experience chronic fatigue and other symptoms that could be associated with AIDS. I wasn't diagnosed until 1 year after that, but as a couple already we had been grappling with our fears of loss, anger, and frustration.

The shifting perspectives occur on myriad levels: individually within both the person with AIDS and his lover, between the partners in the relationship, and among the many people who affect their day-to-day lives. Participants shared with us vast numbers of stories of how AIDS had changed their lives as individuals and couples in both negative as well as positive ways.

PAUL: When you're a couple, you generally move together in one direction and AIDS pulls you apart. You move in separate directions. Dennis is moving in a very separate direction from me. His direction has to do with middle age issues, with his career, with things like that. The direction I'm going in is finishing up my life. I'm moving toward the end of my life and Dennis is moving somewhere else. I want to be able to do things that provide quality in my life, whatever that is, that day. I guess you get very self-centered, because you're dealing with survival of the self.

KENNY: The effect of AIDS on the two of us has been multidimensional. It has affected our sex life. It has destroyed my career which to me had been so important and rewarding. So much of the emotional things, so much joy from my research and out of my interaction with colleagues has stopped. I don't interact with my colleagues, virtually none, now. Phone calls from colleagues and friends have gone down. I sometimes wonder if they are afraid to call because they might think I'm dead. Periodically, I send letters to the editors of various journals to let them know I'm still around.

Developing and Maintaining a Support System

We asked the following question of every couple we interviewed: "What advice would you give to couples who are just beginning to deal with AIDS in their relationship?" The most common response was to urge others "not to go it alone." They reviewed their own personal stories, which emphasized the importance of reaching out to friends and mobilizing a network of supportive care services, including emotional, financial, spiritual, social, and legal services, as well as concrete assistance such as getting a ride to the doctor's office.

MAX: After the diagnosis of AIDS, you have to figure out a wide variety of ways to help yourself, to help your lover. You have to. When David got sick in December, I started going back to church. I was raised a Catholic and church had really never been important in my life. I found something there that works for me. I don't know what it is, I don't care what it is. It's working. I go to a care partners group every week. I go to a therapist on my own—one-on-one, once a week. We are relying on our families, relying on our friends. We learned how to create these support structures. For example, when David got sick, a good friend of our said, "What is my role going to be to do for you?" When David was incapable of bathing himself, I learned to ask for help in bathing him and shaving him. We got a friend of ours who cuts hair to come in. I arranged for people to come in. I arranged for people to come in and give him massages.

STEVE: It took me a while to find the joy and satisfaction that I have now in helping him and doing the things that Larry says he depends on. At first, I had a lot of anger and resentment about his being ill. It was coming out left and right, and as time went on, I began to accept things more and experienced a greater feeling and satisfaction and even joy in helping Larry do things. One of the things that I've had to set up is meal time companions to be with Larry here in the hospital. Our friends come to be here with him for lunch and dinner when I can't be at his bedside. Larry and I had worked out plans for friends bringing in home-cooked meals and in general making these long hospital stays as comfortable as we could humanly make possible.

Our surviving partners warned against the temptation to pull away, the temptation to isolate, the temptation not to tell the truth about the illness.

BOB: I think I've become more open and honest since the diagnosis. I've come out of hiding, as it were, with family and friends, and most people know the complete story now. You know, I don't have to say it's some kind of skin cancer. I tell them directly that I have AIDS. And if they don't like it, you know, it's their problem.

It's the same thing with the gayness. I just feel so much more comfortable with life and living that I don't have to hide anymore. There is a negative side. I go to bed at night sometimes and wonder, if I dare close my eyes, will I get up tomorrow? But I try to focus on the positive and accomplish things and do things that I think are good and worthwhile.

For some, creating a support system was difficult to achieve. We heard stories of "friends who turned their backs," difficulty in locating community services, or a reticence on the part of the couple to reach out for help. John had been living with AIDS himself for almost 3 years when he befriended a man who was rejected by his family when they learned of his diagnosis of AIDS:

JOHN: I met a young man whose mother heard that he has AIDS and she came to see him in the hospital and gave him a $500 check and said, "We don't want to hear from you. You've embarrassed me and your father. If this is the life you want, we don't even want to know if you're dead. So here's the $500." I enjoyed warmth and love from my own family and know how lucky I am, so I brought this young man into my own family. And he needed that.

Developing and maintaining support systems with couples in which both partners have AIDS can be a monumental task that taxes all available resources. However, some of the most inspiring stories and creative approaches to supportive systems were heard from just those couples.

Relationship with the Health Care Family

We asked every couple and surviving partner to share their own experiences with health care, and the shift from independent functioning to the dependencies as they continued to become more ill. For those who had been diagnosed early on in this epidemic, we heard stories more frequently of fear, aversion, and confusion of health care workers that translated into horror stories. They told stories of assistants who wouldn't change bedpans, nurses who interacted with patients with disgust, physicians who wouldn't prescribe drugs, and hospital administrators who were reluctant to hospitalize them. Taken together with the fears of the persons with AIDS themselves, the effect for patients became compounded. One commented, "I feel like a leper of the 20th century, but without a haven or a place to go."

TONY: When Norm got diagnosed with AIDS, his doctor shared the news directly with me and his advice was that I should terminate my relationship with Norm because AIDS was fatal. From that point on, I was excluded by medical staff from any information about his care as if Norm were single and didn't have a lover, and didn't have family. I was devastated and Norm was more upset with that level of treatment than he was with having AIDS.

KENNY: The doctors tried to maintain the sense of absolute control and yet so often with me as an AIDS patient, they didn't have that kind of control at all and they looked for areas and treatment where they could make some kind of semblance of control as if they had absolute power over the disease. I was kind of amused by these attitudes and futile attempts by the medical profession. On the other hand, I was angry and scared that my emotional needs, my human needs, were overlooked completely. It was as if I were just a disease.

For those diagnosed within the last few years, more commonly we heard stories ranging from tolerance to extraordinary efforts of health care workers. For many health professionals, it was a first experience in dealing with such a devastating disease. For many, it was a first experience in dealing on an intimate basis with gay men and male couples.
We heard numerous stories from couples and surviving partners of great compassion and tremendous caring from their health care families and an attitude that profoundly affected the overall well-being of patients, their lovers, friends, and families.

KEITH: Jessie's final placement was at a private hospital here in the city. And the nurses there on the AIDS unit became a family to him in the most personal and warmhearted ways. I was there every day to visit. I wasn't there enough every day. And I finally realized that I didn't have to be there all the time because of the way these nurses were for him and to me. They accepted us as a couple and consulted me in making medical decisions. That was so incredibly important to both of us.

JOHN: The nurse would come in and put the blankets around me. And she'd give me a little hug when I wasn't feeling well and she'd encourage me to sit up and look at that sunset. I realized how lucky I was to see that sunset. She wasn't just concerned with drawing blood or washing me. She really cared about me and that made a difference in the way I felt overall. The nurses here, male and female, have been outstanding.

It was interesting to note the high degree of responsibility persons with AIDS assumed in informing themselves on AIDS evaluation procedures, traditional and nontraditional medical therapies, hospital procedures, sympathetic nurses, and knowledgeable physicians. Information was gathered from a multitude of sources, including media reports, professional journals, articles, and, most commonly, information shared with other persons with AIDS through the development of informal support networks. Many assumed a patient activist stance, which included an attitude of collaborating with the health care team in the management of their own care.

> LEON: When I got diagnosed with AIDS, I acted helpless as if I were a child. I looked to my doctors and nurses to provide the definitive answers about my treatment. I soon realized that some of them were learning too about this disease, and sometimes there were not cut-and-dried answers. I learned to be part of my own health care team. I learned to avoid doctors and technicians with bad attitudes. I asked other patients about medical procedures and treatments. I became very aggressive about making decisions.

Dealing with Mourning and Loss

The most common emotions experienced by the participants were the tremendous sense of grief and the profound feeling of loss. Again, anticipating grieving began early with those who had lost friends and lovers to this disease. For some, the sadness and worry about themselves as a couple began even before the diagnosis of AIDS. Loss often began with not being able to function at one's usual optimal level.

> JIM: Just because I had AIDS doesn't mean the world stops. I continued my career and managing the usual day-to-day issues of living and our relationship. However, a little over a year after the diagnosis I started to get fatigued easily. At first, I fought it and then simply left the office a little early every day. I stopped going to the gym. Then I began to forget little things and get irritated very easily. I got very scared as well as mad. I became very aware of being able to do less and less as the weeks passed.

> PETER: I, too, was frustrated and scared in seeing the changes with Jim. I felt so helpless and at times a little guilty that he was the one stricken. I must confess to some anger and lots of terror at the thoughts of Jim's dying.

Numerous couples told us that in addition to developing a fighting spirit in doing the best they could in battling this disease and supporting each other emotionally, there was what was frequently described as a "push and pull" with the degrees of closeness and distance to the other.

PAT: When Peter was sick I did find myself fleeing. I found myself pulling away, rather than getting more close and I think that had to do with the fears that I had, the irrational thinking that somehow it would be easier for me when Peter dies if I were more distant now. So there is a sense of trying to pull away. Of course, that's crazy because the real part of it is, whatever time we have left, I want it to be very close. But the fear of feeling is what you might feel, so you want to pull away.

Many had made funeral arrangements and made legal wills well in advance of the final steps of the disease process to begin the releasing with their lovers and families.

JOE: Even though I'm the one who is sick, both Jim and I are going through the same crisis, the same terror and anguish. It's important that we talk about everything including my funeral plans and all the legal stuff. That helped us talk about my dying, and his life without me. You know, the dying have work to do. There were lots of things I needed to share with my family and them with me.

In addition to experiencing the hour-to-hour pain of grief, most talked about a need to take a break from talking about AIDS and all of its ensuing sadness.

GEORGE: Both of us talk about the continual losses and changes as Roger gets sicker. But we can't grieve all the time. It can get too overwhelming. Actually one of our rules is that we don't talk about AIDS and our sadness at dinner. That's the time we remember the good memories we've shared over the years together.

Some couples didn't talk much directly to each other about their grief. A few couples thought they could cope best by denying the reality of loss. Most acknowledged the pain of the day-to-day losses of functioning, as well as the ultimate pain of loss by death.

LOUIS: The closer you get to an understanding that you're dying, the more lonely it feels. Once again, you start moving in a direction that's different from your lover. It's very separate from everyone you know. It's isolating and it's lonely. Sometimes in a relationship you want to leave out of choice because you've outgrown each other or you're angry or you have someone else. But this isn't that kind of leaving.

The surviving partners recounted numerous painful stories of the last few days of the lives of their lovers and the more difficult experiences of adjusting to a life in which their spouse was missing.

SAL: I was doing things that needed to be done, but had been putting off for a long time. One afternoon, our delightful nurse said that I had better get into the bedroom because there was something really wrong with him. I walked in and David was making the strangest noise. He had lost control of his voice and when he saw me come in he calmed down. I crawled into bed with David, crawled up around him, and put my arms around him. We started talking about what death is. Over the course of his illness we had talked a lot about reading about death and that when you die you go into the light. Into the light you find friends and loved ones there. I was talking about all the people that we loved that were gone. David quieted down. He slowly slipped away, quietly, in my arms on a Saturday in his bed as he wanted to.

A Final Word

We heard a lot about courage in these interviews with all of our participants. Many couples spoke of the emotional struggles they faced in dealing with the impact of an AIDS diagnosis on their relationship. Yet many shared that the greatest challenge they faced was finding the courage to fight against giving up, and to live every day with compassion, devotion, and an added appreciation of themselves, their lovers, and their families.

References

Kinsey, A. C., Pomeroy, W. B., & Martin, C. E. (1948). *Sexual behavior in the human male*. Philadelphia: W. B. Saunders Company.

Mattison, A. M., & McWhirter, D. P. (1990, May). *Gay couples: Twenty years and beyond*. Paper presented at the Annual Meeting of the American Psychiatric Association, New York.

McWhirter, D. P., & Mattison, A. M. (1984). *The male couple: How relationships develop*. Englewood Cliffs, NJ: Prentice-Hall.

Author Index

Subject Index

Abbott Laboratories, 23
ABC's "20/20," 236, 238, 241
ABC's "Nightline," 231
Abortion
  maternal mortality in Brazil, 198
Abrahms, Donald, 395
Acquired immune deficiency
      syndrome (AIDS)
  in Africa, *See also* HIV infection,
      Africa
    clinical presentation, 190
    euphemisms for, 190
    herbal medicines and, 191
    incidence of compared to U.S.
      and world totals, 189
    prostitution and, 192–3
    sex education and, 193–4
    sociocultural aspects of, 190–3
    traditional healers/doctors and,
      191–2
    underreporting of, 190
  among Blacks
    AIDS/HIV risk compared to
      other ethnic/racial groups, 97,
      104–7
    perception of AIDS risk, 104–5
    predicting the spread of HIV
      problems with theoretical
        models of behavior, 113–4
      problems with statistical
        models of infection rates, 114–
        5
  in Brazil, *See also* HIV infection, in
      Brazil

AIDS National Program, 198–9
AIDS National Reference Center,
      201
    clinical presentation of, 201
    health care issues and, 199–202
    incidence of, 198–200
    National Program for Prevention
      and Control of AIDS, 202, 209
    pharmaceutical industry
      corruption and, 208–9
    other public health crises and,
      197–8
    socioeconomic conditions and,
      197–8, 205–6
    underreporting of, 198
  confidentiality and, 223–5
  Darwinian evolution and, 214–7
  denial and, 239–40
  education about, *See* Public
      education about AIDS
  emotional impact of, *See* Emotional
      impact of AIDS
  fear and, 221, 226–7, 230, 240
  as federal government germ
      warfare, 110, 390
  as a gay disease, 38, 104–5, 179–
      80, 207, 222, 238, 240, 376
  geographical distribution of, 211
  humanizing AIDS, 223–4
  among Hispanics
    AIDS/HIV risk compared to
      other ethnic/racial groups, 97,
      104–7